D1452958

The Resource Handbook of

ELECTRONICS

ELECTRONICS HANDBOOK SERIES

Series Editor:
Jerry C. Whitaker
Technical Press
Morgan Hill, California

PUBLISHED TITLES

AC POWER SYSTEMS HANDBOOK, SECOND EDITION
Jerry C. Whitaker

THE COMMUNICATIONS FACILITY DESIGN HANDBOOK
Jerry C. Whitaker

THE ELECTRONIC PACKAGING HANDBOOK
Glenn R. Blackwell

POWER VACUUM TUBES HANDBOOK, SECOND EDITION
Jerry C. Whitaker

MICROELECTRONICS
Jerry C. Whitaker

SEMICONDUCTOR DEVICES AND CIRCUITS
Jerry C. Whitaker

SIGNAL MEASUREMENT, ANALYSIS, AND TESTING
Jerry C. Whitaker

THERMAL DESIGN OF ELECTRONIC EQUIPMENT
Ralph Remsburg

THE RESOURCE HANDBOOK OF ELECTRONICS
Jerry C. Whitaker

FORTHCOMING TITLES

ELECTRONIC SYSTEMS MAINTENANCE HANDBOOK
Jerry C. Whitaker

The Resource Handbook of

ELECTRONICS

Jerry C. Whitaker
Technical Press
Morgan Hill, California

CRC Press

Boca Raton   London   New York   Washington, D.C.

Library of Congress Cataloging-in-Publication Data

Whitaker, Jerry C.
    The resource handbook of electronics / Jerry C. Whitaker.
        p.   cm.--(The Electronics handbook series)
    Includes bibliographical references and index.
    ISBN 0-8493-8353-6 (alk. paper)
    1. Electonics--Handbooks, manuals, etc. I. Title. II. Series.

TK7825 .W48 2000
621.381--dc21   00-057935

© 2001 by CRC Press LLC

No claim to original U.S. Government works
International Standard Book Number 0-8493-8353-6
Library of Congress Card Number 00-057935
Printed in the United States of America   2 3 4 5 6 7 8 9 0
Printed on acid-free paper

Preface

The hallmark of the CRC Press "Electronics Engineering Series" of books is their depth of coverage on targeted subjects. Even the more general-interest publication of the series—*The Electronics Handbook*—covers the entire realm of electronics in exceptional detail.

This book is a departure from those that have gone before it. *The Resource Handbook of Electronics* is intended to provide quick access to basic information, mostly through figures and tables. For each of the 20-plus chapters, a broad-brush overview is given, followed in most cases by extensive tabular data. *The Resource Handbook of Electronics* is intended for readers who need specific data at their fingertips, accessible in a convenient format.

This book is intended for engineers, technicians, operators, and technical managers involved in the specification, design, installation, operation, maintenance, and management of electronics facilities. The book is designed to be a hands-on pocket guide that holds solutions to specific problems. In this regard, it is a companion publication to *The Electronics Handbook* and the other books in the series. For readers who need extensive background on a given subject, *The Electronics Handbook* and its related works provide the necessary level of detail. For readers who need a broad overview of the subject and essential data relating to it, *The Resource Handbook of Electronics* is the ideal publication.

This book is organized in a logical sequence that begins with fundamental electrical properties and builds to higher levels of sophistication from one chapter to the next. Chapters are devoted to all of the most common components and devices, in addition to higher-level applications of those components.

Among the extensive data contained in *The Resource Handbook of Electronics* are

- **Frequency assignments**—A complete and up-to-date listing of frequencies used by various services in the U.S. and elsewhere

- **Glossary of terms**—An extensive dictionary of electronic terms, including abbreviations and acronyms

- **Conversion factors**—Detailed tables covering all types of conversion requirements in the field of electronics

The Resource Handbook of Electronics is the most detailed publication of its kind. I trust you will find it useful on the job, day in and day out.

Jerry C. Whitaker
Morgan Hill, California

For updated information on this and other engineering books, visit the author's
Internet site
www.technicalpress.com

About the Author

Jerry Whitaker is a technical writer based in Morgan Hill, California, where he operates the consulting firm *Technical Press*. Mr. Whitaker has been involved in various aspects of the communications industry for more than 25 years. He is a Fellow of the Society of Broadcast Engineers and an SBE-certified Professional Broadcast Engineer. He is also a member and Fellow of the Society of Motion Picture and Television Engineers, and a member of the Institute of Electrical and Electronics Engineers. Mr. Whitaker has written and lectured extensively on the topic of electronic systems installation and maintenance.

Mr. Whitaker is the former editorial director and associate publisher of *Broadcast Engineering* and *Video Systems* magazines. He is also a former radio station chief engineer and TV news producer.

Mr. Whitaker is the author of a number of books, including:

- *The Communications Facility Design Handbook*, CRC Press, 2000.

- *Power Vacuum Tubes Handbook*, 2nd edition, CRC Press, 1999.

- *AC Power Systems*, 2nd edition, CRC Press, 1998.

- *DTV: The Revolution in Electronic Imaging*, 2nd edition, McGraw-Hill, 1999.

- Editor-in-Chief, *NAB Engineering Handbook*, 9th edition, National Association of Broadcasters, 1999.

- Editor-in-Chief, *The Electronics Handbook*, CRC Press, 1996.

- Coauthor, *Communications Receivers: Principles and Design*, 2nd edition, McGraw-Hill, 1996.

- *Electronic Displays: Technology, Design, and Applications*, McGraw-Hill, 1994.

- Coeditor, *Standard Handbook of Video and Television Engineering*, 3rd edition, McGraw-Hill, 2000.

- Coeditor, *Information Age Dictionary*, Intertec/Bellcore, 1992.

- *Maintaining Electronic Systems*, CRC Press, 1991.

- *Radio Frequency Transmission Systems: Design and Operation*, McGraw-Hill, 1990.

Mr. Whitaker has twice received a Jesse H. Neal Award *Certificate of Merit* from the Association of Business Publishers for editorial excellence. He also has been recognized as *Educator of the Year* by the Society of Broadcast Engineers.

Acknowledgment

The author wishes to express appreciation to the following contributors for their assistance in the preparation of this book.

K. Blair Benson
E. Stanley Busby
Michael W. Dahlgren
Gene DeSantis
Donald C. McCroskey
C. Robert Paulson

Table of Contents

For baby
Ashley Grace Whitaker
The journey begins...

Fundamental Electrical Properties

1.1   Introduction

The atomic theory of matter specifies that each of the many chemical elements is composed of unique and identifiable particles called atoms. In ancient times only 10 were known in their pure, uncombined form; these were carbon, sulfur, copper, antimony, iron, tin, gold, silver, mercury, and lead. Of the several hundred now identified, less than 50 are found in an uncombined, or chemically free, form on earth.

Each atom consists of a compact nucleus of positively and negatively charged particles (protons and electrons, respectively). Additional electrons travel in well-defined orbits around the nucleus. The electron orbits are grouped in regions called *shells*, and the number of electrons in each orbit increases with the increase in orbit diameter in accordance with quantum-theory laws of physics. The diameter of the outer orbiting path of electrons in an atom is in the order of one-millionth (10^{-6}) millimeter, and the nucleus, one-millionth of that. These typical figures emphasize the minute size of the atom.

1.2   Electrical Fundamentals

The nucleus and the free electrons for an iron atom are shown in the schematic diagram in Figure 1.1. Note that the electrons are spinning in different directions. This rotation creates a magnetic field surrounding each electron. If the number of electrons with positive spins is equal to the number with negative spins, then the net field is zero and the atom exhibits no magnetic field.

In the diagram, although the electrons in the first, second, and fourth shells balance each other, in the third shell five electrons have clockwise positive spins, and one a counterclockwise negative spin, which gives the iron atom in this particular electron configuration a cumulative *magnetic effect*.

The parallel alignment of the electron spins over regions, known as *domains*, containing a large number of atoms. When a magnetic material is in a demagnetized state, the direction of magnetization in the domain is in a random order. Magnetization by an

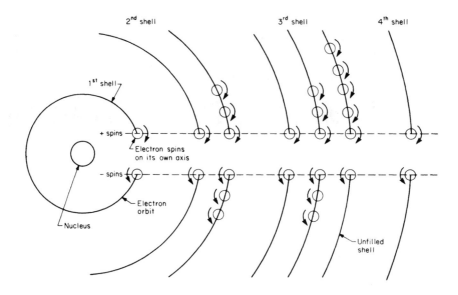

Figure 1.1 Schematic of the iron (Fe) atom.

external field takes place by a change or displacement in the isolation of the domains, with the result that a large number of the atoms are aligned with their charged electrons in parallel.

1.2.1 Conductors and Insulators

In some elements, such as copper, the electrons in the outer shells of the atom are so weakly bound to the nucleus that they can be released by a small electrical force, or voltage. A voltage applied between two points on a length of a metallic conductor produces the flow of an electric current, and an electric field is established around the conductor. The conductivity is a constant for each metal that is unaffected by the current through or the intensity of any external electric field.

In some nonmetallic materials, the free electrons are so tightly bound by forces in the atom that, upon the application of an external voltage, they will not separate from their atom except by an electrical force strong enough to destroy the insulating properties of the material. However, the charges will realign within the structure of their atom. This condition occurs in the insulating material (dielectric) of a capacitor when a voltage is applied to the two conductors encasing the dielectric.

Semiconductors are electronic conducting materials wherein the conductivity is dependent primarily upon impurities in the material. In addition to negative mobile charges of electrons, positive mobile charges are present. These positive charges are called *holes* because each exists as an absence of electrons. Holes (+) and electrons (−), because they are oppositely charged, move in opposite directions in an electric field. The conductivity of semiconductors is highly sensitive to, and increases with, temperature.

1.2.2 Direct Current (dc)

Direct current is defined as a unidirectional current in which there are no significant changes in the current flow. In practice, the term frequently is used to identify a voltage source, in which case variations in the load can result in fluctuations in the current but not in the direction.

Direct current was used in the first systems to distribute electricity for household and industrial power. For safety reasons, and the voltage requirements of lamps and motors, distribution was at the low nominal voltage of 110. The losses in distribution circuits at this voltage seriously restricted the length of transmission lines and the size of the areas that could be covered. Consequently, only a relatively small area could be served by a single generating plant. It was not until the development of alternating-current systems and the voltage transformer that it was feasible to transport high levels of power at relatively low current over long distances for subsequent low-voltage distribution to consumers.

1.2.3 Alternating Current (ac)

Alternating current is defined as a current that reverses direction at a periodic rate. The average value of alternating current over a period of one cycle is equal to zero. The effective value of an alternating current in the supply of energy is measured in terms of the root mean square (rms) value. The rms is the square root of the square of all the values, positive and negative, during a complete cycle, usually a sine wave. Because rms values cannot be added directly, it is necessary to perform an rms addition as shown in the equation:

$$V_{rms\ total} = \sqrt{V_{rms\ 1}{}^2 + V_{rms\ 2}{}^2 + L\ V_{rms\ n}{}^2} \qquad (1.1)$$

As in the definition of direct current, in practice the term frequently is used to identify a voltage source.

The level of a sine-wave alternating current or voltage can be specified by two other methods of measurement in addition to rms. These are *average* and *peak*. A sine-wave signal and the rms and average levels are shown in Figure 1.2. The levels of complex, symmetrical ac signals are specified as the peak level from the axis, as shown in the figure.

1.2.4 Static Electricity

The phenomenon of static electricity and related potential differences concerns configurations of conductors and insulators where no current flows and all electrical

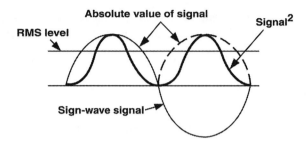

Figure 1.2 Root mean square (rms) measurements. The relationship of rms and average values is shown.

forces are unchanging; hence the term *static*. Nevertheless, static forces are present because of the number of excess electrons or protons in an object. A static charge can be induced by the application of a voltage to an object. A flow of current to or from the object can result from either a breakdown of the surrounding nonconducting material or by the connection of a conductor to the object.

Two basic laws regarding electrons and protons are:

- Like charges exert a repelling force on each other; electrons repel other electrons and protons repel other protons

- Opposite charges attract each other; electrons and protons are attracted to each other

Therefore, if two objects each contain exactly as many electrons as protons in each atom, there is no electrostatic force between the two. On the other hand, if one object is charged with an excess of protons (deficiency of electrons) and the other an excess of electrons, there will be a relatively weak attraction that diminishes rapidly with distance. An attraction also will occur between a neutral and a charged object.

Another fundamental law, developed by Faraday, governing static electricity is that all of the charge of any conductor not carrying a current lies in the surface of the conductor. Thus, any electric fields external to a completely enclosed metal box will not penetrate beyond the surface. Conversely, fields within the box will not exert any force on objects outside the box. The box need not be a solid surface; a conduction cage or grid will suffice. This type of isolation frequently is referred to as a *Faraday shield*.

1.2.5 Noise in Electronic Circuits

Noise has become the standard term for signals that are random and that are combined with the circuit signal to affect the overall performance of a system. As the study of noise has progressed, engineers have come to realize that there are many sources of noise in circuits. The following definitions are commonly used in discussions of circuit noise:

- *White noise*: a signal that has its energy evenly distributed over the entire frequency spectrum, within the frequency range of interest (typically below frequencies in the infrared range). Because *white noise* is totally random, it may seem inappropriate to refer to its frequency range, because it is not really periodic in the ordinary sense. Nevertheless, by examining an oscilloscope trace of white noise, it can be verified that every trace is different, as the noise never repeats itself, and yet each trace looks the same. There is a strong theoretical foundation to represent the frequency content of such signals as covering the frequency spectrum evenly. In this way the impact on other periodic signals can be analyzed. The term white noise arises from the fact that, similar to white light, which has equal amounts of all light frequencies, white noise has equal amounts of noise at all frequencies within circuit operating ranges.

- *Interference*: the name given to any predictable, periodic signal that occurs in an electronic circuit in addition to the signal the circuit is designed to process. This is distinguished from a noise signal by the fact that it occupies a relatively small frequency range, and because it is predictable it can often be filtered out. Usually, interference comes from another electronic system such as an interfering radio source.

- *Thermal noise*: any noise that is generated within a circuit and is temperature-dependent. This signal usually is the result of the influence of temperature directly on the operating characteristics of circuit components, which because of the random motion of molecules as a result of temperature, in turn creates a random fluctuation of the signal being processed.

- *Shot noise*: a type of circuit noise that is not temperature-dependent, and is not white noise in the sense that it tends to diminish at higher frequencies. This noise usually occurs in components whose operation depends on a mean *particle residence time* for the active electrons within the device. The *cutoff frequency* above which noise disappears is closely related to the inverse of this characteristic particle residence time.

1.3 References

1. Whitaker, Jerry C. (ed.), *The Electronics Handbook*, CRC Press, Boca Raton, FL, 1996.

1.4 Bibliography

Benson, K. Blair, and Jerry C. Whitaker, *Television and Audio Handbook for Technicians and Engineers*, McGraw-Hill, New York, NY, 1990.
Benson, K. Blair, *Audio Engineering Handbook*, McGraw-Hill, New York, NY, 1988.
Whitaker, Jerry C., *Television Engineers' Field Manual*, McGraw-Hill, New York, NY, 2000.

1.5 Tabular Data

Table 1.1 Symbols and Terminology for Physical and Chemical Quantities: Classical Mechanics (*From* [1]. *Used with permission.*)

Name	Symbol	Definition	SI unit
mass	m		kg
reduced mass	μ	$\mu = m_1 m_2/(m_1 + m_2)$	kg
density, mass density	ρ	$\rho = m/V$	kg m^{-3}
relative density	d	$d = \rho/\rho^\theta$	1
surface density	ρ_A, ρ_S	$\rho_A = m/A$	kg m^{-2}
specific volume	v	$v = V/M = 1/\rho$	m^3 kg^{-1}
momentum	$\boldsymbol{p}$	$\boldsymbol{p} = m\boldsymbol{v}$	kg ms^{-1}
angular momentum, action	L	$L = \boldsymbol{r} \times \boldsymbol{p}$	J s
moment of inertia	I, J	$I = \sum m_i r_i^2$	kg m^2
force	F	$F = dp/dt = m\boldsymbol{a}$	N
torque, moment of a force	$T, (M)$	$T = \boldsymbol{r} \times \boldsymbol{F}$	N m
energy	E		J
potential energy	E_p, V, Φ	$E_p = -\int \boldsymbol{F} \cdot d\boldsymbol{s}$	J
kinetic energy	E_k, T, K	$E_k = (1/2)mv^2$	J
work	W, w	$W = \int \boldsymbol{F} \cdot d\boldsymbol{s}$	J
Hamilton function	H	$H(q, p)$ $= T(q, p) + V(q)$	J
Lagrange function	L	$L(q, \dot{q})$ $= T(q, \dot{q}) - V(q)$	J
pressure	p, P	$p = F/A$	Pa, N m^{-2}
surface tension	γ, σ	$\gamma = dW/dA$	N m^{-1}, J m^{-2}
weight	$G, (W, P)$	$G = mg$	N
gravitational constant	G	$F = Gm_1 m_2/r^2$	N m^2 kg^{-2}
normal stress	σ	$\sigma = F/A$	Pa
shear stress	τ	$\tau = F/A$	Pa
linear strain, relative elongation	ε, e	$\varepsilon - \Delta l/l$	1
modulus of elasticity, Young's modulus	E	$E = \sigma/\varepsilon$	Pa
shear strain	γ	$\gamma = \Delta x/d$	1
shear modulus	G	$G = \tau/\gamma$	Pa
volume strain, bulk strain	θ	$\theta = \Delta V/V_0$	1
bulk modulus, compression modulus	K	$K = -V_0(dp/dV)$	Pa
viscosity, dynamic viscosity	η, μ	$\tau_{x,z} = \eta(dv_x/dz)$	Pa s
fluidity	ϕ	$\phi = 1/\eta$	m kg^{-1}s
kinematic viscosity	v	$v = \eta/\rho$	m^2 s^{-1}
friction coefficient	$\mu, (f)$	$F_{\text{frict}} = \mu F_{\text{norm}}$	1
power	P	$P = dW/dt$	W
sound energy flux	P, P_a	$P = dE/dt$	W
acoustic factors			
reflection factor	ρ	$\rho = P_r/P_0$	1
acoustic absorption factor	$\alpha_a, (\alpha)$	$\alpha_a = 1 - \rho$	1
transmission factor	τ	$\tau = P_{\text{tr}}/P_0$	1
dissipation factor	δ	$\delta = \alpha_a - \tau$	1

Table 1.2 Symbols and Terminology for Physical and Chemical Quantities: Electricity and Magnetism (*From* [1]. *Used with permission.*)

Name	Symbol	Definition	SI unit
quantity of electricity, electric charge	Q		C
charge density	ρ	$\rho = Q/V$	$C\,m^{-3}$
surface charge density	σ	$\sigma = Q/A$	$C\,m^{-2}$
electric potential	V, ϕ	$V = dW/dQ$	$V, J\,C^{-1}$
electric potential difference	$U, \Delta V, \Delta\phi$	$U = V_2 - V_1$	V
electromotive force	E	$E = \int (F/Q) \cdot ds$	V
electric field strength	E	$E = F/Q = -\text{grad}\,V$	$V\,m^{-1}$
electric flux	Ψ	$\Psi = \int D \cdot dA$	C
electric displacement	D	$D = \varepsilon E$	$C\,m^{-2}$
capacitance	C	$C = Q/U$	$F, C\,V^{-1}$
permittivity	ε	$D = \varepsilon E$	$F\,m^{-1}$
permittivity of vacuum	ε_0	$\varepsilon_0 = \mu_0^{-1} c_0^{-2}$	$F\,m^{-1}$
relative permittivity	ε_r	$\varepsilon_r = \varepsilon/\varepsilon_0$	1
dielectric polarization (dipole moment per volume)	P	$P = D - \varepsilon_0 E$	$C\,m^{-2}$
electric susceptibility	χ_e	$\chi_e = \varepsilon_r - 1$	1
electric dipole moment	p, μ	$p = Qr$	$C\,m$
electric current	I	$I = dQ/dt$	A
electric current density	j, J	$I = \int j \cdot dA$	$A\,m^{-2}$
magnetic flux density, magnetic induction	B	$F = Qv \times B$	T
magnetic flux	Φ	$\Phi = \int B \cdot dA$	Wb
magnetic field strength	H	$B = \mu H$	$A\,M^{-1}$
permeability	μ	$B = \mu H$	$N\,A^{-2}, H\,m^{-1}$
permeability of vacuum	μ_0		$H\,m^{-1}$
relative permeability	μ_r	$\mu_r = \mu/\mu_0$	1
magnetization (magnetic dipole moment per volume)	M	$M = B/\mu_0 - H$	$A\,m^{-1}$
magnetic susceptibility	$\chi, \kappa, (\chi_m)$	$\chi = \mu_r - 1$	1
molar magnetic susceptibility	χ_m	$\chi_m = V_m \chi$	$m^3\,mol^{-1}$
magnetic dipole moment	m, μ	$E_p = -m \cdot B$	$A\,m^2, J\,T^{-1}$
electrical resistance	R	$R = U/I$	Ω
conductance	G	$G = 1/R$	S
loss angle	δ	$\delta = (\pi/2) + \phi_I - \phi_U$	1, rad
reactance	X	$X = (U/I) \sin \delta$	Ω
impedance (complex impedance)	Z	$Z = R + iX$	Ω
admittance (complex admittance)	Y	$Y = 1/Z$	S
susceptance	B	$Y = G + iB$	S
resistivity	ρ	$\rho = E/j$	$\Omega\,m$
conductivity	κ, γ, σ	$\kappa = 1/\rho$	$S\,m^{-1}$
self-inductance	L	$E = -L(dI/dt)$	H
mutual inductance	M, L_{12}	$E_1 = L_{12}(dI_2/dt)$	H
magnetic vector potential	A	$B = \nabla \times A$	$Wb\,m^{-1}$
Poynting vector	S	$S = E \times H$	$W\,m^{-2}$

Table 1.3 Symbols and Terminology for Physical and Chemical Quantities: Electromagnetic Radiation (*From* [1]. *Used with permission.*)

Name	Symbol	Definition	SI unit
wavelength	λ		m
speed of light in vacuum	c_0		$\mathrm{m\,s^{-1}}$
in a medium	c	$c = c_0/n$	$\mathrm{m\,s^{-1}}$
wavenumber in vacuum	$\tilde{v}$	$\tilde{v} = v/c_0 = 1/n\lambda$	$\mathrm{m^{-1}}$
wavenumber (in a medium)	σ	$\sigma = 1/\lambda$	$\mathrm{m^{-1}}$
frequency	v	$v = c/\lambda$	Hz
circular frequency,	ω	$\omega = 2\pi v$	$\mathrm{s^{-1}}$, $\mathrm{rad\,s^{-1}}$
pulsatance			
refractive index	n	$n = c_0/c$	1
Planck constant	h		J s
Planck constant/2π	$\hbar$	$\hbar = h/2\pi$	J s
radiant energy	Q, W		J
radiant energy density	ρ, w	$\rho = Q/V$	$\mathrm{J\,m^{-3}}$
spectral radiant energy density			
in terms of frequency	ρ_v, w_v	$\rho_v = d\rho/dv$	$\mathrm{J\,m^{-3}Hz^{-1}}$
in terms of wavenumber	$\rho_{\tilde{v}}, w_{\tilde{v}}$	$\rho_{\tilde{v}} = d\rho/d\tilde{v}$	$\mathrm{J\,m^{-2}}$
in terms of wavelength	ρ_λ, w_λ	$\rho_\lambda = d\rho/d\lambda$	$\mathrm{J\,m^{-4}}$
Einstein transition probabilities			
spontaneous emission	A_{nm}	$dN_n/dt = -A_{nm}N_n$	$\mathrm{s^{-1}}$
stimulated emission	B_{nm}	$dN_n/dt = -\rho_{\tilde{v}}(\tilde{v}_{nm}) \times B_{nm}N_n$	$\mathrm{s\,kg^{-1}}$
stimulated absorption	B_{mn}	$dN_n/dt = \rho_{\tilde{v}}(\tilde{v}_{nm})B_{mn}N_m$	$\mathrm{s\,kg^{-1}}$
radiant power,	Φ, P	$\Phi = dQ/dt$	W
radiant energy per time			
radiant intensity	I	$I = d\Phi/d\Omega$	$\mathrm{W\,sr^{-1}}$
radiant exitance	M	$M = d\Phi/dA_{\text{source}}$	$\mathrm{W\,m^{-2}}$
(emitted radiant flux)			
irradiance	$E, (I)$	$E = d\Phi/dA$	$\mathrm{W\,m^{-2}}$
(radiant flux received)			
emittance	ε	$\varepsilon = M/M_{\text{bb}}$	1
Stefan-Boltzman constant	σ	$M_{\text{bb}} = \sigma T^4$	$\mathrm{W\,m^{-2}\,K^{-4}}$
first radiation constant	c_1	$c_1 = 2\pi hc_0^2$	$\mathrm{W\,m^2}$
second radiation constant	c_2	$c_2 = hc_0/k$	K m
transmittance,	τ, T	$\tau = \Phi_{\text{tr}}/\Phi_0$	1
transmission factor			
absorptance,	α	$\alpha = \Phi_{\text{abs}}/\Phi_0$	1
absorption factor			
reflectance,	ρ	$\rho = \Phi_{\text{refl}}/\Phi_0$	1
reflection factor			
(decadic) absorbance	A	$A = \lg(1 - \alpha_i)$	1
napierian absorbance	B	$B = \ln(1 - \alpha_i)$	1
absorption coefficient			
(linear) decadic	a, K	$a = A/l$	$\mathrm{m^{-1}}$
(linear) napierian	α	$\alpha = B/l$	$\mathrm{m^{-1}}$
molar (decadic)	ε	$\varepsilon = a/c = A/cl$	$\mathrm{m^2\,mol^{-1}}$
molar napierian	κ	$\kappa = \alpha/c = B/cl$	$\mathrm{m^2\,mol^{-1}}$
absorption index	k	$k = \alpha/4\pi\tilde{v}$	1
complex refractive index	$\hat{n}$	$\hat{n} = n + ik$	1
molar refraction	R, R_m	$R = \dfrac{n^2 - 1}{n^2 + 2}V_m$	$\mathrm{m^3\,mol^{-1}}$
angle of optical rotation	α		1, rad

Table 1.4 Symbols and Terminology for Physical and Chemical Quantities: Solid State (*From* [1]. *Used with permission.*)

Name	Symbol	Definition	SI unit
lattice vector	R, R_0		m
fundamental translation vectors for the crystal lattice	$a_1; a_2; a_3,$ $a; b; c$	$R = n_1 a_1 + n_2 a_2 + n_3 a_3$	m
(circular) reciprocal lattice vector	G	$G \cdot R = 2\pi m$	m^{-1}
(circular) fundamental translation vectors for the reciprocal lattice	$b_1; b_2; b_3,$ $a^*; b^*; c^*$	$a_i \cdot b_k = 2\pi \delta_{ik}$	m^{-1}
lattice plane spacing	d		m
Bragg angle	θ	$n\lambda = 2d \sin\theta$	1, rad
order of reflection	n		1
order parameters			
short range	σ		1
long range	s		1
Burgers vector	b		m
particle position vector	r, R_j		m
equilibrium position vector of an ion	R_0		m
equilibrium position vector of an ion	R_0		m
displacement vector of an ion	u	$u = R - R_0$	m
Debye–Waller factor	B, D		1
Debye circular wavenumber	q_D		m^{-1}
Debye circular frequency	ω_D		s^{-1}
Grüneisen parameter	γ, Γ	$\gamma = \alpha V / \kappa C_V$	1
Madelung constant	$\alpha, \mathcal{M}$	$E_{coul} = \dfrac{\alpha N_A z_+ z_- e^2}{4\pi \varepsilon_0 R_0} \cdot$	1
density of states	N_E	$N_E = dN(E)/dE$	$J^{-1}\,m^{-3}$
(spectral) density of vibrational modes	N_ω, g	$N_\omega = dN(\omega)/d\omega$	$s\,m^{-3}$
resistivity tensor	ρ_{ik}	$E = \rho \cdot j$	$\Omega\,m$
conductivity tensor	σ_{ik}	$\sigma = \rho^{-1}$	$S\,m^{-1}$
thermal conductivity tensor	λ_{ik}	$J_q = -\lambda \cdot \mathrm{grad}\, T$	$W\,m^{-1}\,K^{-1}$
residual resistivity	ρ_R		$\Omega\,m$
relaxation time	τ	$\tau = l/v_F$	s
Lorenz coefficient	L	$L = \lambda/\sigma T$	$V^2\,K^{-2}$
Hall coefficient	A_H, R_H	$E = \rho \cdot j + R_H(B \times j)$	$m^3\,C^{-1}$
thermoelectric force	E		V
Peltier coefficient	Π		V
Thomson coefficient	$\mu, (\tau)$		$V\,K^{-1}$
work function	Φ	$\Phi = E_\infty - E_F$	J
number density, number concentration	$n, (p)$		m^{-3}
gap energy	E_g		J
donor ionization energy	E_d		J
acceptor ionization energy	E_a		J
Fermi energy	E_F, ε_F		J
circular wave vector, propagation vector	k, q	$k = 2\pi/\lambda$	m^{-1}
Bloch function	$u_k(r)$	$\psi(r) = u_k(r)\exp(ik \cdot r)$	$m^{-3/2}$
charge density of electrons	ρ	$\rho(r) = -e\psi^*(r)\psi(r)$	$C\,m^{-3}$
effective mass	m^*		kg
mobility	μ	$\mu = v_{drift}/E$	$m^2\,V^{-1}\,s^{-1}$
mobility ratio	b	$b = \mu_n/\mu_p$	1
diffusion coefficient	D	$dN/dt = -DA(dn/dx)$	$m^2\,s^{-1}$
diffusion length	L	$L = \sqrt{D\tau}$	m
characteristic (Weiss) temperature	ϕ, ϕ_W		K
Curie temperature	T_C		K
Néel temperature	T_N		K

Table 1.5 Total Elongation at Failure of Selected Polymers (*From* [1]. *Used with permission.*)

Polymer	Elongation
ABS	5–20
Acrylic	2–7
Epoxy	4.4
HDPE	700–1000
Nylon, type 6	30–100
Nylon 6/6	15–300
Phenolic	0.4–0.8
Polyacetal	25
Polycarbonate	110
Polyester	300
Polypropylene	100–600
PTFE	250–350

Table 1.6 Tensile Strength of Selected Wrought Aluminum Alloys (*From* [1]. *Used with permission.*)

Alloy	Temper	TS (MPa)
1050	0	76
1050	H16	130
2024	0	185
2024	T361	495
3003	0	110
3003	H16	180
5050	0	145
5050	H34	195
6061	0	125
6061	T6, T651	310
7075	0	230
7075	T6, T651	570

Table 1.7 Density of Selected Materials, Mg/m^3 (*From* [1]. *Used with permission.*)

Metal		Ceramic		Glass		Polymer	
Ag	10.50	Al_2O_3	3.97–3.986	SiO_2	2.20	ABS	1.05–1.07
Al	2.7	BN (cub)	3.49	SiO_2 10 wt% Na_2O	2.291	Acrylic	1.17–1.19
Au	19.28	BeO	3.01–3.03	SiO_2 19.55 wt% Na_2O	2.383	Epoxy	1.80–2.00
Co	8.8	MgO	3.581	SiO_2 29.20 wt% Na_2O	2.459	HDPE	0.96
Cr	7.19	SiC(hex)	3.217	SiO_2 39.66 wt% Na_2O	2.521	Nylon, type 6	1.12–1.14
Cu	8.93	Si_3N_4 (α)	3.184	SiO_2 39.0 wt% CaO	2.746	Nylon 6/6	1.13–1.15
Fe	7.87	Si_3N_4 (β)	3.187			Phenolic	1.32–1.46
Ni	8.91	TiO_2 (rutile)	4.25			Polyacetal	1.425
Pb	11.34	UO_2	10.949–10.97			Polycarbonate	1.2
Pt	21.44	ZrO_2 (CaO)	5.5			Polyester	1.31
Ti	4.51	Al_2O_3 MgO	3.580			Polystyrene	1.04
W	19.25	$3Al_2O_3$ $2SiO_2$	2.6–3.26			PTFE	2.1–2.3

Table 1.8 Dielectric Constants of Ceramics (*From* [1]. *Used with permission.*)

Material	Dielectric constant, 10^6 Hz	Dielectric strength V/mil	Volume resistivity $\Omega \cdot$ cm (23°C)	Loss factor[a]
Alumina	4.5–8.4	40–160	10^{11}–10^{14}	0.0002–0.01
Corderite	4.5–5.4	40–250	10^{12}–10^{14}	0.004–0.012
Forsterite	6.2	240	10^{14}	0.0004
Porcelain (dry process)	6.0–8.0	40–240	10^{12}–10^{14}	0.0003–0.02
Porcelain (wet process)	6.0–7.0	90–400	10^{12}–10^{14}	0.006–0.01
Porcelain, zircon	7.1–10.5	250–400	10^{13}–10^{15}	0.0002–0.008
Steatite	5.5–7.5	200–400	10^{13}–10^{15}	0.0002–0.004
Titanates (Ba, Sr, Ca, Mg, and Pb)	15–12.000	50–300	10^8–10^{15}	0.0001–0.02
Titanium dioxide	14–110	100–210	10^{13}–10^{18}	0.0002–0.005

[a]Power factor × dielectric constant equals loss factor.

Table 1.9 Dielectric Constants of Glass (*From* [1]. *Used with permission.*)

Type	Dielectric constant at 100 MHz (20°C)	Volume resistivity (350°C M $\Omega \cdot$ cm)	Loss factor[a]
Corning 0010	6.32	10	0.015
Corning 0080	6.75	0.13	0.058
Corning 0120	6.65	100	0.012
Pyrex 1710	6.00	2,500	0.025
Pyrex 3320	4.71	–	0.019
Pyrex 7040	4.65	80	0.013
Pyrex 7050	4.77	16	0.017
Pyrex 7052	5.07	25	0.019
Pyrex 7060	4.70	13	0.018
Pyrex 7070	4.00	1,300	0.0048
Vycor 7230	3.83	–	0.0061
Pyrex 7720	4.50	16	0.014
Pyrex 7740	5.00	4	0.040
Pyrex 7750	4.28	50	0.011
Pyrex 7760	4.50	50	0.0081
Vycor 7900	3.9	130	0.0023
Vycor 7910	3.8	1,600	0.00091
Vycor 7911	3.8	4,000	0.00072
Corning 8870	9.5	5,000	0.0085
G.E. Clear (silica glass)	3.81	4,000–30,000	0.00038
Quartz (fused)	3.75–4.1 (1 MHz)	–	0.0002 1 MHz

[a]Power factor × dielectric constant equals loss factor.

Table 1.10 Dielectric Constants of Solids in the Temperature Range 17–22°C (*From*[1]. *Used with permission.*)

Material	Freq., Hz	Dielectric constant	Material	Freq., Hz	Dielectric Constant
Acetamide	4×10^8	4.0	Phenanthrene	4×10^8	2.80
Acetanilide	–	2.9	Phenol (10°C)	4×10^8	4.3
Acetic acid (2°C)	4×10^8	4.1	Phosphorus, red	10^8	4.1
Aluminum oleate	4×10^8	2.40	Phosphorus, yellow	10^8	3.6
Ammonium bromide	10^8	7.1	Potassium aluminum		
Ammonium chloride	10^8	7.0	sulfate	10^6	3.8
Antimony trichloride	10^8	5.34	Potassium carbonate		
Apatite ⊥ optic axis	3×10^8	9.50	(15°C)	10^8	5.6
Apatite ‖ optic axis	3×10^8	7.41	Potassium chlorate	6×10^7	5.1
Asphalt	$<3 \times 10^6$	2.68	Potassium chloride	10^4	5.03
Barium chloride (anhyd.)	6×10^7	11.4	Potassium chromate	6×10^7	7.3
Barium chloride (2H$_2$O)	6×10^7	9.4	Potassium iodide	6×10^7	5.6
Barium nitrate	6×10^7	5.9	Potassium nitrate	6×10^7	5.0
Barium sulfate(15°C)	10^8	11.4	Potassium sulfate	6×10^7	5.9
Beryl ⊥ optic axis	10^4	7.02	Quartz ⊥ optic axis	3×10^7	4.34
Beryl ‖ optic axis	10^4	6.08	Quartz ‖ optic axis	3×10^7	4.27
Calcite ⊥ optic axis	10^4	8.5	Resorcinol	4×10^8	3.2
Calcite ‖ optic axis	10^4	8.0	Ruby⊥ optic axis	10^4	13.27
Calcium carbonate	10^6	6.14	Ruby ‖ optic axis	10^4	11.28
Calcium fluoride	10^4	7.36	Rutile ⊥ optic axis	10^8	86
Calcium sulfate (2H$_2$O)	10^4	5.66	Rutile ‖ optic axis	10^8	170
Cassiterite ⊥ optic axis	10^{12}	23.4	Selenium	10^8	6.6
Cassiterite ‖ optic axis	10^{12}	24	Silver bromide	10^6	12.2
d-Cocaine	5×10^8	3.10	Silver chloride	10^6	11.2
Cupric oleate	4×10^8	2.80	Silver cyanide	10^6	5.6
Cupric oxide (15°C)	10^8	18.1	Smithsonite ⊥ optic axis	10^{12}	9.3
Cupric sulfate (anhyd.)	6×10^7	10.3			
Cupric sulfate (5H$_2$O)	6×10^7	7.8	Smithsonite ‖ optic axis	10^{10}	9.4
Diamond	10^8	5.5			
Diphenylymethane	4×10^8	2.7	Sodium carbonate (anhvd.)	6×10^7	8.4
Dolomite ⊥ optic axis	10^8	8.0	Sodium carbonate	6×10^7	5.3
Dolomite ‖ optic axis	10^8	6.8	(10H$_2$O)		
Ferrous oxide (15°C)	10^8	14.2	Sodium chloride	10^4	6.12
Iodine	10^8	4	Sodium nitrate	–	5.2
Lead acetate	10^6	2.6	Sodium oleate	4×10^8	2.75
Lead carbonate (15°C)	10^8	18.6	Sodium perchlorate	6×10^7	5.4
Lead chloride	10^6	4.2	Sucrose (mean)	3×10^8	3.32
Lead monoxide (15°C)	10^8	25.9	Sulfur (mean)	–	4.0
Lead nitrate	6×10^7	37.7	Thallium chloride	10^6	46.9
Lead oleate	4×10^8	3.27	p-Toluidine	4×10^8	3.0
Lead sulfate	10^6	14.3	Tourmaline ⊥ optic axis	10^4	7.10
Lead sulfide (15°)	10^6	17.9			
Malachite (mean)	10^{12}	7.2	Tourmaline ‖ optic axis	10^4	6.3
Mercuric chloride	10^6	3.2			
Mercurous chloride	10^6	9.4	Urea	4×10^8	3.5
Naphthalene	4×10^8	2.52	Zircon ⊥, ‖	10^8	12

International Standards and Constants

2.1   Introduction

Standardization usually starts within a company as a way to reduce costs associated with parts stocking, design drawings, training, and retraining of personnel. The next level might be a cooperative agreement between firms making similar equipment to use standardized dimensions, parts, and components. Competition, trade secrets, and the *NIH factor* (not invented here) often generate an atmosphere that prevents such an understanding. Enter the professional engineering society, which promises a forum for discussion between users and engineers while downplaying the commercial and business aspects.

2.2   The History of Modern Standards

In 1836, the U.S. Congress authorized the Office of Weights and Measures (OWM) for the primary purpose of ensuring uniformity in custom house dealings. The Treasury Department was charged with its operation. As advancements in science and technology fueled the industrial revolution, it was apparent that standardization of hardware and test methods was necessary to promote commercial development and to compete successfully with the rest of the world. The industrial revolution in the 1830s introduced the need for interchangeable parts and hardware. Economical manufacture of transportation equipment, tools, weapons, and other machinery was possible only with mechanical standardization.

By the late 1800s professional organizations of mechanical, electrical, chemical, and other engineers were founded with this aim in mind. The Institute of Electrical Engineers developed standards between 1890 and 1910 based on the practices of the major electrical manufacturers of the time. Such activities were not within the purview of the OWM, so there was no government involvement during this period. It took the pressures of war production in 1918 to cause the formation of the American Engineering

Standards Committee (AESC) to coordinate the activities of various industry and engineering societies. This group became the American Standards Association (ASA) in 1928.

Parallel developments would occur worldwide. The International Bureau of Weights and Measures was founded in 1875, the International Electrotechnical Commission (IEC) in 1904, and the International Federation of Standardizing Bodies (ISA) in 1926. Following World War II (1946) this group was reorganized as the International Standards Organization (ISO) comprised of the ASA and the standardizing bodies of 25 other countries. Present participation is approximately 55 countries and 145 technical committees. The stated mission of the ISO is *to facilitate the internationalization and unification of industrial standards*.

The International Telecommunications Union (ITU) was founded in 1865 for the purpose of coordinating and interfacing telegraphic communications worldwide. Today, its member countries develop regulations and voluntary recommendations, and provide coordination of telecommunications development. A sub-group, the International Radio Consultative Committee (CCIR) (which no longer exists under this name), is concerned with certain transmission standards and the compatible use of the frequency spectrum, including geostationary satellite orbit assignments. Standardized transmission formats to allow interchange of communications over national boundaries are the purview of this committee. Because these standards involve international treaties, negotiations are channeled through the U.S. State Department.

2.2.1 American National Standards Institute (ANSI)

ANSI coordinates policies to promote procedures, guidelines, and the consistency of standards development. Due process procedures ensure that participation is open to all persons who are materially affected by the activities without domination by a particular group. Written procedures are available to ensure that consistent methods are used for standards developments and appeals. Today, there are more than 1000 members who support the U.S. voluntary standardization system as members of the ANSI federation. This support keeps the Institute financially sound and the system free of government control.

The functions of ANSI include: (1) serving as a clearinghouse on standards development and supplying standards-related publications and information, and (2) the following business development issues:

- Provides national and international standards information necessary to market products worldwide.

- Offers American National Standards that assist companies in reducing operating and purchasing costs, thereby assuring product quality and safety.

- Offers an opportunity to voice opinion through representation on numerous technical advisory groups, councils, and boards.

- Furnishes national and international recognition of standards for credibility and force in domestic commerce and world trade.

- Provides a path to influence and comment on the development of standards in the international arena.

Prospective standards must be submitted by an ANSI accredited standards developer. There are three methods which may be used:

- **Accredited organization method**. This approach is most often used by associations and societies having an interest in developing standards. Participation is open to all interested parties as well as members of the association or society. The standards developer must fashion its own operating procedures, which must meet the general requirements of the ANSI procedures.

- **Accredited standards committee method**. Standing committees of directly and materially affected interests develop documents and establish consensus in support of the document. This method is most often used when a standard affects a broad range of diverse interests or where multiple associations or societies with similar interests exist. These committees are administered by a *secretariat*, an organization that assumes the responsibility for providing compliance with the pertinent operating procedures. The committee can develop its own operating procedures consistent with ANSI requirements, or it can adopt standard ANSI procedures.

- **Accredited canvass method**. This approach is used by smaller trade associations or societies that have documented current industry practices and desire that these standards be recognized nationally. Generally, these developers are responsible for less than five standards. The developer identifies those who are directly and materially affected by the activity in question and conducts a letter ballot *canvass* of those interests to determine consensus. Developers must use standard ANSI procedures.

Note that all methods must fulfill the basic requirements of public review, voting, consideration, and disposition of all views and objections, and an appeals mechanism.

The introduction of new technologies or changes in the direction of industry groups or engineering societies may require a mediating body to assign responsibility for a developing standard to the proper group. The Joint Committee for Intersociety Coordination (JCIC) operates under ANSI to fulfill this need.

2.2.2 Professional Society Engineering Committees

The engineering groups that collate and coordinate activities that are eventually presented to standardization bodies encourage participation from all concerned parties. Meetings are often scheduled in connection with technical conferences to promote greater participation. Other necessary meetings are usually scheduled in geographical locations of the greatest activity in the field. There are no charges or dues to be a member or to attend the meetings. An interest in these activities can still be served by reading the reports from these groups in the appropriate professional journals. These wheels may seem to grind exceedingly slowly at times, but the adoption of standards that may have to endure for 50 years or more should not be taken lightly.

2.3 References

1. Whitaker, Jerry C. (ed.), *The Electronics Handbook*, CRC Press, Boca Raton, FL, 1996.

2.4 Bibliography

Whitaker, Jerry C., and K. Blair Benson (eds.), *Standard Handbook of Video and Television Engineering*, McGraw-Hill, New York, NY, 2000.

2.5 Tabular Data

Table 2.1 Common Standard Units

Name	Symbol	Quantity
ampere	A	electric current
ampere per meter	A/m	magnetic field strength
ampere per square meter	A/m^2	current density
becquerel	Bg	activity (of a radionuclide)
candela	cd	luminous intensity
coulomb	C	electric charge
coulomb per kilogram	C/kg	exposure (x and gamma rays)
coulomb per sq. meter	C/m^2	electric flux density
cubic meter	m^3	volume
cubic meter per kilogram	m^3/kg	specific volume
degree Celsius	°C	Celsius temperature
farad	F	capacitance
farad per meter	F/m	permittivity
henry	H	inductance
henry per meter	H/m	permeability
hertz	Hz	frequency
joule	J	energy, work, quantity of heat
joule per cubic meter	J/m^3	energy density
joule per kelvin	J/K	heat capacity
joule per kilogram K	J/(kg•K)	specific heat capacity
joule per mole	J/mol	molar energy
kelvin	K	thermodynamic temperature
kilogram	kg	mass
kilogram per cubic meter	kg/m^3	density, mass density
lumen	lm	luminous flux
lux	lx	luminance

Table 2.1 Common Standard Units (continued)

Name	Symbol	Quantity
meter	m	length
meter per second	m/s	speed, velocity
meter per second sq.	m/s^2	acceleration
mole	mol	amount of substance
newton	N	force
newton per meter	N/m	surface tension
ohm	Ω	electrical resistance
pascal	Pa	pressure, stress
pascal second	Pa•s	dynamic viscosity
radian	rad	plane angle
radian per second	rad/s	angular velocity
radian per second squared	rad/s^2	angular acceleration
second	s	time
siemens	S	electrical conductance
square meter	m^2	area
steradian	sr	solid angle
tesla	T	magnetic flux density
volt	V	electrical potential
volt per meter	V/m	electric field strength
watt	W	power, radiant flux
watt per meter kelvin	W/(m•K)	thermal conductivity
watt per square meter	W/m^2	heat (power) flux density
weber	Wb	magnetic flux

Table 2.2 Standard Prefixes

Multiple	Prefix	Symbol
10^{18}	exa	E
10^{15}	peta	P
10^{12}	tera	T
10^{9}	giga	G
10^{6}	mega	M
10^{3}	kilo	k
10^{2}	hecto	h
10	deka	da
10^{-1}	deci	d
10^{-2}	centi	c
10^{-3}	milli	m
10^{-6}	micro	μ
10^{-9}	nano	n
10^{-12}	pico	p
10^{-15}	femto	f
10^{-18}	atto	a

Table 2.3 Common Standard Units for Electrical Work

Unit	Symbol
centimeter	cm
cubic centimeter	cm^3
cubic meter per second	m^3/s
gigahertz	GHz
gram	g
kilohertz	kHz
kilohm	kΩ
kilojoule	kJ
kilometer	km
kilovolt	kV
kilovoltampere	kVA
kilowatt	kW
megahertz	MHz
megavolt	MV
megawatt	MW
megohm	MΩ
microampere	μA
microfarad	μF
microgram	μg
microhenry	μH
microsecond	μs
microwatt	μW
milliampere	mA
milligram	mg
millihenry	mH
millimeter	mm
millisecond	ms
millivolt	mV
milliwatt	mW
nanoampere	nA
nanofarad	nF
nanometer	nm
nanosecond	ns
nanowatt	nW
picoampere	pA
picofarad	pF
picosecond	ps
picowatt	pW

Table 2.4 Names and Symbols for the SI Base Units (*From* [1]. *Used with permission.*)

Physical quantity	Name of SI unit	Symbol for SI unit
length	meter	m
mass	kilogram	kg
time	second	s
electric current	ampere	A
thermodynamic temperature	kelvin	K
amount of substance	mole	mol
luminous intensity	candela	cd

Table 2.5 Units in Use Together with the SI (These units are not part of the SI, but it is recognized that they will continue to be used in appropriate contexts. *From* [1]. *Used with permission.*)

Physical quantity	Name of unit	Symbol for unit	Value in SI units
time	minute	min	60 s
time	hour	h	3600 s
time	day	d	86 400 s
plane angle	degree	°	$(\pi/180)$ rad
plane angle	minute	′	$(\pi/10\ 800)$ rad
plane angle	second	″	$(\pi/648\ 000)$ rad
length	ångstrom[a]	Å	10^{-10} m
area	barn	b	10^{-28} m^2
volume	litre	l, L	$dm^3 = 10^{-3}\,m^3$
mass	tonne	t	$Mg = 10^3\,kg$
pressure	bar[a]	bar	$10^5\,Pa = 10^5\,N\,m^{-2}$
energy	electronvolt[b]	eV $(= e \times V)$	$\approx 1.60218 \times 10^{-19}$ J
mass	unified atomic mass unit[b,c]	u $(= m_a(^{12}C)/12)$	$\approx 1.66054 \times 10^{-27}$ kg

[a]The ångstrom and the bar are approved by CIPM for temporary use with SI units, until CIPM makes a further recommendation. However, they should not be introduced where they are not used at present.

[b]The values of these units in terms of the corresponding SI units are not exact, since they depend on the values of the physical constants e (for the electronvolt) and N_A (for the unified atomic mass unit), which are determined by experiment.

[c]The unified atomic mass unit is also sometimes called the dalton, with symbol Da, although the name and symbol have not been approved by CGPM.

Table 2.6 Derived Units with Special Names and Symbols (*From* [1]. *Used with permission.*)

Physical quantity	Name of SI unit	Symbol for SI unit	Expression in terms of SI base units
frequency[a]	hertz	Hz	s^{-1}
force	newton	N	$m\,kg\,s^{-2}$
pressure, stress	pascal	Pa	$N\,m^{-2} = m^{-1}\,kg\,s^{-2}$
energy, work, heat	joule	J	$N\,m = m^2\,kg\,s^{-2}$
power, radiant flux	watt	W	$J\,s^{-1} = m^2\,kg\,s^{-3}$
electric charge	coulomb	C	$A\,s$
electric potential, electromotive force	volt	V	$J\,C^{-1} = m^2\,kg\,s^{-3}\,A^{-1}$
electric resistance	ohm	Ω	$V\,A^{-1} = m^2\,kg\,s^{-3}\,A^{-2}$
electric conductance	siemens	S	$\Omega^{-1} = m^{-2}\,kg^{-1}\,s^3\,A^2$
electric capacitance	farad	F	$C\,V^{-1} = m^{-2}\,kg^{-1}\,s^4\,A^2$
magnetic flux density	tesla	T	$V\,s\,m^{-2} = kg\,s^{-2}\,A^{-1}$
magnetic flux	weber	Wb	$V\,s = m^2\,kg\,s^{-2}\,A^{-1}$
inductance	henry	H	$V\,A^{-1}\,s = m^2\,kg\,s^{-2}\,A^{-2}$
Celsius temperature[b]	degree Celsius	°C	K
luminous flux	lumen	lm	$cd\,sr$
illuminance	lux	lx	$cd\,sr\,m^{-2}$
activity (radioactive)	becquerel	Bq	s^{-1}
absorbed dose (of radiation)	gray	Gy	$J\,kg^{-1} = m^2\,s^{-2}$
dose equivalent (dose equivalent index)	sievert	Sv	$J\,kg^{-1} = m^2\,s^{-2}$
plane angle	radian	rad	$1 = m\,m^{-1}$
solid angle	steradian	sr	$1 = m^2\,m^{-2}$

[a]For radial (circular) frequency and for angular velocity the unit rad s^{-1}, or simply s^{-1}, should be used, and this may not be simplified to Hz. The unit Hz should be used only for frequency in the sense of cycles per second.

[b]The Celsius temperature θ is defined by the equation:

$$\theta/°C = T/K - 273.15$$

The SI unit of Celsius temperature interval is the degree Celsius, °C, which is equal to the kelvin, K. °C should be treated as a single symbol, with no space between the ° sign and the letter C. (The symbol °K and the symbol ° should no longer be used.)

Table 2.7 The Greek Alphabet (*From* [1]. *Used with permission.*)

Greek letter		Greek name	English equivalent	Greek letter		Greek name	English equivalent
A	α	Alpha	a	N	ν	Nu	n
B	β	Beta	b	Ξ	ξ	Xi	x
Γ	γ	Gamma	g	O	o	Omicron	ŏ
Δ	δ	Delta	d	Π	π	Pi	p
E	ϵ	Epsilon	ĕ	P	ρ	Rho	r
Z	ζ	Zeta	z	Σ	σ ς	Sigma	s
H	η	Eta	ē	T	τ	Tau	t
Θ	θ ϑ	Theta	th	Υ	υ	Upsilon	u
I	ι	Iota	i	Φ	ϕ φ	Phi	ph
K	κ	Kappa	k	X	χ	Chi	ch
Λ	λ	Lambda	l	Ψ	ψ	Psi	ps
M	μ	Mu	m	Ω	ω	Omega	ō

Table 2.8 Constants (*From* [1]. *Used with permission.*)

π Constants

π =	3.14159	26535	89793	23846	26433	83279	50288	41971	69399	37511
$1/\pi$ =	0.31830	98861	83790	67153	77675	26745	02872	40689	19291	48091
π^2 =	9.8690	44010	89358	61883	44909	99876	15113	53136	99407	24079
$\log_e \pi$ =	1.14472	98858	49400	17414	34273	51353	05871	16472	94812	91531
$\log_{10} \pi$ =	0.49714	98726	94133	85435	12682	88290	89887	36516	78324	38044
$\log_{10} \sqrt{2\pi}$ =	0.39908	99341	79057	52478	25035	91507	69595	02099	34102	92128

Constants Involving e

e =	2.71828	18284	59045	23536	02874	71352	66249	77572	47093	69996
$1/e$ =	0.36787	94411	71442	32159	55237	70161	46086	74458	11131	03177
e^2 =	7.38905	60989	30650	22723	04274	60575	00781	31803	15570	55185
$M = \log_{10} e$ =	0.43429	44819	03251	82765	11289	18916	60508	22943	97005	80367
$1/M = \log_e 10$ =	2.30258	50929	94045	68401	79914	54684	36420	76011	01488	62877
$\log_{10} M$ =	9.63778	43113	00536	78912	29674	98645	−10			

Numerical Constants

$\sqrt{2}$ =	1.41421	35623	73095	04880	16887	24209	69807	85696	71875	37695
$\sqrt[3]{2}$ =	1.25992	10498	94873	16476	72106	07278	22835	05702	51464	70151
$\log_e 2$ =	0.69314	71805	59945	30941	72321	21458	17656	80755	00134	36026
$\log_{10} 2$ =	0.30102	99956	63981	19521	37388	94724	49302	67881	89881	46211
$\sqrt{3}$ =	1.73205	08075	68877	29352	74463	41505	87236	69428	05253	81039
$\sqrt[3]{3}$ =	1.44224	95703	07408	38232	16383	10780	10958	83918	69253	49935
$\log_e 3$ =	1.09861	22886	68109	69139	52452	36922	52570	46474	90557	82275
$\log_{10} 3$ =	0.47712	12547	19662	43729	50279	03255	11530	92001	28864	19070

Electromagnetic Spectrum

3.1   Introduction

The usable spectrum of electromagnetic-radiation frequencies extends over a range from below 100 Hz for power distribution to 1020 for the shortest X-rays. The lower frequencies are used primarily for terrestrial broadcasting and communications. The higher frequencies include visible and near-visible infrared and ultraviolet light, and X-rays.

3.1.1   Operating Frequency Bands

The standard frequency band designations are listed in Tables 3.1 and 3.2. Alternate and more detailed subdivision of the VHF, UHF, SHF, and EHF bands are given in Tables 3.3 and 3.4.

Low-End Spectrum Frequencies (1 to 1000 Hz)

Electric power is transmitted by wire but not by radiation at 50 and 60 Hz, and in some limited areas, at 25 Hz. Aircraft use 400-Hz power in order to reduce the weight of iron in generators and transformers. The restricted bandwidth that would be available for communication channels is generally inadequate for voice or data transmission, although some use has been made of communication over power distribution circuits using modulated carrier frequencies.

Low-End Radio Frequencies (1000 to 100 kHz)

These low frequencies are used for very long distance radio-telegraphic communication where extreme reliability is required and where high-power and long antennas can be erected. The primary bands of interest for radio communications are given in Table 3.5.

Table 3.1 Standardized Frequency Bands (*From* [1]. *Used with permission.*)

Extremely low-frequency (ELF) band:	30 Hz up to 300 Hz	(10 Mm down to 1 Mm)
Voice-frequency (VF) band:	300 Hz up to 3 kHz	(1 Mm down to 100 km)
Very low-frequency (VLF) band:	3 kHz up to 30 kHz	(100 km down to 10 km)
Low-frequency (LF) band:	30 kHz up to 300 kHz	(10 km down to 1 km)
Medium-frequency (MF) band:	300 kHz up to 3 MHz	(1 km down to 100 m)
High-frequency (HF) band:	3 MHz up to 30 MHz	(100 m down to 10 m)
Very high-frequency (VHF) band:	30 MHz up to 300 MHz	(10 m down to 1 m)
Ultra high-frequency (UHF) band:	300 MHz up to 3 GHz	(1 m down to 10 cm)
Super high-frequency (SHF) band:	3 GHz up to 30 GHz	(1 cm down to 1 cm)
Extremely high-frequency (EHF) band:	30 GHz up to 300 GHz	(1 cm down to 1 mm)

Table 3.2 Standardized Frequency Bands at 1 GHz and Above (*From* [1]. *Used with permission.*)

L band:	1 GHz up to 2 GHz	(30 cm down to 15 cm)
S band:	2 GHz up to 4 GHz	(15 cm down to 7.5 cm)
C band:	4 GHz up to 8 GHz	(7.5 cm down to 3.75 cm)
X band:	8 GHz up to 12 GHz	(3.75 cm down to 2.5 cm)
Ku band:	12 GHz up to 18 GHz	(2.5 cm down to 1.67 cm)
K band:	18 GHz up to 26.5 GHz	(1.67 cm down to 1.13 cm)
Ka band:	26.5 GHz up to 40 GHz	(1.13 cm down to 7.5 mm)
Q band:	32 GHz up to 50 GHz	(9.38 mm down to 6 mm)
U band:	40 GHz up to 60 GHz	(7.5 mm down to 5 mm)
V band:	50 GHz up to 75 GHz	(6 mm down to 4 mm)
W band:	75 GHz up to 100 GHz	(4 mm down to 3.33 mm)

Medium-Frequency Radio (20 kHz to 2 MHz)

The low-frequency portion of the band is used for around-the-clock communication services over moderately long distances and where adequate power is available to overcome the high level of atmospheric noise. The upper portion is used for AM radio, although the strong and quite variable *sky wave* occurring during the night results in substandard quality and severe fading at times. The greatest use is for AM broadcasting, in addition to fixed and mobile service, LORAN ship and aircraft navigation, and amateur radio communication.

High-Frequency Radio (2 to 30 MHz)

This band provides reliable medium-range coverage during daylight and, when the transmission path is in total darkness, worldwide long-distance service, although the

Table 3.3 Detailed Subdivision of the UHF, SHF, and EHF Bands (*From* [1]. *Used with permission.*)

L band:	1.12 GHz up to 1.7 GHz	(26.8 cm down to 17.6 cm)
LS band:	1.7 GHz up to 2.6 GHz	(17.6 cm down to 11.5 cm)
S band:	2.6 GHz up to 3.95 GHz	(11.5 cm down to 7.59 cm)
C(G) band:	3.95 GHz up to 5.85 GHz	(7.59 cm down to 5.13 cm)
XN(J, XC) band:	5.85 GHz up to 8.2 GHz	(5.13 cm down to 3.66 cm)
XB(H, BL) band:	7.05 GHz up to 10 GHz	(4.26 cm down to 3 cm)
X band:	8.2 GHz up to 12.4 GHz	(3.66 cm down to 2.42 cm)
Ku(P) band:	12.4 GHz up to 18 GHz	(2.42 cm down to 1.67 cm)
K band:	18 GHz up to 26.5 GHz	(1.67 cm down to 1.13 cm)
V(R, Ka) band:	26.5 GHz up to 40 GHz	(1.13 cm down to 7.5 mm)
Q(V) band:	33 GHz up to 50 GHz	(9.09 mm down to 6 mm)
M(W) band:	50 GHz up to 75 GHz	(6 mm down to 4 mm)
E(Y) band:	60 GHz up to 90 GHz	(5 mm down to 3.33 mm)
F(N) band:	90 GHz up to 140 GHz	(3.33 mm down to 2.14 mm)
G(A) band:	140 GHz up to 220 GHz	(2.14 mm down to 1.36 mm)
R band:	220 GHz up to 325 GHz	(1.36 mm down to 0.923 mm)

Table 3.4 Subdivision of the VHF, UHF, SHF Lower Part of the EHF Band (*From* [1]. *Used with permission.*)

A band:	100 MHz up to 250 MHz	(3 m down to 1.2 m)
B band:	250 MHz up to 500 MHz	(1.2 m down to 60 cm)
C band:	500 MHz up to 1 GHz	(60 cm down to 30 cm)
D band:	1 GHz up to 2 GHz	(30 cm down to 15 cm)
E band:	2 GHz up to 3 GHz	(15 cm down to 10 cm)
F band:	3 GHz up to 4 GHz	(10 cm down to 7.5 cm)
G band:	4 GHz up to 6 GHz	(7.5 cm down to 5 cm)
H band:	6 GHz up to 8 GHz	(5 cm down to 3.75 cm)
I band:	8 GHz up to 10 GHz	(3.75 cm down to 3 cm)
J band:	10 GHz up to 20 GHz	(3 cm down to 1.5 cm)
K band:	20 GHz up to 40 GHz	(1.5 cm down to 7.5 mm)
L band:	40 GHz up to 60 GHz	(7.5 mm down to 5 mm)
M band:	60 GHz up to 100 GHz	(5 mm down to 3 mm)

reliability and signal quality of the latter is dependent to a large degree upon ionospheric conditions and related long-term variations in sun-spot activity affecting sky-wave propagation. The primary applications include broadcasting, fixed and mobile services, telemetering, and amateur transmissions.

Table 3.5 Radio Frequency Bands (*From* [1]. *Used with permission.*)

Longwave broadcasting band:	150–290 kHz
AM broadcasting band:	550–1640 kHz (1.640 MHz) (107 Channels, 10-kHz separation)
International broadcasting band:	3–30 MHz
Shortwave broadcasting band:	5.95–26.1 MHz (8 bands)
VHF television (channels 2–4):	54–72 MHz
VHF television (channels 5–6):	76–88 MHz
FM broadcasting band:	88–108 MHz
VHF television (channels 7–13):	174–216 MHz
UHF television (channels 14–83):	470–890 MHz

Very High and Ultrahigh Frequencies (30 MHz to 3 GHz)

VHF and UHF bands, because of the greater channel bandwidth possible, can provide transmission of a large amount of information, either as television detail or data communication. Furthermore, the shorter wavelengths permit the use of highly directional parabolic or multielement antennas. Reliable long-distance communication is provided using high-power *tropospheric scatter* techniques. The multitude of uses include, in addition to television, fixed and mobile communication services, amateur radio, radio astronomy, satellite communication, telemetering, and radar.

Microwaves (3 to 300 GHz)

At these frequencies, many transmission characteristics are similar to those used for shorter optical waves, which limit the distances covered to line of sight. Typical uses include television relay, satellite, radar, and wide-band information services. (See Tables 3.6 and 3.7.)

Infrared, Visible, and Ultraviolet Light

The portion of the spectrum visible to the eye covers the gamut of transmitted colors ranging from red, through yellow, green, cyan, and blue. It is bracketed by infrared on the low-frequency side and ultraviolet (UV) on the high side. Infrared signals are used in a variety of consumer and industrial equipments for remote controls and sensor circuits in security systems. The most common use of UV waves is for excitation of phosphors to produce visible illumination.

X-Rays

Medical and biological examination techniques and industrial and security inspection systems are the best-known applications of X-rays. X-rays in the higher-frequency range are classified as *hard X-rays* or *gamma rays*. Exposure to X-rays for long periods can result in serious irreversible damage to living cells or organisms.

Table 3.6 Applications in the Microwave Bands (*From* [1]. *Used with permission.*)

Aeronavigation:	0.96–1.215 GHz
Global positioning system (GPS) down link:	1.2276 GHz
Military communications (COM)/radar:	1.35–1.40 GHz
Miscellaneous COM/radar:	1.40–1.71 GHz
L-band telemetry:	1.435–1.535 GHz
GPS downlink:	1.57 GHz
Military COM (troposcatter/telemetry):	1.71–1.85 GHz
Commercial COM and private line of sight (LOS):	1.85–2.20 GHz
Microwave ovens:	2.45 GHz
Commercial COM/radar:	2.45–2.69 GHz
Instructional television:	2.50–2.69 GHz
Military radar (airport surveillance):	2.70–2.90 GHz
Maritime navigation radar:	2.90–3.10 GHz
Miscellaneous radars:	2.90–3.70 GHz
Commercial C-band satellite (SAT) COM downlink:	3.70–4.20 GHz
Radar altimeter:	4.20–4.40 GHz
Military COM (troposcatter):	4.40–4.99 GHz
Commercial microwave landing system:	5.00–5.25 GHz
Miscellaneous radars:	5.25–5.925 GHz
C-band weather radar:	5.35–5.47 GHz
Commercial C-band SAT COM uplink:	5.925–6.425 GHz
Commercial COM:	6.425–7.125 GHz
Mobile television links:	6.875–7.125 GHz
Military LOS COM:	7.125–7.25 GHz
Military SAT COM downlink:	7.25–7.75 GHz
Military LOS COM:	7.75–7.9 GHz
Military SAT COM uplink:	7.90–8.40 GHz
Miscellaneous radars:	8.50–10.55 GHz
Precision approach radar:	9.00–9.20 GHz
X-band weather radar (and maritime navigation radar):	9.30–9.50 GHz
Police radar:	10.525 GHz
Commercial mobile COM [LOS and electronic news gathering (ENG)]:	10.55–10.68 GHz
Common carrier LOS COM:	10.70–11.70 GHz
Commercial COM:	10.70–13.25 GHz
Commercial Ku-band SAT COM downlink:	11.70–12.20 GHz
Direct broadcast satellite (DBS) downlink and private LOS COM:	12.20–12.70 GHz
ENG and LOS COM:	12.75–13.25 GHz
Miscellaneous radars and SAT COM:	13.25–14.00 GHz
Commercial Ku-band SAT COM uplink:	14.00–14.50 GHz
Military COM (LOS, mobile, and Tactical):	14.50–15.35 GHz
Aeronavigation:	15.40–15.70 GHz
Miscellaneous radars:	15.70–17.70 GHz
DBS uplink:	17.30–17.80 GHz

Table 3.6 Applications in the Microwave Bands (continued)

Common carrier LOS COM:	17.70–19.70 GHz
Commercial COM (SAT COM and LOS):	17.70–20.20 GHz
Private LOS COM:	18.36–19.04 GHz
Military SAT COM:	20.20–21.20 GHz
Miscellaneous COM:	21.20–24.00 GHz
Police radar:	24.15 GHz
Navigation radar:	24.25–25.25 GHz
Military COM:	25.25–27.50 GHz
Commercial COM:	27.50–30.00 GHz
Military SAT COM:	30.00–31.00 GHz
Commercial COM:	31.00–31.20 GHz
Navigation radar:	31.80–33.40 GHz
Miscellaneous radars:	33.40–36.00 GHz
Military COM:	36.00–38.60 GHz
Commercial COM:	38.60–40.00 GHz

3.2 Radio Wave Propagation

To visualize a radio wave, consider the image of a sine wave being traced across the screen of an oscilloscope [2]. As the image is traced, it sweeps across the screen at a specified rate, constantly changing amplitude and phase with relation to its starting point at the left side of the screen. Consider the left side of the screen to be the antenna, the horizontal axis to be distance instead of time, and the sweep speed to be the speed of light, or at least very close to the speed of light, and the propagation of the radio wave is visualized. To be correct, the traveling, or propagating, radio wave is really a wavefront, as it comprises an electric field component and an orthogonal magnetic field component. The distance between wave crests is defined as the *wavelength* and is calculated by,

$$\lambda = \frac{c}{f} \tag{3.1}$$

where:
λ = wavelength, m
c = the speed of light, approximately 2.998×10^8 m/s
f = frequency, Hz

At any point in space far away from the antenna, on the order of 10 wavelengths or 10 times the aperture of the antenna to avoid *near-field effects*, the electric and magnetic fields will be orthogonal and remain constant in amplitude and phase in relation to any other point in space. The polarization of the radio wave is defined by the polarization of the electric field, horizontal if parallel to the Earth's surface and vertical if perpendicu-

Table 3.7 Satellite Frequency Allocations (*From* [1]. *Used with permission.*)

Band	Uplink	Downlink	Satellite Service
VHF		0.137–0.138	Mobile
VHF	0.3120–0.315	0.387–0.390	Mobile
L-Band		1.492–1.525	Mobile
	1.610–1.6138		Mobile, Radio Astronomy
	1.613.8–1.6265	1.6138–1.6265	Mobile LEO
	1.6265–1.6605	1.525–1.545	Mobile
		1.575	Global Positioning System
		1.227	GPS
S-Band	1.980–2.010	2.170–2.200	MSS. Available Jan. 1, 2000
	(1.980–1.990)		(Available in U.S. in 2005)
	2.110–2.120	2.290–2.300	Deep-space research
		2.4835–2.500	Mobile
C-Band	5.85–7.075	3.4–4.2	Fixed (FSS)
	7.250–7.300	4.5–4.8	FSS
X-Band	7.9–8.4	7.25–7.75	FSS
Ku-Band	12.75–13.25	10.7–12.2	FSS
	14.0–14.8	12.2–12.7	Direct Broadcast (BSS) (U.S.)
Ka-Band		17.3–17.7	FSS (BSS in U.S.)
			22.55–23.55 Intersatellite
			24.45–24.75 Intersatellite
			25.25–27.5 Intersatellite
	27–31	17–21	FSS
Q	42.5–43.5, 47.2–50.2	37.5–40.5	FSS, MSS
	50.4–51.4		Fixed
		40.5–42.5	Broadcast Satellite
V	54.24–58.2		Intersatellite
	59–64		Intersatellite

Sources: Final Acts of the World Administrative Radio Conference (WARC-92), Malaga–Torremolinos, 1992; 1995 World Radiocommunication Conference (WRC-95). Also, see Gagliardi, R.M. 1991. *Satellite Communications,* van Nostrand Reinhold, New York. Note that allocations are not always global and may differ from region to region in all or subsets of the allocated bands.

lar to it. Typically, polarization can be determined by the orientation of the antenna radiating elements.

An *isotropic antenna* is one that radiates equally in all directions. To state this another way, it has a gain of unity.

If this isotropic antenna is located in an absolute vacuum and excited with a given amount of power at some frequency, as time progresses the radiated power must be equally distributed along the surface of an ever expanding sphere surrounding the isotropic antenna. The power density at any given point on the surface of this imaginary sphere is simply the radiated power divided by the surface area of the sphere, that is:

$$P_d = \frac{P_t}{4\pi D^2} \tag{3.2}$$

where:
P_d = power density, W/m^2
D = distance from antenna, m
P_t = radiated power, W

Because power and voltage, in this case power density and electric field strength, are related by impedance, it is possible to determine the electric field strength as a function of distance given that the impedance of free space is taken to be approximately 377 Ω

$$E = \sqrt{Z P_d} = 5.48 \frac{\sqrt{P_t}}{D} \tag{3.3}$$

where E is the electric field strength in volts per meter.
Converting to units of kilowatts for power, the equation becomes

$$E = 173 \frac{\sqrt{P_{t(kW)}}}{D} \text{ V/m} \tag{3.4}$$

which is the form in which the equation is usually seen. Because a half-wave dipole has a gain of 2.15 dB over that of an isotropic radiator (dBi), the equation for the electric field strength from a half-wave dipole is

$$E = 222 \frac{\sqrt{P_{t(kW)}}}{D} \text{ V/m} \tag{3.5}$$

From these equations it is evident that, for a given radiated power, the electric field strength decreases linearly with the distance from the antenna, and power density decreases as the square of the distance from the antenna.

3.2.1 Free Space Path Loss

A typical problem in the design of a radio frequency communications system requires the calculation of the power available at the output terminals of the receive antenna [2]. Although the gain or loss characteristics of the equipment at the receiver and transmitter sites can be ascertained from manufacturer's data, the effective loss between the two antennas must be stated in a way that allows for the characterization of the transmission path between the antennas. The ratio of the power radiated by the transmit antenna to the power available at the receive antenna is known as the *path loss* and is usually expressed in decibels. The minimum loss on any given path occurs between two antennas when there are no intervening obstructions and no ground losses. In such a case when the receive and transmit antennas are *isotropic*, the path loss is known as *free space path loss*.

If the transmission path is between isotropic antennas, then the power received by the receive antenna is the power density at the receive antenna multiplied by the effective area of the antenna and is expressed as

$$P_r = \frac{P_t A}{4\pi D^2}$$

(3.6)

where A is the effective area of the receive antenna in square meters.

The effective area of an isotropic antenna is defined as $\lambda^2/4\pi$. Note that an isotropic antenna is not a point source, but has a defined area; this is often a misunderstood concept. As a result, the received power is

$$P_r = \frac{P_t}{4\pi D^2} \cdot \frac{\lambda^2}{4\pi} = P_t \left(\frac{\lambda}{4\pi D}\right)^2$$

(3.7)

The term $(\lambda^2/4\pi D)^2$ is the free space path loss. Expressed in decibels with appropriate constants included for consistency of units, the resulting equation for free space path loss, written in terms of frequency, becomes

$$L_{fs} = 32.5 + 20\log D + 20\log f$$

(3.8)

where:
D = distance, km
f = frequency, MHz

The equation for the received power along a path with no obstacles and long enough to be free from any near-field antenna effects, such as that in Figure 3.1, then becomes

$$P_r = P_t - L_t + G_t - L_{fs} + G_r - L_r$$

(3.9)

where:
P_r = received power, dB
P_t = transmitted power, dB
L_t = transmission line loss, dB
G_t = gain of transmit antenna referenced to an isotropic antenna, dBi
L_{fs} = free space path loss, dB
G_r = gain of receive antenna, dBi
L_r = line loss of receiver downlead, dB

It should be pointed out that the only frequency-dependent term in the equation for free space path loss occurs in the expression for the power received by an isotropic antenna. This is a function of the antenna area and, as stated previously, the area of an isotropic radiator is defined in terms of wavelength. As a result, the calculated field

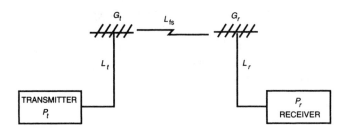

Figure 3.1 Path loss variables. (*From* [2]. *Used with permission.*)

strength at a given distance from sources with equal radiated powers but on frequencies separated by one octave will be identical, but the free space path loss equation will show 6-dB additional loss for the higher frequency path. To view this another way, for the two paths to have the same calculated loss, the antennas for both paths must have equal effective areas. An antenna with a constant area has higher gain at higher frequencies. As a result, to achieve the same total path loss over these two paths, the higher frequency path requires a higher gain antenna, but the required effective areas of the antennas for the two paths are equal. The most important concept to remember is that the resultant field strength and power density at a given distance for a given radiated power are the same regardless of frequency, as long as the path approximates a free space path, but that the free space path loss increases by 6 dB for a doubling of frequency or distance.

The representation of the radio wave path in Figure 3.1 and the previous discussion have only considered a direct path between the receiver and transmitter. In reality, there are two major modes of propagation: the *skywave* and the *groundwave*.

The skywave refers to propagation via the ionosphere, which consists of several layers of ionized particles in the Earth's atmosphere from approximately 50 to several hundred kilometers in altitude. Some frequencies will be reflected by the ionosphere resulting in potentially long-distance propagation.

Groundwave propagation consists of two components, the space wave and the surface wave. The space wave also has two components known as the *direct path* and the *reflected path*. The direct path is the commonly depicted line-of-sight path that has been previously discussed. The reflected path is that path which ends at the receiver by way of reflection from the ground or some other object. Note that there may be multiple reflected paths. The surface wave is that portion of the wavefront that interacts with and travels along the surface of the Earth. The surface wave is commonly incorrectly called the *groundwave*.

3.3 References

1. Whitaker, Jerry C. (ed.), *The Electronics Handbook*, CRC Press, Boca Raton, FL, 1996.

2. Straub, Gerhard J., "Radio Wave Propagation," in *The Electronics Handbook*, Jerry C. Whitaker (ed.), CRC Press, Boca Raton, FL, pp. 131–1332, 1996.

3.4 Bibliography

Whitaker, Jerry C., and K. Blair Benson (eds.), *Standard Handbook of Video and Television Engineering*, McGraw-Hill, New York, NY, 2000.

3.5 Tabular Data

Table 3.8 Power Conversion Factors (decibels to watts)

dBm	dBw	Watts	Multiple Prefix	
+150	+120	1,000,000,000,000	10^{12}	1 Terawatt
+140	+110	100,000,000,000	10^{11}	100 Gigawatts
+130	+100	10,000,000,000	10^{10}	10 Gigawatts
+120	+90	1,000,000,000	10^{9}	1 Gigawatt
+110	+80	100,000,000	10^{8}	100 Megawatts
+100	+70	10,000,000	10^{7}	10 Megawatts
+90	+60	1,000,000	10^{6}	1 Megawatt
+80	+50	100,000	10^{5}	100 Kilowatts
+70	+40	10,000	10^{4}	10 Kilowatts
+60	+30	1,000	10^{3}	1 Kilowatt
+50	+20	100	10^{2}	1 Hectrowatt
+40	+10	10	10	1 Decawatt
+30	0	1	1	1 Watt
+20	−10	0.1	10^{-1}	1 Deciwatt
+10	−20	0.01	10^{-2}	1 Centiwatt
0	−30	0.001	10^{-3}	1 Milliwatt
−10	−40	0.0001	10^{-4}	100 Microwatts
−20	−50	0.00001	10^{-5}	10 Microwatts
−30	−60	0.000,001	10^{-6}	1 Microwatt
−40	−70	0.0,000,001	10^{-7}	100 Nanowatts
−50	−80	0.00,000,001	10^{-8}	10 Nanowatts
−60	−90	0.000,000,001	10^{-9}	1 Nanowatt
−70	−100	0.0,000,000,001	10^{-10}	100 Picowatts
−80	−110	0.00,000,000,001	10^{-11}	10 Picowatts
−90	−120	0.000,000,000,001	10^{-12}	1 Picowatt

Table 3.9 Relationships of Voltage Standing Wave Ratio and Key Operating Parameters

SWR	Reflection Coefficient	Return Loss	Power Ratio	Percent Reflected
1.01:1	0.0050	46.1 dB	0.00002	0.002
1.02:1	0.0099	40.1 dB	0.00010	0.010
1.04:1	0.0196	34.2 dB	0.00038	0.038
1.06:1	0.0291	30.7 dB	0.00085	0.085
1.08:1	0.0385	28.3 dB	0.00148	0.148
1.10:1	0.0476	26.4 dB	0.00227	0.227
1.20:1	0.0909	20.8 dB	0.00826	0.826
1.30:1	0.1304	17.7 dB	0.01701	1.7
1.40:1	0.1667	15.6 dB	0.02778	2.8
1.50:1	0.2000	14.0 dB	0.04000	4.0
1.60:1	0.2308	12.7 dB	0.05325	5.3
1.70:1	0.2593	11.7 dB	0.06722	6.7
1.80:1	0.2857	10.9 dB	0.08163	8.2
1.90:1	0.3103	10.2 dB	0.09631	9.6
2.00:1	0.3333	9.5 dB	0.11111	11.1
2.20:1	0.3750	8.5 dB	0.14063	14.1
2.40:1	0.4118	7.7 dB	0.16955	17.0
2.60:1	0.4444	7.0 dB	0.19753	19.8
2.80:1	0.4737	6.5 dB	0.22438	22.4
3.00:1	0.5000	6.0 dB	0.25000	25.0
3.50:1	0.5556	5.1 dB	0.30864	30.9
4.00:1	0.6000	4.4 dB	0.36000	36.0
4.50:1	0.6364	3.9 dB	0.40496	40.5
5.00:1	0.6667	3.5 dB	0.44444	44.4
6.00:1	0.7143	2.9 dB	0.51020	51.0
7.00:1	0.7500	2.5 dB	0.56250	56.3
8.00:1	0.7778	2.2 dB	0.60494	60.5
9.00:1	0.8000	1.9 dB	0.64000	64.0
10.00:1	0.8182	1.7 dB	0.66942	66.9
15.00:1	0.8750	1.2 dB	0.76563	76.6
20.00:1	0.9048	0.9 dB	0.81859	81.9
30.00:1	0.9355	0.6 dB	0.87513	97.5
40.00:1	0.9512	0.4 dB	0.90482	90.5
50.00:1	0.9608	0.3 dB	0.92311	92.3

Frequency Assignment and Allocations

4.1　Introduction

The Communications Act of 1934, as amended, provides for the regulation of interstate and foreign commerce in communication by wire or radio in the U.S[1]. This Act is printed in Title 47 of the U.S. Code, beginning with Section 151. The primary treaties and other international agreements in force relating to radiocommunication and to which the U.S. is a party are as follows:

- The International Telecommunication Convention, signed at Nairobi on November 6, 1982. The U.S. deposited its instrument of ratification on January 7, 1986.

- The Radio Regulations annexed to the International Telecommunication Convention, signed at Geneva on December 6, 1979 and entered into force with respect to the U.S. on January 1, 1982.

- The United States-Canada Agreement relating to the Coordination and Use of Radio Frequencies above 30 MHz, effected by an exchange of notes at Ottawa on October 24, 1962. A revision to the Technical Annex to the Agreement, made in October 1964 at Washington, was effected by an exchange of notes signed by the U.S. on June 16, 1965, and by Canada on June 24, 1965. The revision entered into force on June 24, 1965. A revision to this Agreement to add *Arrangement E* (Arrangement between the Department of Communications of Canada and the National Telecommunications and Information Administration and the Federal Communications Commission of the U.S. concerning the use of the 406.1 to 430

1　This chapter is based on: *Manual of Regulations and Procedures for Federal Radio Frequency Management*, September 1995 edition, revisions for September 1996, January and May 1997, NTIA, Washington, D.C.

MHz band in Canada-U.S. border areas) was effected by an exchange of notes signed by the U.S. on February 26, 1982, and Canada on April 7, 1982.

4.1.1 The International Telecommunication Union (ITU)

The International Telecommunication Union (IT) is the international body responsible for international frequency allocations, worldwide telecommunications standards, and telecommunication development activities. At this writing, 185 countries were members of the ITU. The broad functions of the ITU are the regulation, coordination, and development of international telecommunications. The U.S. is an active member of the ITU and its work is considered critical to the interest of the United States.

The ITU is the oldest of the intergovernmental organizations that have become specialized agencies within the United Nations. The ITU was born with the spread of one of the great inventions of the 19th century, the telegraph, which crossed national frontiers to link major cities in Europe. International action was essential to establish an international telegraph network. It was necessary to reach agreement on the technical systems to be used, on uniform methods of handling messages, and on the collection of charges. A procedure of international accounting had to be set up.

First came bilateral understanding between bordering countries, then international agreement between regional groups of countries, ending in an inter-European association. Extra-European countries were progressively drawn in, and a truly international organization came into being. In 1865 the International Telegraph Union was created in Paris by the first International Telegraph Convention. The member countries agreed to a set of basic telegraph service regulations. These were modified later as a result of practical operating experience. At Vienna, in 1868, a permanent international bureau was created and established in Berne.

The international telephone service came much later and its progress was much slower. It was not until 1927, when radio provided the means to carry the human voice across the ocean from continent to continent, that this service became world-wide; nevertheless, in 1885, in Berlin, the first provisions concerning the international telephone service were drawn up.

When, at the end of the 19th century, wireless (radiotelegraphy) became practicable, it was seen at once to be an invaluable complement of telegraphy by wire and cable, since radio alone could provide telecommunication between land and ships at sea. The first International Radiotelegraph Convention was signed in Berlin in 1906 by twenty-nine countries. Nearly two decades later, in 1924 and 1925, at Conferences in Paris, the International Telephone Consultative Committee (CCIF) and the International Telegraph Consultative Committee (CCIT) were established. This was followed by the 1927 International Radiotelegraph Conference in Washington, D.C. in 1927, which was attended by 80 countries. It was a historical milestone in the development of radio because it was at this Conference that the Table of Frequency Allocations was first devised and the International Radio Consultative Committee (CCIR) was formed.

In 1932, two Plenipotentiary Conferences were held in Madrid: a Telegraph and Telephone Conference and a Radiotelegraph Conference. On that occasion, the two existing Conventions were amalgamated in a single International Telecommunication

Convention, and the countries that signed and acceded to it renamed the Union the International Telecommunication Union (ITU) to indicate its broader scope. Four sets of Regulations were annexed to the Convention: telegraph, telephone, radio, and additional radio regulations.

A Plenipotentiary Conference met in Atlantic City, N.J., in 1947 to revise the Madrid Convention. It introduced important changes in the organization of the Union. The International Frequency Registration Board (IFRB) and the Administrative Council were created. Also, the ITU became the specialized agency within the United Nations in the sphere of telecommunications, and its headquarters was transferred from Berne to Geneva.

The Union remained essentially unchanged until 1992, when an Additional Plenipotentiary Conference in Geneva extensively restructured the ITU. The Nice Constitution and Convention of 1989, which had not been ratified, was used as the general model for the 1992 Conference. The CCIR, IFRB, and World Administrative Radio Conference (WARC) functions were incorporated into the Radiocommunication Sector (ITU-R); the CCITT and Telecommunication Conference functions were incorporated into the Telecommunication Standardization Sector (ITU-T); development activities were incorporated into the Telecommunication Development Sector (ITU-D); and the Secretariats were combined into one General Secretariat.

Purposes of the Union

The purposes of the Union are as follows:

- To promote the development and efficient operation of telecommunication facilities, in order to improve the efficiency of telecommunication services, their usefulness, and their general availability to the public

- Promote and offer technical assistance to developing countries in the field of telecommunications, to promote the mobilization of the human and financial resources needed to develop telecommunications, and to promote the extension of the benefits of new telecommunications technologies to people everywhere

- Promote, at the international level, the adoption of a broader approach to the issues of telecommunications in the global information economy and society

While the principal facilities of the ITU are in Geneva adjacent to the grounds of the United Nations, the Union also has a number of regional and sub-regional offices.

Structure of the Union

The ITU Constitution states that the Union shall comprise:

- The Plenipotentiary Conference, which is the supreme authority of the Union

- The Council, which acts on behalf of the Plenipotentiary Conference

- World conferences on international telecommunications

- The Radiocommunication Sector, including world and regional radiocommunication conferences, radiocommunication assemblies, and the Radio Regulations Board
- The Telecommunication Standardization Sector, including world telecommunication standardization conferences
- The Telecommunication Development Sector, including world and regional telecommunication development conferences
- The General Secretariat

4.1.2 The Federal Communications Commission (FCC)

Congress, through adoption of the Communications Act of 1934, created the Federal Communications Commission (FCC) as an independent regulatory agency. Section I of the Act specifies that the FCC was created, "For the purpose of regulation of interstate and foreign commerce in communication by wire and radio so as to make available, so far as possible, to all the people of the United States a rapid, efficient, nationwide, and worldwide wire and radio communication service with adequate facilities at reasonable charges, for the purpose of the national defense, for the purpose of promoting the safety of life and property through the use of wire and radio communication, and for the purpose of securing a more effective execution of this policy by centralizing authority heretofore granted by law to several agencies and by granting additional authority with respect to interstate and foreign commerce in wire and radio communication."

The FCC is directed by five Commissioners appointed by the President, by and with the advice and consent of the Senate, for staggered five-year terms. No more than three can be members of the same political party. The President designates one Commissioner as Chairman. The Commissioners make their decisions collectively by formal vote although authority to act on routine matters is normally delegated to the staff.

The staff of the FCC performs day-to-day functions of the agency, including license and application processing, drafting of rulemaking items, enforcing rules and regulations, and formulating policy.

The Commission reorganized itself in 1995 to establish two new bureaus—Wireless Telecommunications and International—to reflect the changes in the industries it regulates. The staff is divided along functional lines into six operating bureaus and 10 remote offices.

4.2 National Table of Frequency Allocations

The National Table of Frequency Allocations is comprised of the U.S. Government Table of Frequency Allocations and the FCC Table of Frequency Allocations. The National Table indicates the normal national frequency allocation planning and the degree of conformity with the ITU table. When required in the national interest and consistent with national rights, as well as obligations undertaken by the U.S. to other countries that may be affected, additional uses of frequencies in any band may be authorized to meet service needs other than those provided for in the National Table.

Specific exceptions to the National Table of Frequency Allocations are as follows:

- A government frequency assignment may be authorized in a non-government band, as an exception, provided: *a*) the assignment is coordinated with the FCC, and *b*) no harmful interference will be caused to the service rendered by non-government stations, present or future.

- A non-government frequency assignment may be authorized in a government band, as an exception, provided: *a*) the assignment is coordinated with the IRAC, and *b*) no harmful interference will be caused to the service rendered by government stations, present or future.

In the case of bands shared by government and non-government services, frequency assignments therein are subject to coordination between the IRAC and the FCC, and no priority is recognized unless the terms of such priority are specifically defined in the National Table of Frequency Allocations or unless they are subject to mutually agreed arrangements in specific cases.

4.2.1   U.S. Government Table of Frequency Allocations

The U.S. Government Table of Frequency Allocations is used as a guide in the assignment of radio frequencies to government radio stations in the United States and Possessions. Exceptions to the table may be made by the IRAC after careful consideration to avoid harmful interference and to ensure compliance with the ITU radio regulations.

For the use of frequencies by government radio stations outside the U.S., government agencies are guided insofar as practicable by the ITU Table of Frequency Allocations and, where applicable, by the authority of the host government. Maximum practicable effort should be made to avoid the possibility of harmful interference to other authorized U.S. operations. If harmful interference is considered likely, it is incumbent upon the agency conducting the operation to coordinate with other U.S. users.

Application of the U.S. Government Table is subject to the recognition that:

- Below 25000 kHz the table is only applicable in the assignment of frequencies after September 5, 1961.

- Under Article 38 of the International Telecommunication Convention, administrations "retain their entire freedom with regard to military radio installations of their army, naval and air forces."

- Under No. 342 of the ITU Radio Regulations, administrations may assign frequencies in derogation of the ITU Table of Frequency Allocations "on the express condition that harmful interference shall not be caused to services carried on by stations operating in accordance with the provisions of the Convention and of these Regulations."

Some frequency assignments below 25000 kHz that were made before September 5, 1961, are not in conformity with the government table. Because of the exception mentioned previously, the status of these assignments can be determined only on a case-by-case basis. With this exception, the rules pertaining to the relative status between radio services are as follows:

- Primary and permitted services have equal rights, except that, in the preparation of frequency plans, the primary service, as compared with the permitted service, has prior choice of frequencies.

- Secondary services are on a non-interference basis to the primary and permitted services. Stations of a secondary service: (*a*) must not cause harmful interference to stations of primary or permitted services to which frequencies are already assigned or to which frequencies may be assigned at a later date; (*b*) cannot claim protection from harmful interference from stations of a primary or permitted service to which frequencies are already assigned or may be assigned at a later date; (*c*) can claim protection, however, from harmful interference from stations of the same or other secondary service(s) to which frequencies may be assigned at a later date.

Important definitions for terms used in the table include the following:

- *Additional allocation*, where a band is indicated in a footnote of the table as "also allocated" to a service in an area smaller than a *region*, or in a particular country. For example, an allocation that is added in this area or in this country to the service or services which are indicated in the table.

- *Alternative allocation*, where a band is indicated in a footnote of the table as "allocated" to one or more services in an area smaller than a region, or in a particular country. For example, an allocation that replaces, in this area or in this country, the allocation indicated in the table.

- *Different category of service*, where the allocation category (primary, permitted, or secondary) of the service in the table is changed. For example, the table reflects the allocation as Fixed, Mobile, and RADIOLOCATION; the category of these services are changed by the footnote to FIXED, MOBILE, and Radiolocation.

- An *allocation* or a footnote to the government table denoting relative status between radio services automatically applies to each assignment in the band to which the footnote or allocation pertains, unless at the time of a particular frequency assignment action a different provision is decided upon for the assignment concerned.

- A *priority note* reflecting the same provisions as an allocation or an applicable footnote to the government table is redundant and is not applied to frequency assignments.

An assignment that is in conformity with the service allocation (as amplified by pertinent footnotes) for the band in which it is contained takes precedence over assignments therein that are not in conformity unless, at the time of the frequency assignment action, a different provision is decided upon.

Where in the table a band is indicated as allocated to more than one service, such services are listed in the following order:

- *Primary services*, the names of which are printed in all capital letters (example: FIXED)

- *Permitted services*, the names of which are printed in "capitals between oblique strokes" (example: /RADIOLOCATION/)

- *Secondary services*, the names of which are printed in "normal characters" (example: Mobile)

Other details of the table include the following:

- The columns to the right of the double line show the national provisions; those to the left show the provisions of the ITU Table of Frequency Allocations.

- Column 1 indicates the national band limits.

- Column 2 indicates the government allocation, including all "US" and "G" footnotes considered to be applicable to the government nationally. Where the allocated service is followed by a function in parentheses, e.g., SPACE (space-to-Earth), the allocation is limited to the function shown.

- Column 3 indicates the non-government allocation including all "US" footnotes, and certain "NG" footnotes as contained in Part 2 of the FCC Rules and Regulations. Where the allocated service is followed by a function in parentheses, e.g., SPACE (space-to-Earth), the allocation is limited to the function shown. These data have been included in the Government Table for information purposes only.

- Column 4 contains such remarks as serve to amplify the government and non-government allocations or point up understanding between the FCC and IRAC/NTIA in respect thereof.

- The international footnotes shown in the columns to the left of the double line are applicable only in the relationships between the U.S. and other countries. An international footnote is applicable to the U.S. Table of Allocations if the number also appears in Columns 2 and 3 of the U.S. table. The international footnote is then applicable to both government and non-government use.

The texts of footnotes in the table are listed in numerical order at the end of the table, in sections headed Government Footnotes, U.S. Footnotes, International Footnotes, and NG Footnotes. Because of space limitations, the footnotes are not included in this chapter. The complete set of footnotes is available from the National Telecommunications and Information Administration, Washington, D.C. (www.ntia.doc.gov).

The U.S. Government Table of Frequency Allocations is given on the following pages.

TABLES OF FREQUENCY ALLOCATIONS

INTERNATIONAL			Band kHz	UNITED STATES		Remarks
Region 1 kHz	Region 2 kHz	Region 3 kHz		Government Allocation	Non-Govt. Allocation	
Below 9 (Not Allocated) 444 445			Below 9	(Not Allocated) 444 445	(Not Allocated) 444 445	
9-14 RADIONAVIGATION			9-14	RADIONAVIGATION US18 US294	RADIONAVIGATION US18 US294	
14-19.95 FIXED MARITIME MOBILE 448 446 447	MARITIME MOBILE 448		14-19.95	FIXED MARITIME MOBILE US294 448	Fixed US294 448	
19.95-20.05 STANDARD FREQUENCY AND TIME SIGNAL (20 kHz)		STANDARD FREQUENCY AND TIME SIGNAL (20 kHz)	19.95-20.05	STANDARD FREQUENCY AND TIME SIGNAL (20 kHz) US294	STANDARD FREQUENCY AND TIME SIGNAL (20 kHz) US294	FCC Rules and Regulations make no provisions for the licensing of standard frequency stations.
20.05-70 FIXED MARITIME MOBILE 448 447 449	MARITIME MOBILE 448		20.05-59	FIXED MARITIME MOBILE US294 448	FIXED US294 448	
			59-61	STANDARD FREQUENCY AND TIME SIGNAL (60 kHz) US294	STANDARD FREQUENCY AND TIME SIGNAL (60 kHz) US294	FCC Rules and Regulations make no provisions for the licensing of standard frequency stations.
			61-70	FIXED MARITIME MOBILE US294 448	FIXED US294 448	
70-72 RADIONAVIGATION 451	70-90 FIXED MARITIME MOBILE 448 MARITIME RADIONAVIGATION 451 Radiolocation 452	70-72 RADIONAVIGATION 451 Fixed Maritime Mobile 448 / 450	70-90	FIXED MARITIME MOBILE Radiolocation US294 448 451	FIXED Radiolocation US294 448 451	

TABLES OF FREQUENCY ALLOCATIONS

INTERNATIONAL			Band kHz	UNITED STATES		Remarks
Region 1 kHz	Region 2 kHz	Region 3 kHz		Government Allocation	Non-Govt. Allocation	
72-84 FIXED MARITIME MOBILE 448 RADIONAVIGATION 451 447		72-84 FIXED MARITIME MOBILE 448 RADIONAVIGATION 451				
84-86 RADIONAVIGATION 451 454		84-86 RADIONAVIGATION 451 Fixed Maritime Mobile 448 450				
86-90 FIXED MARITIME MOBILE 448 RADIONAVIGATION 451 447		86-90 FIXED MARITIME MOBILE 448 RADIONAVIGATION 451				
90-110 RADIONAVIGATION 453 Fixed 453A 454	RADIONAVIGATION 453 453A 454		90-110	RADIONAVIGATION US18 US104 US294 453	RADIONAVIGATION US18 US104 US294 453	
110-112 FIXED MARITIME MOBILE RADIONAVIGATION 451 454	110-130 FIXED MARITIME MOBILE MARITIME RADIONAVIGATION 451 Radiolocation 452 454	110-112 FIXED MARITIME MOBILE RADIONAVIGATION 451	110-130	FIXED MARITIME MOBILE Radiolocation US294 451 454	FIXED MARITIME MOBILE Radiolocation US294 451 454	
112-115 RADIONAVIGATION 451		112-117.6 RADIONAVIGATION 451 Fixed Maritime Mobile 454 455				

TABLES OF FREQUENCY ALLOCATIONS

INTERNATIONAL				UNITED STATES		
Region 1 kHz	Region 2 kHz	Region 3 kHz	Band kHz	Government Allocation	Non-Govt. Allocation	Remarks
115-117.6 RADIONAVIGATION 451 Fixed Maritime Mobile 454 456						
117.6-126 FIXED MARITIME MOBILE RADIONAVIGATION 451 454		117.6-126 FIXED MARITIME MOBILE RADIONAVIGATION 451 454				
126-129 RADIONAVIGATION 451		126-129 RADIONAVIGATION 451 Fixed Maritime Mobile 454 455				
129-130 FIXED MARITIME MOBILE RADIONAVIGATION 451 454		129-130 FIXED MARITIME MOBILE RADIONAVIGATION 451 454				
130-148.5 MARITIME MOBILE /FIXED/ 454 457	130-160 FIXED MARITIME MOBILE 454	130-160 FIXED MARITIME MOBILE RADIONAVIGATION 454	130-160	FIXED MARITIME MOBILE US294 454	FIXED MARITIME MOBILE US294 454	
148.5-255 BROADCASTING 460 461 462	160-190 FIXED 459	160-190 FIXED Aeronautical Radionavigation	160-190	FIXED MARITIME MOBILE US294 459	FIXED US294 459	
	190-200 AERONAUTICAL RADIONAVIGATION	190-200 AERONAUTICAL RADIONAVIGATION	190-200	AERONAUTICAL RADIONAVIGATION US18 US226 US294	AERONAUTICAL RADIONAVIGATION US18 US226 US294	

TABLES OF FREQUENCY ALLOCATIONS

INTERNATIONAL			UNITED STATES			
Region 1 kHz	Region 2 kHz	Region 3 kHz	Band kHz	Government Allocation	Non-Govt. Allocation	Remarks
255-283.5 BROADCASTING /AERONAUTICAL RADIONAVIGATION- /463 462 464	200-275 AERONAUTICAL RADIONAVIGATION Aeronautical Mobile	200-285 AERONAUTICAL RADIONAVIGATION Aeronautical Mobile	200-275	AERONAUTICAL RADIONAVIGATION Aeronautical Mobile US18 US294	AERONAUTICAL RADIONAVIGATION Aeronautical Mobile US18 US294	
	275-285 AERONAUTICAL RADIONAVIGATION Aeronautical Mobile Maritime Radionavigation (radiobeacons)		275-285	AERONAUTICAL RADIONAVIGATION Aeronautical Mobile Maritime Radionavigation (radiobeacons) US18 US294	AERONAUTICAL RADIONAVIGATION Aeronautical Mobile Maritime Radionavigation (radiobeacons) US18 US294	
283.5-315 MARITIME RADIONAVIGATION (radiobeacons) 466 /AERONAUTICAL RADIONAVIGATION/ 465 466A	285-315 MARITIME RADIONAVIGATION/ (radiobeacons) 466 /AERONAUTICAL RADIONAVIGATION/	285-315 MARITIME RADIONAVIGATION (radiobeacons) 466 /AERONAUTICAL RADIONAVIGATION/	285-325	MARITIME RADIONAVIGATION (radiobeacons) Aeronautical Radionavigation (Radiobeacons) US18 US294 G121 466	MARITIME RADIONAVIGATION (radiobeacons) Aeronautical Radionavigation (Radiobeacons) US18 US294 466	
315-325 AERONAUTICAL RADIONAVIGATION Maritime Radionavigation (radiobeacons) 466 465 467	315-325 MARITIME RADIONAVIGATION 466 Aeronautical Radionavigation	315-325 AERONAUTICAL RADIONAVIGATION MARITIME RADIONAVIGATION (radiobeacons) 466				
325-405 AERONAUTICAL RADIONAVIGATION 465	325-335 AERONAUTICAL RADIONAVIGATION Aeronautical Mobile Maritime Radionavigation (radiobeacons)	325-405 AERONAUTICAL RADIONAVIGATION Aeronautical Mobile	325-335	AERONAUTICAL RADIONAVIGATION (radiobeacons) Aeronautical Mobile Maritime Radionavigation (radiobeacons) US18 US294	AERONAUTICAL RADIONAVIGATION (radiobeacons) Aeronautical Mobile Maritime Radionavigation (radiobeacons) US18 US294	
335-405 AERONAUTICAL RADIONAVIGATION Aeronautical Mobile	335-405 AERONAUTICAL RADIONAVIGATION Aeronautical Mobile		335-405	AERONAUTICAL RADIONAVIGATION (radiobeacons) Aeronautical Mobile US18 US294	AERONAUTICAL RADIONAVIGATION (radiobeacons) Aeronautical Mobile US18 US294	

TABLES OF FREQUENCY ALLOCATIONS

INTERNATIONAL			Band	UNITED STATES		Remarks
Region 1 kHz	Region 2 kHz	Region 3 kHz	kHz	Government Allocation	Non-Govt. Allocation	
405-415 RADIONAVIGATION 468 465	405-415 RADIONAVIGATION 468 Aeronautical Mobile		405-415	RADIONAVIGATION Aeronautical Mobile US18 US294 468	RADIONAVIGATION Aeronautical Mobile US18 US294 468	
415-435 AERONAUTICAL RADIONAVIGATION /MARITIME MOBILE- /470 465	415-495 MARITIME MOBILE 470 Aeronautical Radionavigation 469 469A 471 472A	AERONAUTICAL Radionavigation 470A	415-435	AERONAUTICAL RADIONAVIGATION MARITIME MOBILE US294 469A 470	AERONAUTICAL RADIONAVIGATION MARITIME MOBILE US294 469A 470	
435-495 MARITIME MOBILE 470 Aeronautical Radionavigation 465 471 472A			435-495	MARITIME MOBILE Aeronautical Radionavigation US231 US294 470 471 472A	MARITIME MOBILE US231 US294 470 471 472A	The frequency 480 kHz is available to low power Government Coast stations for the calibration of ship direction finders on the condition that harmful interference is not caused to the maritime mobile service.
495-505 MOBILE (distress and calling) 472	495-505 MOBILE (distress and calling)		495-505	MOBILE (distress and calling) 472	MOBILE (distress and calling) 472	500 kHz distress and calling
505-526.5 MARITIME MOBILE 470 /AERONAUTICAL RADIONAVIGATION/ 465 471 474 476	505-510 MARITIME MOBILE 470 471	505-526.5 MARITIME MOBILE 470 /AERONAUTICAL RADIONAVIGATION/ 474	505-510	MARITIME MOBILE 470 471	MARITIME MOBILE 470 471	
	510-525 MOBILE 474 AERONAUTICAL RADIONAVIGATION 471	AERONAUTICAL Mobile Land Mobile 471	510-525	AERONAUTICAL RADIONAVIGATION (radiobeacons) MARITIME MOBILE (Ships Only) US14 US18 US225 474	AERONAUTICAL RADIONAVIGATION (radiobeacons) MARITIME MOBILE (Ships Only) US14 US18 US225 474	518 kHz is used for international NAVTEX in the maritime mobile service.

TABLES OF FREQUENCY ALLOCATIONS

INTERNATIONAL			Band kHz	UNITED STATES		Remarks
Region 1 kHz	Region 2 kHz	Region 3 kHz		Government Allocation	Non-Govt. Allocation	
	525-535 BROADCASTING 477 AERONAUTICAL RADIONAVIGATION		525-535	AERONAUTICAL RADIONAVIGATION (radiobeacons) MOBILE US18 US221 US239	AERONAUTICAL RADIONAVIGATION (radiobeacons) MOBILE US18 US221 US239	530 kHz Travelers Information Service
526.5-1606.5 BROADCASTING 478		526.5-535 BROADCASTING Mobile 479				
	535-1605 BROADCASTING	535-1606.5 BROADCASTING	535-1605		BROADCASTING NG128	
	1605-1625 BROADCASTING 480		1605-1615	MOBILE US221 480 G127	MOBILE US221 480	1610 kHz Travelers Information Systems
1606.5-1625 MARITIME MOBILE 480A /FIXED/ /LAND MOBILE/ 483 484	480A	1606.5-1800 FIXED MOBILE RADIOLOCATION RADIONAVIGATION 482	1615-1625	US237 US299 480	BROADCASTING US237 US299 480	Broadcasting implementation is subject to decisions of a future Region 2 Administrative Radio Conference.
	1625-1705 BROADCASTING 480 /FIXED/ /MOBILE/ Radiolocation 480A		1625-1705	Radiolocation US238 US299 480	BROADCASTING Radiolocation US238 US299 480	
1635-1800 MARITIME MOBILE 480A /FIXED/ /LAND MOBILE/ 483 484 488	1705-1800 FIXED MOBILE RADIOLOCATION AERONAUTICAL RADIONAVIGATION		1705-1800	FIXED MOBILE RADIOLOCATION US240	FIXED MOBILE RADIOLOCATION US240	

TABLES OF FREQUENCY ALLOCATIONS

INTERNATIONAL			Band kHz	UNITED STATES		Remarks
Region 1 kHz	Region 2 kHz	Region 3 kHz		Government Allocation	Non-Govt. Allocation	
1800-1810 RADIOLOCATION 487 485 486	1800-1850 AMATEUR	1800-2000 AMATEUR FIXED MOBILE except aero-nautical mobile RADIONAVIGATION Radiolocation 489	1800-1900		AMATEUR	
1810-1850 AMATEUR 490 491 492 493						
1850-2000 FIXED MOBILE except aero-nautical mobile RADIOLOCATION RADIONAVIGATION 484 488 495	1850-2000 AMATEUR FIXED MOBILE except aero-nautical mobile RADIOLOCATION RADIONAVIGATION 494		1900-2000	RADIOLOCATION US290	RADIOLOCATION US290	
2000-2025 FIXED MOBILE except aero-nautical mobile (R) 484 495	2000-2065 FIXED MOBILE		2000-2065	FIXED MOBILE	MARITIME MOBILE NG19	2003 kHz, intership frequency on the Great Lakes.
2025-2045 FIXED MOBILE except aero-nautical mobile (R) Meteorological Aids 496 484 495						
2045-2160 MARITIME MOBILE /FIXED/ /LAND MOBILE/ 483 484	2065-2107 MARITIME MOBILE 497 498		2065-2107 2065-2068.5	MARITIME MOBILE Ship and coast (telephony) 497	MARITIME MOBILE Ship and coast (telephony) 497	

TABLES OF FREQUENCY ALLOCATIONS

INTERNATIONAL			Band kHz	UNITED STATES		Remarks
Region 1 kHz	Region 2 kHz	Region 3 kHz		Government Allocation	Non-Govt. Allocation	
			2068.5-2078.5	Ship (Wide-band telegraphy, facsimile and space transmission systems) US296	Ship (Wide-band telegraphy, facsimile and space transmission systems) US296	
			2078.5-2089.5	Ship and coast (telephony) 497	Ship and coast (telephony) 497	
			2089.5-2092.5	Ship (Calling, telegraphy)	Ship (Calling, telegraphy)	
			2092.5-2107	Ship and coast (telephony) 497	Ship and coast (telephony) 497	
2107-2170 FIXED MOBILE	2107-2170 FIXED MOBILE		2107-2170	FIXED MOBILE	FIXED LAND MOBILE MARITIME MOBILE NG19	
2160-2170 RADIOLOCATION 487						
485 486 499						
2170-2173.5 MARITIME MOBILE			2170-2173.5	MARITIME MOBILE (Telephony)	MARITIME MOBILE (Telephony)	
2173.5-2190.5 MOBILE (distress and calling)			2173.5-2190.5	MOBILE (distress and calling) US279 500 500A 500B 501	MOBILE (distress and calling) US279 500 500A 500B 501	2182 kHz Distress and Calling
500 500A 500B 501						
2190.5-2194 MARITIME MOBILE			2190.5-2194	MARITIME MOBILE (Telephony)	MARITIME MOBILE (Telephony)	
2194-2300 FIXED MOBILE except aero-nautical mobile (R)	2194-2300 FIXED MOBILE		2194-2495	FIXED MOBILE	FIXED LAND MOBILE MARITIME MOBILE NG19	
484 495 502	502					

TABLES OF FREQUENCY ALLOCATIONS

INTERNATIONAL			Band kHz	UNITED STATES		Remarks
Region 1 kHz	Region 2 kHz	Region 3 kHz		Government Allocation	Non-Govt. Allocation	
2300-2498 FIXED MOBILE except aeronautical mobile (R) BROADCASTING 503 495 484 495 504	2300-2495 FIXED MOBILE BROADCASTING 503					
2498-2501 STANDARD FREQUEN-CY AND TIME SIGNAL (2500 kHz)	2495-2501 STANDARD FREQUENCY AND TIME SIGNAL (2500 kHz) 495		2495-2505	STANDARD FREQUENCY AND TIME SIGNAL (2500 kHz) G106	STANDARD FREQUENCY AND TIME SIGNAL (2500 kHz)	FCC Rules and Regulations make no provisions for licensing of standard frequency stations.
2501-2502 STANDARD FREQUENCY AND TIME SIGNAL Space Research	2502-2505 STANDARD FREQUENCY AND TIME					
2502-2625 FIXED MOBILE except aeronautical mobile (R) 484 495 504	2505-2850 FIXED MOBILE		2505-2850	FIXED MOBILE US285	FIXED LAND MOBILE MARITIME MOBILE US285	2635 kHz and 2638 kHz intership frequencies 2738 kHz intership frequency except in Gulf of Mexico 2830 kHz intership frequency in Gulf of Mexico
2625-2650 MARITIME MOBILE MARITIME RADIONAVIGATION 484						
2650-2850 FIXED MOBILE except aeronautical mobile (R) 484 495						
2850-3025 AERONAUTICAL MOBILE (R) 501 505			2850-3025	AERONAUTICAL MOBILE (R) US283 501 505	AERONAUTICAL MOBILE (R) US283 501 505	

TABLES OF FREQUENCY ALLOCATIONS

INTERNATIONAL				UNITED STATES		
Region 1 kHz	Region 2 kHz	Region 3 kHz	Band kHz	Government Allocation	Non-Govt. Allocation	Remarks
3025-3155 AERONAUTICAL MOBILE (OR)			3025-3155	AERONAUTICAL MOBILE (OR)	AERONAUTICAL MOBILE (OR)	Operation in the (OR) bands by Non-Government stations shall be authorized only by special arrangements between the FCC and the IRAC.
3155-3200 FIXED MOBILE except aeronautical mobile (R) 506 507			3155-3230	FIXED MOBILE except aeronautical mobile (R)	FIXED MOBILE except aeronautical mobile (R)	
3200-3230 FIXED MOBILE except aeronautical mobile (R) BROADCASTING 503 506						
3230-3400 FIXED MOBILE except aeronautical mobile BROADCASTING 503 506 508			3230-3400	FIXED MOBILE except aeronautical mobile Radiolocation	FIXED MOBILE except aeronautical mobile Radiolocation	
3400-3500 AERONAUTICAL MOBILE (R)			3400-3500	AERONAUTICAL MOBILE (R) US283	AERONAUTICAL MOBILE (R) US283	
3500-3800 AMATEUR 510 FIXED MOBILE except aeronautical mobile 484	3500-3750 AMATEUR 510 509 511 3750-4000 AMATEUR 510 FIXED	3500-3900 AMATEUR 510 FIXED MOBILE	3500-4000	510	AMATEUR 510	
3800-3900 FIXED AERONAUTICAL MOBILE (OR) LAND MOBILE 511 512 514 515						

TABLES OF FREQUENCY ALLOCATIONS

INTERNATIONAL			Band kHz	UNITED STATES		Remarks
Region 1 kHz	Region 2 kHz	Region 3 kHz		Government Allocation	Non-Govt. Allocation	
3900-3950 AERONAUTICAL MOBILE (OR) 513		3900-3950 AERONAUTICAL MOBILE BROADCASTING				
3950-4000 FIXED BROADCASTING		3950-4000 FIXED BROADCASTING 516				
4000-4063 FIXED MARITIME MOBILE 517 516			4000-4063	MARITIME MOBILE US236	MARITIME MOBILE US236	See Section 4.3.13 for use.
4063-4438 MARITIME MOBILE 500A 500B 520 520A 520B 518 519			4063-4438	MARITIME MOBILE	MARITIME MOBILE	See Annex H for Maritime Mobile channel use.
			4063-4065	Ship stations, oceanographic data transmission	Ship stations, oceanographic data transmission	
			4065-4146	Ship stations, telephony, duplex operation 520	Ship stations, telephony, duplex operation 520	
			4146-4152	Ship and coast stations, telephony simplex operation US82	Ship and coast stations, telephony simplex operation US82	
			4152-4172	Ship stations, wide-band telegraphy, facsimile and special transmission systems US296	Ship stations, wide-band telegraphy, facsimile and special transmission systems US296	

TABLES OF FREQUENCY ALLOCATIONS

Region 1 kHz	Region 2 kHz	Region 3 kHz	Band kHz	Government Allocation	Non-Govt. Allocation	Remarks
			4172-4181.75	Ship stations, narrow-band direct-printing telegraphy and data transmission systems (paired frequencies) 500B	Ship stations, narrow-band direct-printing telegraphy and data transmission systems (paired frequencies) 500B	
			4181.75-4186.75	Ship stations, A1A Morse telegraphy, calling	Ship stations, A1A Morse telegraphy, calling	
			4186.75-4202.25	Ship stations, A1A Morse telegraphy, working	Ship stations, A1A Morse telegraphy, working	
			4202.25-4207.25	Ship stations, narrow-band direct-printing telegraphy and A1A Morse telegraphy, working (non-paired frequencies)	Ship stations, narrow-band direct-printing telegraphy and A1A Morse telegraphy, working (non-paired frequencies)	
			4207.25-4209.25	Ship stations, digital selective calling 500A	Ship stations, digital selective calling 500A	
			4209.25-4219.25	Coast stations, narrow-band direct-printing telegraph data transmission systems (paired frequencies) 520B	Coast stations, narrow-band direct-printing telegraph data transmission systems (paired frequencies) 520B	
			4219.25-4221	Coast stations, digital selective calling	Coast stations, digital selective calling	

TABLES OF FREQUENCY ALLOCATIONS

INTERNATIONAL Region 1 kHz	Region 2 kHz	Region 3 kHz	Band kHz	UNITED STATES Government Allocation	Non-Govt. Allocation	Remarks
			4221-4351	Coast stations, wide-band and A1A Morse telegraphy, facsimile, special and data transmission systems and direct-printing telegraphy systems	Coast stations, wide-band and A1A Morse telegraphy, facsimile, special and data transmission systems and direct-printing telegraphy systems	
			4351-4438	Coast stations, telephony, duplex operation	Coast stations, telephony, duplex operation	
4438-4650 FIXED MOBILE except aeronautical mobile (R)	4438-4650 FIXED MOBILE except aeronautical mobile (R)	4438-4650 FIXED MOBILE except aeronautical mobile	4438-4650	FIXED MOBILE except aero-nautical mobile (R)	FIXED MOBILE except aero-nautical mobile (R)	
4650-4700 AERONAUTICAL MOBILE (R)			4650-4700	AERONAUTICAL MOBILE (R) US282 US283	AERONAUTICAL MOBILE (R) US282 US283	
4700-4750 AERONAUTICAL MOBILE (OR)			4700-4750	AERONAUTICAL MOBILE (OR)	AERONAUTICAL MOBILE (OR)	Operations in the (OR) bands by Non-Government stations shall be authorized only by special arrangements between the FCC and the IRAC.
4750-4850 FIXED AERONAUTICAL MOBILE (OR) LAND MOBILE BROADCASTING 503	4750-4850 FIXED MOBILE except aero-nautical mobile (R) BROADCASTING 503	4750-4850 FIXED BROADCASTING 503 Land Mobile	4750-4850	FIXED MOBILE except aero-nautical mobile (R)	FIXED MOBILE except aero-nautical mobile (R)	
4850-4995 FIXED LAND MOBILE BROADCASTING 503			4850-4995	FIXED MOBILE	FIXED	

TABLES OF FREQUENCY ALLOCATIONS

INTERNATIONAL Region 1 kHz	Region 2 kHz	Region 3 kHz	Band kHz	UNITED STATES Government Allocation	Non-Govt. Allocation	Remarks
4995-5003 STANDARD FREQUENCY AND TIME SIGNAL (5000 kHz) 5003-5005 STANDARD FREQUENCY AND TIME SIGNAL Space Research			4995-5005	STANDARD FREQUENCY AND TIME SIGNAL (5000 kHz) G106	STANDARD FREQUENCY AND TIME SIGNAL (5000 kHz)	FCC Rules and Regulations make no provisions for the licensing of standard frequency stations.
5005-5060 FIXED BROADCASTING 503			5005-5060	FIXED	FIXED	
5060-5250 FIXED Mobile except aeronautical mobile 521 5250-5450 FIXED MOBILE except aeronautical mobile			5060-5450	FIXED Mobile except aeronautical mobile US212	FIXED Mobile except aeronautical mobile US212	
5450-5480 FIXED AERONAUTICAL MOBILE (OR) LAND MOBILE 501 505	5450-5480 AERONAUTICAL MOBILE (R)	5450-5480 FIXED AERONAUTICAL MOBILE (OR) LAND MOBILE	5450-5680	AERONAUTICAL MOBILE (R) US283 501 505	AERONAUTICAL MOBILE (R) US283 501 505	
5480-5680 AERONAUTICAL MOBILE (R) 501 505						
5680-5730 AERONAUTICAL MOBILE (OR) 501 505			5680-5730	AERONAUTICAL MOBILE (OR) 501 505	AERONAUTICAL MOBILE (OR) 501 505	Operation in the (OR) bands by Non-Government stations shall be authorized only by special arrangements between the FCC and the IRAC.
5730-5900 FIXED LAND MOBILE	5730-5900 FIXED MOBILE except aeronautical mobile (R)	5730-5900 FIXED Mobile except aeronautical mobile (R)	5730-5950	FIXED MOBILE except aeronautical mobile (R)	FIXED MOBILE except aeronautical mobile (R)	
5900-5950 BROADCASTING 521A 521B 521C						

TABLES OF FREQUENCY ALLOCATIONS

INTERNATIONAL			Band kHz	UNITED STATES		Remarks
Region 1 kHz	Region 2 kHz	Region 3 kHz		Government Allocation	Non-Govt. Allocation	
5950-6200 BROADCASTING			5950-6200	BROADCASTING	BROADCASTING	
6200-6525 MARITIME MOBILE 500A 500B 520 520B 522			6200-6525	MARITIME MOBILE	MARITIME MOBILE	See Annex H for Maritime Mobile channel use.
			6200-6224	Ship stations, telephony, duplex operation 520	Ship stations, telephony, duplex operation 520	
			6224-6233	Ship and coast stations, telephony, simplex operation US82	Ship and coast stations, telephony, simplex operation US82	
			6233-6261	Ship stations, wide-band telegraphy, facsimile and special transmission systems US296	Ship stations, wide-band telegraphy, facsimile and special transmission systems US296	
			6261-6262.75	Ship stations, oceanographic data transmission	Ship stations, oceanographic data transmission	
			6262.75-6275.75	Ship stations, narrow-band direct-printing telegraphy and data transmission systems (paired frequencies) 500B	Ship stations, narrow-band direct-printing telegraphy and data transmission systems (paired frequencies) 500B	
			6275.75-6280.75	Ship stations, A1A Morse telegraphy, calling	Ship stations, A1A Morse telegraphy, calling	

TABLES OF FREQUENCY ALLOCATIONS

| INTERNATIONAL | | | Band | UNITED STATES | | Remarks |
Region 1 kHz	Region 2 kHz	Region 3 kHz	kHz	Government Allocation	Non-Govt. Allocation	
			6280.75-6284.75	Ship stations, narrow-band direct-printing telegraphy and data transmission systems (paired frequencies)	Ship stations, narrow-band direct-printing telegraphy and data transmission systems (paired frequencies)	
			6284.75-6300.25	Ship stations, A1A Morse telegraphy, working	Ship stations, A1A Morse telegraphy, working	
			6300.25-6311.75	Ship stations, narrow-band direct-printing telegraph and A1A Morse telegraphy, working (non-paired frequencies)	Ship stations, narrow-band direct-printing telegraph and A1A Morse telegraphy, working (non-paired frequencies)	
			6311.75-6313.75	Ship stations, digital selective calling 500A	Ship stations, digital selective calling 500A	
			6313.75-6330.75	Coast stations, narrow-band direct-printing telegraphy and data transmission systems (paired frequencies) 520B	Coast stations, narrow-band direct-printing telegraphy and data transmission systems (paired frequencies) 520B	
			6330.75-6332.5	Coast stations, digital selective calling	Coast stations, digital selective calling	

TABLES OF FREQUENCY ALLOCATIONS

INTERNATIONAL			Band kHz	UNITED STATES		Remarks
Region 1 kHz	Region 2 kHz	Region 3 kHz		Government Allocation	Non-Govt. Allocation	
			6332.5-6501	Coast stations, wide-band and A1A Morse telegraphy, facsimile, special and data transmission systems and direct-printing telegraphy systems	Coast stations, wide-band and A1A Morse telegraphy, facsimile, special and data transmission systems and direct-printing telegraphy systems	
			6501-6525	Coast stations, telephony, duplex operation	Coast stations, telephony, duplex operation	
6525-6685 AERONAUTICAL MOBILE (R)			6525-6685	AERONAUTICAL MOBILE (R) US283	AERONAUTICAL MOBILE (R) US283	
6685-6765 AERONAUTICAL MOBILE (OR)			6685-6765	AERONAUTICAL MOBILE (OR)	AERONAUTICAL MOBILE (OR)	Operation in the (OR) bands by Non-Government stations shall be authorized only by special arrangement between the FCC and the IRAC.
6765-7000 FIXED Land Mobile 525 524			6765-7000	FIXED Mobile 524	FIXED Mobile 524	ISM 6780 ± 15 kHz
7000-7100 AMATEUR 510 AMATEUR-SATELLITE 526 527			7000-7100	510	AMATEUR AMATEUR-SATELLITE 510	
7100-7300 BROADCASTING 528	7100-7300 AMATEUR 510 528	7100-7300 BROADCASTING	7100-7300	510 528	AMATEUR 510 528	
7300-7350 BROADCASTING 521A 521B 528A			7300-8100	FIXED Mobile	FIXED Mobile	

TABLES OF FREQUENCY ALLOCATIONS

INTERNATIONAL			UNITED STATES		
Region 1 kHz	Region 2 kHz	Region 3 kHz	Band kHz	Government Allocation	Non-Govt. Allocation	Remarks
7350-8100 FIXED Land Mobile 529						
8100-8195 FIXED MARITIME MOBILE			8100-8195	MARITIME MOBILE US236	MARITIME MOBILE US236	See Section 4.3.13 for use.
8195-8815 MARITIME MOBILE 500A 500B 520B 529A 501			8195-8815 8195-8294	MARITIME MOBILE Ship stations, telephony, duplex operation 529A	MARITIME MOBILE Ship stations, telephony, duplex operation 529A	See Annex H for Maritime Mobile channel use.
			8294-8300	Ship and Coast stations, telephony, simplex operation US82	Ship and Coast stations, telephony, simplex operation US82	
			8300-8340	Ship stations, wide-band telegraphy, facsimile, and special transmission systems US296	Ship stations, wide-band telegraphy, facsimile, and special transmission systems US296	
			8340-8341.75	Ship stations, oceanographic data transmission	Ship stations, oceanographic data transmission	
			8341.75-8365.75	Ship stations, A1A Morse telegraphy, working 501	Ship stations, A1A Morse telegraphy, working 501	
			8365.75-8370.75	Ship stations, A1A Morse telegraphy, calling	Ship stations, A1A Morse telegraphy, calling	

TABLES OF FREQUENCY ALLOCATIONS

INTERNATIONAL			Band kHz	UNITED STATES		Remarks
Region 1 kHz	Region 2 kHz	Region 3 kHz		Government Allocation	Non-Govt. Allocation	
			8370.75-8376.25	Ship stations, A1A Morse telegraphy, working	Ship stations, A1A Morse telegraphy, working	
			8376.25-8396.25	Ship stations, narrow-band direct-printing telegraphy and data transmission systems (paired frequencies) 500B	Ship stations, narrow-band direct-printing telegraphy and data transmission systems (paired frequencies) 500B	
			8396.25-8414.25	Ship stations, narrow-band direct-printing telegraphy and A1A Morse telegraphy, working (non-paired frequencies) 500A	Ship stations, narrow-band direct-printing telegraphy and A1A Morse telegraphy, working (non-paired frequencies) 500A	
			8414.25-8416.25	Ship stations, digital selective calling 500A	Ship stations, digital selective calling 500A	
			8416.25-8436.25	Coast stations, narrow-band direct-printing telegraphy and data transmission systems (paired frequencies) 520B	Coast stations, narrow-band direct-printing telegraphy and data transmission systems (paired frequencies) 520B	
			8436.25-8438	Coast stations, digital selective calling	Coast stations, digital selective calling	

TABLES OF FREQUENCY ALLOCATIONS

INTERNATIONAL			Band kHz	UNITED STATES		Remarks
Region 1 kHz	Region 2 kHz	Region 3 kHz		Government Allocation	Non-Govt. Allocation	
			8438-8707	Coast stations, wide-band and A1A Morse telegraphy, facsimile, special and data transmission systems and direct-printing telegraphy systems	Coast stations, wide-band and A1A Morse telegraphy, facsimile, special and data transmission systems and direct-printing telegraphy systems	
			8707-8815	Coast stations, telephony, duplex operation	Coast stations, telephony, duplex operation	
8815-8965 AERONAUTICAL MOBILE (R)			8815-8965	AERONAUTICAL MOBILE (R)	AERONAUTICAL MOBILE (R)	
8965-9040 AERONAUTICAL MOBILE (OR)			8965-9040	AERONAUTICAL MOBILE (OR)	AERONAUTICAL MOBILE (OR)	Operation in the (OR) bands by Non-Government stations shall be authorized only by special arrangements between the FCC and the IRAC.
9040-9400 FIXED			9040-9500	FIXED	FIXED	
9400-9500 BROADCASTING 521A 521B 529B						
9500-9900 BROADCASTING 530 531			9500-9900	BROADCASTING US235	BROADCASTING US235	
9900-9995 FIXED			9900-9995	FIXED	FIXED	
9995-10003 STANDARD FREQUENCY AND TIME SIGNAL (10000 kHz) 501			9995-10005	STANDARD FREQUENCY AND TIME SIGNAL (10000 kHz) 501 G106	STANDARD FREQUENCY AND TIME SIGNAL (10000 kHz) 501	FCC Rules and Regulations make no provisions for the licensing of standard frequency stations.

TABLES OF FREQUENCY ALLOCATIONS

INTERNATIONAL			Band kHz	UNITED STATES		Remarks
Region 1 kHz	Region 2 kHz	Region 3 kHz		Government Allocation	Non-Govt. Allocation	
10003-10005 STANDARD FREQUENCY AND TIME SIGNAL Space Research 501						
10005-10100 AERONAUTICAL MOBILE (R) 501			10005-10100	AERONAUTICAL MOBILE (R) US283 501	AERONAUTICAL MOBILE (R) US283 501	
10100-10150 FIXED Amateur 510			10100-10150	US247 510	AMATEUR US247 510	
10150-11175 FIXED Mobile except aeronautical mobile (R)			10150-11175	FIXED Mobile except aero-nautical mobile (R)	FIXED Mobile except aero-nautical mobile (R)	
11175-11275 AERONAUTICAL MOBILE (OR)			11175-11275	AERONAUTICAL MOBILE (OR) US283	AERONAUTICAL MOBILE (OR)	Operation in the (OR) bands by Non-Government stations shall be authorized only by special ar-rangement between the FCC and the IRAC.
11275-11400 AERONAUTICAL MOBILE (R)			11275-11400	AERONAUTICAL MOBILE (R) US283	AERONAUTICAL MOBILE (R) US283	
11400-11600 FIXED			11400-11650	FIXED	FIXED	
11600-11650 BROADCASTING 521A 521B 529B						
11650-12050 BROADCASTING 530 531			11650-12050	BROADCASTING US235	BROADCASTING US235	
12050-12100 BROADCASTING 521A 521B 529B			12050-12230	FIXED	FIXED	
12100-12230 FIXED						

TABLES OF FREQUENCY ALLOCATIONS

INTERNATIONAL			Band	UNITED STATES		Remarks
Region 1 kHz	Region 2 kHz	Region 3 kHz	kHz	Government Allocation	Non-Govt. Allocation	
12230-13200 MARITIME MOBILE 500A 500B 529A			12230-13200	MARITIME MOBILE	MARITIME MOBILE	See Annex H for Maritime Mobile channel use.
			12230-12353	Ship stations, telephony, duplex operation 529A	Ship stations, telephony, duplex operation 529A	
			12353-12368	Ship and Coast stations, telephony, simplex operation US82	Ship and Coast stations, telephony, simplex operation US82	
			12368-12420	Ship stations, wide-band telegraphy, facsimile and special transmission systems US296	Ship stations, wide-band telegraphy, facsimile and special transmission systems US296	
			12420-12421.75	Ship stations, oceanographic data transmission	Ship stations, oceanographic data transmission	
			12421.75-12476.75	Ship stations, A1A Morse telegraphy, working	Ship stations, A1A Morse telegraphy, working	
			12476.75-12549.75	Ship stations, narrow-band direct-printing telegraphy and data transmission systems (paired frequencies) 500B	Ship stations, narrow-band direct-printing telegraphy and data transmission systems (paired frequencies) 500B	
			12549.75-12554.75	Ship stations, A1A Morse telegraphy, calling	Ship stations, A1A Morse telegraphy, calling	

TABLES OF FREQUENCY ALLOCATIONS

INTERNATIONAL			UNITED STATES		
Region 1 kHz	Region 2 kHz	Region 3 kHz	Band kHz	Government Allocation	Non-Govt. Allocation	Remarks
			12554.75- 12559.75	Ship stations, narrow-band direct-printing telegraphy and data transmis- sion systems (paired frequen- cies)	Ship stations, narrow-band direct-printing telegraphy and data transmission systems (paired frequencies)	
			12559.75- 12576.75	Ship stations, narrow-band direct-printing telegraphy and A1A Morse teleg- raphy, working (non-paired fre- quencies)	Ship stations, narrow-band direct-printing telegraphy and A1A Morse teleg- raphy, working (non-paired frequencies)	
			12576.75- 12578.75	Ship stations, digital selective call- ing 500A	Ship stations, digital selective calling 500A	
			12578.75- 12656.75	Coast stations, narrow-band direct-printing telegraphy and data transmis- sion systems (paired frequen- cies) 520B	Coast stations, narrow-band direct-printing telegraphy and data transmission systems (paired frequencies) 520B	
			12656.75- 12658.5	Coast stations, digital selective call- ing	Coast stations, digital selective calling	

TABLES OF FREQUENCY ALLOCATIONS

INTERNATIONAL			Band kHz	UNITED STATES		Remarks
Region 1 kHz	Region 2 kHz	Region 3 kHz		Government Allocation	Non-Govt. Allocation	
			12658.5-13077	Coast stations, wide-band and A1A Morse telegraphy, facsimile, special and data transmission systems and direct-printing telegraphy systems	Coast stations, wide-band and A1A Morse telegraphy, facsimile, special and data transmission systems and direct-printing telegraphy systems	
			13077-13200	Coast stations, telephony, duplex operation	Coast stations, telephony, duplex operation	
13200-13260 AERONAUTICAL MOBILE (OR)			13200-13260	AERONAUTICAL MOBILE (OR) US283	AERONAUTICAL MOBILE (OR)	Operation in the (OR) bands by Non-Government stations shall be authorized only by special arrangement between the FCC and the IRAC.
13260-13360 AERONAUTICAL MOBILE (R)			13260-13360	AERONAUTICAL MOBILE (R) US283	AERONAUTICAL MOBILE (R) US283	
13360-13410 FIXED RADIO ASTRONOMY 533			13360-13410	RADIO ASTRONOMY 533 G115	RADIO ASTRONOMY 533	
13410-13570 FIXED Mobile except aeronautical mobile (R) 534			13410-13600	FIXED Mobile except aeronautical mobile (R) 534	FIXED 534	ISM 13560 ± 7 kHz
13570-13600 FIXED Mobile except aeronautical mobile (R) 534A						
13600-13800 BROADCASTING 521A 521B			13600-13800	BROADCASTING US235	BROADCASTING US235	
13800 BROADCASTING 531						

TABLES OF FREQUENCY ALLOCATIONS

INTERNATIONAL				UNITED STATES		
Region 1 kHz	Region 2 kHz	Region 3 kHz	Band kHz	Government Allocation	Non-Govt. Allocation	Remarks
13800-13870 BROADCASTING 521A 521B 534A	BROADCASTING 521A 521B 534A		13800-14000	FIXED Mobile except aeronautical mobile (R)	FIXED	
13870-14000 FIXED Mobile except aeronautical mobile (R)						
14000-14250 AMATEUR 510 AMATEUR-SATELLITE			14000-14250	510	AMATEUR AMATEUR-SATELLITE 510	
14250-14350 AMATEUR 510 535			14250-14350	510	AMATEUR 510	
14350-14990 FIXED Mobile except aeronautical mobile (R)			14350-14990	FIXED Mobile except aeronautical mobile (R)	FIXED	
14990-15005 STANDARD FREQUENCY AND TIME SIGNAL (15 000 kHz) 501			14990-15010	STANDARD FREQUENCY AND TIME SIGNAL (15 000 kHz) 501 G106	STANDARD FREQUENCY AND TIME SIGNAL (15 000 kHz) 501	FCC Rules and Regulations make no provisions for the licensing of standard frequency stations.
15005-15010 STANDARD FREQUENCY AND TIME SIGNAL Space Research						
15010-15100 AERONAUTICAL MOBILE (OR)			15010-15100	AERONAUTICAL MOBILE (OR)	AERONAUTICAL MOBILE (OR)	Operation in the (OR) bands by Non-Government stations shall be authorized only by special arrangement between the FCC and the IRAC.
15100-15600 BROADCASTING 531			15100-15600	BROADCASTING US235	BROADCASTING US235	
15600-15800 BROADCASTING 521A 521B 529B			15600-16360	FIXED	FIXED	

TABLES OF FREQUENCY ALLOCATIONS

INTERNATIONAL			Band kHz	UNITED STATES		Remarks
Region 1 kHz	Region 2 kHz	Region 3 kHz		Government Allocation	Non-Govt. Allocation	
15800-16360 FIXED 536						
16360-17410 MARITIME MOBILE 500A 500B 520B 529A						
			16360-17410	MARITIME MOBILE	MARITIME MOBILE	See Annex H for Maritime Mobile channel use.
			16360-16528	Ship stations, telephony, duplex operation 529A	Ship stations, telephony, duplex operation 529A	
			16528-16549	Ship and Coast stations, telephony, simplex operation US82	Ship and Coast stations, telephony, simplex operation US82	
			16549-16617	Ship stations, wide-band telegraphy, facsimile, and special transmission systems US296	Ship stations, wide-band telegraphy, facsimile, and special transmission systems US296	
			16617-16618.75	Ship stations, oceanographic data transmission	Ship stations, oceanographic data transmission	
			16618.75-16683.25	Ship stations, A1A Morse telegraphy, working	Ship stations, A1A Morse telegraphy, working	
			16683.25-16733.75	Ship stations, narrow-band direct-printing telegraphy and data transmission systems (paired frequencies) 500B	Ship stations, narrow-band direct-printing telegraphy and data transmission systems (paired frequencies) 500B	

TABLES OF FREQUENCY ALLOCATIONS

INTERNATIONAL			UNITED STATES		
Region 1 kHz	Region 2 kHz	Region 3 kHz	Band kHz	Government Allocation	Non-Govt. Allocation	Remarks
			16733.75- 16738.75	Ship stations, A1A Morse telegraphy, calling	Ship stations, A1A Morse telegraphy, calling	
			16738.75- 16784.75	Ship stations, narrow-band direct-printing telegraphy and data transmis- sion systems (paired fre- quencies)	Ship stations, narrow-band direct-printing telegraphy and data transmission systems (paired frequencies)	
			16784.75- 16804.25	Ship stations, narrow-band direct-printing and A1A Morse telegraphy, working (non- paired frequencies)	Ship stations, narrow-band direct-printing and A1A Morse telegraphy, working (non- paired frequencies)	
			16804.25- 16806.25	Ship stations, digital selective call- ing 500A	Ship stations, digital selective calling 500A	
			16806.25- 16902.75	Coast stations, narrow-band direct-printing telegraph and data transmis- sion systems (paired fre- quencies) 520B	Coast stations, narrow-band direct-printing telegraph and data transmission systems (paired frequencies) 520B	
			16902.75- 16904.5	Coast stations, digital selective call- ing	Coast stations, digital selective calling	

TABLES OF FREQUENCY ALLOCATIONS

INTERNATIONAL			Band	UNITED STATES		Remarks
Region 1 kHz	Region 2 kHz	Region 3 kHz	kHz	Government Allocation	Non-Govt. Allocation	
			16904.5-17242	Coast stations, wide-band and A1A Morse telegraphy, facsimile, special and data transmission systems and direct-printing telegraphy systems	Coast stations, wide-band and A1A Morse telegraphy, facsimile, special and data transmission systems and direct-printing telegraphy systems	
			17242-17410	Coast stations, telephony, duplex operation	Coast stations, telephony, duplex operation	
17410-17480 FIXED			17410-17550	FIXED	FIXED	
17480-17550 BROADCASTING 521A 521B 529B						
17550-17900 BROADCASTING 531			17550-17900	BROADCASTING US235	BROADCASTING US235	
17900-17970 AERONAUTICAL MOBILE (R)			17900-17970	AERONAUTICAL MOBILE (R) US283	AERONAUTICAL MOBILE (R) US283	
17970-18030 AERONAUTICAL MOBILE (OR)			17970-18030	AERONAUTICAL MOBILE (OR)	AERONAUTICAL MOBILE (OR)	Operation in the (OR) bands by Non-Government stations shall be authorized only by special arrangement between the FCC and the IRAC.
18030-18052 FIXED			18030-18068	FIXED	FIXED	
18052-18068 FIXED Space Research						

TABLES OF FREQUENCY ALLOCATIONS

INTERNATIONAL			UNITED STATES			
Region 1 kHz	Region 2 kHz	Region 3 kHz	Band kHz	Government Allocation	Non-Govt. Allocation	Remarks
18068-18168 AMATEUR 510 AMATEUR-SATELLITE 538			18068-18168	510	AMATEUR AMATEUR-SATELLITE 510	
18168-18780 FIXED Mobile except aeronautical mobile			18168-18780	FIXED Mobile	FIXED Mobile	
18780-18900 MARITIME MOBILE			18780-18900	MARITIME MOBILE	MARITIME MOBILE	See Annex H for Maritime Mobile channel use.
			18780-18825	Ship stations, telephony, duplex operation	Ship stations, telephony, duplex operation	
			18825-18846	Ship and Coast stations, telephony, simplex operation US82	Ship and Coast stations, telephony, simplex operation US82	
			18846-18870	Ship stations, wide-band telegraphy, facsimile, and special transmission systems US296	Ship stations, wide-band telegraphy, facsimile, and special transmission systems US296	
			18870-18892.75	Ship stations, narrow-band direct-printing telegraphy and data transmission systems (paired frequencies)	Ship stations, narrow-band direct-printing telegraphy and data transmission systems (paired frequencies)	
			18892.75-18898.25	Ship stations, narrow-band direct-printing telegraphy and A1A Morse telegraphy, working (non-paired frequencies)	Ship stations, narrow-band direct-printing telegraphy and A1A Morse telegraphy, working (non-paired frequencies)	

TABLES OF FREQUENCY ALLOCATIONS

INTERNATIONAL			Band kHz	UNITED STATES		Remarks
Region 1 kHz	Region 2 kHz	Region 3 kHz		Government Allocation	Non-Govt. Allocation	
			18898.25-18899.75	Ship stations, digital selective calling	Ship stations, digital selective calling	
18900-19020 BROADCASTING 521A 521B 529B			18900-19680	FIXED	FIXED	
19020-19680 FIXED						
19680-19800 MARITIME MOBILE 520B			19680-19800	MARITIME MOBILE	MARITIME MOBILE	See Annex H for Maritime Mobile channel use.
			19680.25-19703.25	Coast stations, narrow-band direct-printing telegraphy and data transmission systems (paired frequencies) 520B	Coast stations, narrow-band direct-printing telegraphy and data transmission systems (paired frequencies) 520B	
			19703.25-19705	Coast stations, digital selective calling	Coast stations, digital selective calling	
			19705-19755	Coast stations, wide-band and A1A Morse telegraphy, facsimile, special and data transmission systems and direct-printing telegraphy systems	Coast stations, wide-band and A1A Morse telegraphy, facsimile, special and data transmission systems and direct-printing telegraphy systems	
			19755-19800	Coast stations, telephony, duplex operation	Coast stations, telephony, duplex operation	
19800-19990 FIXED			19800-19950	FIXED	FIXED	

TABLES OF FREQUENCY ALLOCATIONS

INTERNATIONAL			Band	UNITED STATES		Remarks
Region 1 kHz	Region 2 kHz	Region 3 kHz	kHz	Government Allocation	Non-Govt. Allocation	
19990-19995 STANDARD FREQUENCY AND TIME SIGNAL Space Research 501			19990-20010	STANDARD FREQUENCY AND TIME SIGNAL (20 000 kHz) 501 G106	STANDARD FREQUENCY AND TIME SIGNAL (20 000 kHz) 501	FCC Rules and Regulations make no provisions for the licensing of standard frequency stations.
19995-20010 STANDARD FREQUENCY AND TIME SIGNAL (20 000 kHz) 501						
20010-21000 FIXED Mobile 501			20010-21000	FIXED Mobile	FIXED	
21000-21450 AMATEUR 510 AMATEUR-SATELLITE			21000-21450	510	AMATEUR AMATEUR-SATELLITE 510	
21450-21850 BROADCASTING 531			21450-21850	BROADCASTING US235	BROADCASTING US235	
21850-21870 FIXED 539			21850-21924	FIXED	FIXED	
21870-21924 AERONAUTICAL FIXED						
21924-22000 AERONAUTICAL MOBILE (R)			21924-22000	AERONAUTICAL MOBILE (R)	AERONAUTICAL MOBILE (R)	
22000-22855 MARITIME MOBILE 520B 540			22000-22855	MARITIME MOBILE	MARITIME MOBILE	See Annex H for Maritime Mobile channel use.
			22000-22159	Ship stations, telephony, duplex operation	Ship stations, telephony, duplex operation	
			22159-22180	Ship and Coast stations, telephony, simplex operation US82	Ship and Coast stations, telephony, simplex operation US82	

TABLES OF FREQUENCY ALLOCATIONS

INTERNATIONAL			Band kHz	UNITED STATES		Remarks
Region 1 kHz	Region 2 kHz	Region 3 kHz		Government Allocation	Non-Govt. Allocation	
			22180-22240	Ship stations, wide-band telegraphy, facsimile and special transmission systems US296	Ship stations, wide-band telegraphy, facsimile and special transmission systems US296	
			22240-22241.75	Ship stations, oceanographic data transmission	Ship stations, oceanographic data transmission	
			22241.75-22279.25	Ship stations, A1A Morse telegraphy, working	Ship stations, A1A Morse telegraphy, working	
			22279.25-22284.25	Ship stations, A1A Morse telegraphy, calling	Ship stations, A1A Morse telegraphy, calling	
			22284.25-22351.75	Ship stations, narrow-band direct-printing telegraphy and data transmission systems, working (paired frequencies)	Ship stations, narrow-band direct-printing telegraphy and data transmission systems, working (paired frequencies)	
			22351.75-22374.25	Ship stations, narrow-band direct-printing telegraphy and A1A Morse telegraphy, working (non-paired frequencies)	Ship stations, narrow-band direct-printing telegraphy and A1A Morse telegraphy, working (non-paired frequencies)	
			22374.25-22375.75	Ship stations, digital selective calling	Ship stations, digital selective calling	

TABLES OF FREQUENCY ALLOCATIONS

INTERNATIONAL				UNITED STATES		
Region 1 kHz	Region 2 kHz	Region 3 kHz	Band kHz	Government Allocation	Non-Govt. Allocation	Remarks
			22375.75-22443.75	Coast stations, narrow-band direct-printing telegraphy and data transmission systems (paired frequencies) 520B	Coast stations, narrow-band direct-printing telegraphy and data transmission systems (paired frequencies) 520B	
			22443.75-22445.5	Coast stations, digital selective calling	Coast stations, digital selective calling	
			22445.5-22696	Coast stations, wide-band and A1A Morse telegraphy, facsimile, special and data transmission and direct-printing telegraphy systems	Coast stations, wide-band and A1A Morse telegraphy, facsimile, special and data transmission and direct-printing telegraphy systems	
			22696-22855	Coast stations, telephony, duplex operation	Coast stations, telephony, duplex operation	
22855-23000 FIXED 540			22855-23000	FIXED	FIXED	
23000-23200 FIXED Mobile except aeronautical mobile (R) 540			23000-23200	FIXED Mobile except aero-nautical mobile (R)	FIXED	
23200-23350 AERONAUTICAL FIXED AERONAUTICAL MOBILE (OR)			23200-23350	AERONAUTICAL MOBILE (OR)	AERONAUTICAL MOBILE (OR)	Operation in the (OR) bands by Non-Government stations shall be authorized only by special ar-rangement between the FCC and the IRAC.

TABLES OF FREQUENCY ALLOCATIONS

INTERNATIONAL			Band kHz	UNITED STATES		Remarks
Region 1 kHz	Region 2 kHz	Region 3 kHz		Government Allocation	Non-Govt. Allocation	
23350-24000 FIXED MOBILE except aeronautical mobile 541 542			23350-24890	FIXED MOBILE except aero-nautical mobile	FIXED	
24000-24890 FIXED LAND MOBILE 542						
24890-24990 AMATEUR 510 AMATEUR-SATELLITE 542			24890-24990	510	AMATEUR AMATEUR-SATELLITE 510	
24990-25005 STANDARD FREQUENCY AND TIME SIGNAL (25000 kHz)			24990-25010	STANDARD FREQUENCY AND TIME SIGNAL (25000 kHz) G106	STANDARD FREQUENCY AND TIME SIGNAL (25000 kHz)	
25005-25010 STANDARD FREQUENCY AND TIME SIGNAL Space Research						
25010-25070 FIXED MOBILE except aeronautical mobile			25010-25070		LAND MOBILE NG112	25.02-25.06 kHz Industrial
25070-25210 MARITIME MOBILE			25070-25210	MARITIME MOBILE US281	MARITIME MOBILE US281 NG112	See Annex H for Maritime Mobile channel use.
			25070-25100	Ship stations, telephony, duplex operation	Ship stations, telephony, duplex operation	
			25100-25121	Ship and Coast stations, telephony, simplex operation US82	Ship and Coast stations, telephony, simplex operation US82	

TABLES OF FREQUENCY ALLOCATIONS

INTERNATIONAL			Band	UNITED STATES		Remarks
Region 1 kHz	Region 2 kHz	Region 3 kHz	kHz	Government Allocation	Non-Govt. Allocation	
25210-25550 FIXED MOBILE except aeronautical mobile			25121-25161.25	Ship stations, wide-band telegraphy, facsimile, and special transmission systems US296	Ship stations, wide-band telegraphy, facsimile, and special transmission systems US296	
			25161.25-25171.25	Ship stations, A1A Morse telegraphy, working	Ship stations, A1A Morse telegraphy, working	
			25171.25-25172.75	Ship stations, A1A Morse telegraphy, calling	Ship stations, A1A Morse telegraphy, calling	
			25172.75-25192.75	Ship stations, narrow-band direct-printing telegraphy and data transmission systems (paired frequencies)	Ship stations, narrow-band direct-printing telegraphy and data transmission systems (paired frequencies)	
			25192.75-25208.25	Ship stations, narrow-band direct-printing telegraphy and data transmission systems (non-paired frequencies)	Ship stations, narrow-band direct-printing telegraphy and data transmission systems (non-paired frequencies)	
			25208.25-25210	Ship stations, digital selective calling	Ship stations, digital selective calling	
			25210-25330		LAND MOBILE	25.12-25.32 kHz Industrial
			25330-25550	FIXED MOBILE except aeronautical mobile		

TABLES OF FREQUENCY ALLOCATIONS

INTERNATIONAL			Band kHz	UNITED STATES		Remarks
Region 1 kHz	Region 2 kHz	Region 3 kHz		Government Allocation	Non-Govt. Allocation	
25550-25670 RADIO ASTRONOMY 545			25550-25670	RADIO ASTRONOMY US74 545	RADIO ASTRONOMY US74 545	
25670-26100 BROADCASTING			25670-26100	BROADCASTING US25	BROADCASTING US25	International broadcasting
26100-26175 MARITIME MOBILE 520B			26100-26175 26120.25-26120.75	MARITIME MOBILE Coast stations, narrow-band direct-printing telegraphy and data transmission systems (paired frequencies) 520B	MARITIME MOBILE Coast stations, narrow-band direct-printing telegraphy and data transmission systems (paired frequencies) 520B	See Annex H for Maritime Mobile channel use.
			26120.75-26122.5	Coast stations, digital selective calling	Coast stations, digital selective calling	
			26122.5-26145	Coast stations, wide-band and A1A Morse telegraphy, facsimile, special and data transmission systems and direct-printing telegraphy systems	Coast stations, wide-band and A1A Morse telegraphy, facsimile, special and data transmission systems and direct-printing telegraphy systems	
			26145-26175	Coast stations, telephony, duplex operation	Coast stations, telephony, duplex operation	
26175-27500 FIXED MOBILE except aeronautical mobile 546			26175-26480		LAND MOBILE	
			26480-26950	FIXED MOBILE except aero-nautical mobile US10	US10	

TABLES OF FREQUENCY ALLOCATIONS

INTERNATIONAL				UNITED STATES		
Region 1 kHz	Region 2 kHz	Region 3 kHz	Band kHz	Government Allocation	Non-Govt. Allocation	Remarks
			26950-26960		FIXED 546	26.955 kHz International fixed public
				546		
			26960-27230		MOBILE except aero- nautical mobile	ISM 27120 ± 160 kHz
				546	546	
			27230-27410		FIXED MOBILE except aero- nautical mobile	ISM 27120 ± 160 kHz
				546	546	Personal
						Public Safety
						Industrial
						Land Transportation

TABLES OF FREQUENCY ALLOCATIONS

| INTERNATIONAL | | | Band MHz | UNITED STATES | | Remarks |
Region 1 MHz	Region 2 MHz	Region 3 MHz		Government Allocation	Non-Govt. Allocation	
27.5-28 METEOROLOGICAL AIDS FIXED MOBILE			27.41-27.54		LAND MOBILE	Industrial
			27.54-28	FIXED MOBILE US298	US298	
28-29.7 AMATEUR AMATEUR-SATELLITE			28-29.7		AMATEUR AMATEUR-SATELLITE	
29.7-30.005 FIXED MOBILE			29.7-29.8		LAND MOBILE	Industrial
			29.8-29.89		FIXED	29.81-29.88 MHz Aeronautical fixed International fixed public
			29.89-29.91	FIXED MOBILE		See Section 4.3.6 of the NTIA Manual for Channeling Plan.
			29.91-30		FIXED	29.92-29.99 MHz Aeronautical fixed International fixed public
30.005-30.01 SPACE OPERATION (satellite identification) FIXED MOBILE SPACE RESEARCH			30-30.56	FIXED MOBILE		See Section 4.3.6 of the NTIA Manual for Channeling Plan.
30.01-37.5 FIXED MOBILE			30.56-32		LAND MOBILE NG124	Industrial Land Transportation Public Safety

TABLES OF FREQUENCY ALLOCATIONS

INTERNATIONAL			Band MHz	UNITED STATES		Remarks
Region 1 MHz	Region 2 MHz	Region 3 MHz		Government Allocation	Non-Govt. Allocation	
			32-33	FIXED MOBILE		See Section 4.3.6 of the NTIA Manual for Channeling Plan.
			33-34		LAND MOBILE NG124	33.00-33.01 MHz Land Transportation 33.01-33.11 MHz Public Safety 33.11-33.41 MHz Industrial 33.41-34 MHz Public Safety
			34-35	FIXED MOBILE		See Section 4.3.6 of the NTIA Manual for Channeling Plan.
			35-36		LAND MOBILE NG124	35.00-35.19 MHz Industrial 35.19-35.69 MHz Domestic Public Industrial Public Safety 35.69-36.00 MHz Industrial
			36-37	FIXED MOBILE US220	US220	See Section 4.3.6 of the NTIA Manual for Channeling Plan.
			37-37.5		LAND MOBILE NG124	37.00-37.01 MHz Industrial 37.01-37.43 MHz Public Safety 37.43-37.5 MHz Industrial

TABLES OF FREQUENCY ALLOCATIONS

INTERNATIONAL			UNITED STATES			
Region 1 MHz	Region 2 MHz	Region 3 MHz	Band MHz	Government Allocation	Non-Govt. Allocation	Remarks
37.5-38.25 FIXED MOBILE Radio Astronomy			37.5-38	Radio Astronomy	LAND MOBILE Radio Astronomy	37.50-37.89 MHz Industrial
547				547	547 NG59 NG124	37.89-38.00 Public Safety
			38-38.25	FIXED MOBILE RADIO ASTRONOMY	RADIO ASTRONOMY	See Section 4.3.6 of the NTIA Manual for Channeling Plan.
				US81 547	US81 547	
38.25-39.986 FIXED MOBILE			38.25-39	FIXED MOBILE		See Section 4.3.6 of the NTIA Manual for Channeling Plan.
			39-40		LAND MOBILE	Public Safety
39.986-40.02 FIXED MOBILE Space Research					NG124	
40.02-40.98 FIXED MOBILE			40-42	FIXED MOBILE	US210 US220 548	See Section 4.3.6 of the NTIA Manual for Channeling Plan.
548				US210 US220 548		ISM 40.68 ± 0.02 MHz
40.98-41.015 FIXED MOBILE Space Research						
549 550						
41.015-44 FIXED MOBILE			42-46.6		LAND MOBILE	42.00-42.95 MHz Public Safety 42.95-43.19 MHz Industrial 43.19-43.69 MHz Domestic Public Industrial Public Safety 43.69-44.61 MHz Land Transportation 44.61-46.60 MHz Public Safety
549 550					NG124 NG141	

TABLES OF FREQUENCY ALLOCATIONS

INTERNATIONAL			Band MHz	UNITED STATES		Remarks
Region 1 MHz	Region 2 MHz	Region 3 MHz		Government Allocation	Non-Govt. Allocation	
44-47 FIXED MOBILE 552	47-50 FIXED MOBILE	47-50 FIXED MOBILE BROADCASTING	46.6-47	FIXED MOBILE		See Section 4.3.6 of the NTIA Manual for Channeling Plan.
47-68 BROADCASTING 553 554 555 559 561			47-49.6		LAND MOBILE NG124	47.00-47.43 MHz Public safety 47.43-47.69 MHz Public Safety Industrial 47.69-49.60 Industrial
			49.6-50	FIXED MOBILE		See Section 4.3.6 of the NTIA Manual for Channeling Plan.
	50-54 AMATEUR 556 557 558 560		50-54		AMATEUR	
	54-68 BROADCASTING Fixed Mobile 562	54-68 FIXED MOBILE BROADCASTING	54-72		BROADCASTING NG128 NG149	Television broadcasting
68-74.8 FIXED MOBILE except aero-nautical mobile 564 565 567 568 571	68-72 BROADCASTING Fixed Mobile 563	68-74.8 FIXED MOBILE 566 568 571				
	72-73 FIXED MOBILE		72-73		FIXED MOBILE NG3 NG49 NG56	72.02-72.98 MHz Operational Fixed

TABLES OF FREQUENCY ALLOCATIONS

INTERNATIONAL			Band MHz	UNITED STATES		Remarks
Region 1 MHz	Region 2 MHz	Region 3 MHz		Government Allocation	Non-Govt. Allocation	
	73-74.6 RADIO ASTRONOMY 570		73-74.6	RADIO ASTRONOMY US74	RADIO ASTRONOMY US74	
	74.6-74.8 FIXED MOBILE	74.6-74.8 FIXED MOBILE	74.6-74.8	FIXED MOBILE US273 572	FIXED MOBILE US273 572	
74.8-75.2 AERONAUTICAL RADIONAVIGATION 572 572A	571		74.8-75.2	AERONAUTICAL RADIONAVIGATION 572	AERONAUTICAL RADIONAVIGATION 572	75 MHz Marker beacons.
75.2-87.5 FIXED MOBILE except aeronautical mobile 571	75.2-75.4 FIXED MOBILE		75.2-75.4	FIXED MOBILE US273 572	FIXED MOBILE US273 572	
565 571 575 578	75.4-76 FIXED MOBILE	75.4-87 FIXED MOBILE 573 574 577 579	75.4-76		FIXED MOBILE NG3 NG49 NG56	75.42-75.98 MHz Operational Fixed
	76-88 BROADCASTING Fixed Mobile 576	87-100 FIXED MOBILE BROADCASTING 580	76-88	576	BROADCASTING 576 NG128 NG129 NG149	Television broadcasting
87.5-100 BROADCASTING 581	88-100 BROADCASTING		88-108	US93	BROADCASTING US93 NG2 NG128 NG129	FM broadcasting
100-108 BROADCASTING 584 585 586 587 588 589						
108-117.975 AERONAUTICAL RADIONAVIGATION 590A			108-117.975	AERONAUTICAL RADIONAVIGATION G126 US93	AERONAUTICAL RADIONAVIGATION US93	

TABLES OF FREQUENCY ALLOCATIONS

INTERNATIONAL			Band MHz	UNITED STATES		Remarks
Region 1 MHz	Region 2 MHz	Region 3 MHz		Government Allocation	Non-Govt. Allocation	
117.975-136 AERONAUTICAL MOBILE (R) 501 591 592 593 594			117.975-121.9375	AERONAUTICAL MOBILE (R) US26 US28 501 591 592 593	AERONAUTICAL MOBILE (R) US26 US28 501 591 592 593	
			121.9375-123.0875	US30 US31 US33 US80 US102 US213 591	AERONAUTICAL MOBILE US30 US31 US33 US80 US102 US213 591	Private aircraft
			123.0875-123.5875	AERONAUTICAL MOBILE US32 US33 US112 591 593	AERONAUTICAL MOBILE US32 US33 US112 591 593	123.1 MHz for SAR Scene-of Action communication (See Section 7.5.4 of the NTIA Manual)
			123.5875-128.8125	AERONAUTICAL MOBILE (R) US26 591	AERONAUTICAL MOBILE (R) US26 591	
			128.8125-132.0125	591	AERONAUTICAL MOBILE (R) 591	
			132.0125-136.00	AERONAUTICAL MOBILE (R) US26 591	AERONAUTICAL MOBILE (R) US26 591	
136-137 AERONAUTICAL MOBILE (R) Fixed Mobile except aeronautical mobile (R) 591 594A 595			136-137	US244 591	AERONAUTICAL MOBILE (R) US244 591	

TABLES OF FREQUENCY ALLOCATIONS

INTERNATIONAL			UNITED STATES			
Region 1 MHz	Region 2 MHz	Region 3 MHz	Band MHz	Government Allocation	Non-Govt. Allocation	Remarks
137-137.025 SPACE OPERATION (space-to-Earth) METEOROLOGICAL-SATELLITE (space-to-Earth) MOBILE-SATELLITE (space-to-Earth) 599B SPACE RESEARCH (space-to-Earth) Fixed Mobile except aeronautical mobile(R) 596 597 598 599 599A			137- 137.025	SPACE OPERATION (space-to-Earth) METEOROLOGICAL SATELLITE (space-to-Earth) SPACE RESEARCH (space-to-Earth) MOBILE-SATELLITE (space-to-Earth) US319 US320 599B US318 599A	SPACE OPERATION (space-to-Earth) METEOROLOGICAL SATELLITE (space-to-Earth) SPACE RESEARCH (space-to-Earth) MOBILE-SATELLITE (space-to-Earth) US319 US320 599B US318 599A	
137.025-137.175 SPACE OPERATION (space-to-Earth) METEOROLOGICAL-SATELLITE (space-to-Earth) SPACE RESEARCH (space-to-Earth) Fixed Mobile-Satellite (space-to-Earth) 599B Mobile except aeronautical mobile (R) 596 597 598 599 599A			137.025- 137.175	SPACE OPERATION (space-to-Earth) METEOROLOGICAL SATELLITE (space-to-earth) SPACE RESEARCH (space-to-Earth) Mobile-Satellite (space-to-Earth) US319 US320 599B US318 599A	SPACE OPERATION (space-to-Earth) METEOROLOGICAL- SATELLITE (space-to-Earth) SPACE RESEARCH (space-to-Earth) Mobile-Satellite (space-to-Earth) US319 US320 599B US318 599A	

TABLES OF FREQUENCY ALLOCATIONS

INTERNATIONAL			Band MHz	UNITED STATES		Remarks
Region 1 MHz	Region 2 MHz	Region 3 MHz		Government Allocation	Non-Govt. Allocation	
137.175-137.825 SPACE OPERATION (space-to-Earth) METEOROLOGICAL-SATELLITE (space-to-Earth) MOBILE-SATELLITE (space-to-Earth) 599B SPACE RESEARCH (space-to-Earth) Fixed Mobile except aeronautical mobile (R) 596 597 598 599 599A			137.175-137.825	SPACE OPERATION (space-to-Earth) METEOROLOGICAL SATELLITE (space-to-Earth) SPACE RESEARCH (space-to-Earth) MOBILE-SATELLITE (space-to-Earth) US319 US320 599B US318 599A	SPACE OPERATION (space-to-Earth) METEOROLOGICAL SATELLITE (space-to-Earth) SPACE RESEARCH (space-to-Earth) MOBILE-SATELLITE (space-to-Earth) US319 US320 599B US318 599A	
137.825-138 SPACE OPERATION (space-to-Earth) METEOROLOGICAL-SATELLITE (space-to-Earth) SPACE RESEARCH (space-to-Earth) Fixed Mobile-Satellite (space-to-Earth) 599B Mobile except aeronautical mobile (R) 596 597 598 599 599A			137.825-138	SPACE OPERATION (space-to-Earth) METEOROLOGICAL-SATELLITE (space-to-Earth) SPACE RESEARCH (space-to-Earth) Mobile-Satellite (space-to-Earth) US319 US320 599B US318 599A	SPACE OPERATION (space-to-Earth) METEOROLOGICAL-SATELLITE (space-to-Earth) SPACE RESEARCH (space-to-Earth) Mobile-Satellite (space-to-Earth) US319 US320 599B US318 599A	
138-143.6 AERONAUTICAL MOBILE (OR) 600 601 602 604	138-143.6 FIXED MOBILE /RADIOLOCATION/ Space Research (space-to-Earth) 599 603	138-143.6 FIXED MOBILE Space Research (space-to-Earth) 599 603	138-144	FIXED MOBILE US10 G30	US319 US320 US318 599A US10	
143.6-143.65 AERONAUTICAL MOBILE (OR) SPACE RESEARCH (space-to-Earth) 601 602 604	143.6-143.65 FIXED MOBILE SPACE RESEARCH (space-to-Earth) /RADIOLOCATION/	143.6-143.65 FIXED MOBILE SPACE RESEARCH (space-to-Earth) 599 603				

TABLES OF FREQUENCY ALLOCATIONS

INTERNATIONAL			Band MHz	UNITED STATES		Remarks
Region 1 MHz	Region 2 MHz	Region 3 MHz		Government Allocation	Non-Govt. Allocation	
143.65-144 AERONAUTICAL MOBILE (OR)	143.65-144 FIXED MOBILE /RADIOLOCATION/ Space Research (space-to-Earth)	143.65-144 FIXED MOBILE Space Research (space-to-Earth) 599 603				
600 601 602 604						
144-146 AMATEUR 510 AMATEUR-SATELLITE			144-146		AMATEUR AMATEUR-SATELLITE	
605 606				510	510	
146-148 FIXED MOBILE except aeronautical mobile (R)	146-148 AMATEUR 607	146-148 AMATEUR FIXED MOBILE 607	146-148		AMATEUR	
148-149.9 FIXED MOBILE except aeronautical mobile (R) MOBILE-SATELLITE (Earth-to-space) MOBILE-SATELLITE (space-to-Earth) 599B 608 608A 608C	148-149.9 FIXED MOBILE MOBILE-SATELLITE (Earth-to-space) 599B 608 608A 608C	MOBILE-SATELLITE (Earth-to-space)	148-149.9	FIXED MOBILE MOBILE-SATELLITE (Earth-to-space) 599B US319 US320 US323 US325 608 608A US10 G30	MOBILE-SATELLITE (Earth-to-space) 599B US319 US320 US323 US325 608 608A US10	

TABLES OF FREQUENCY ALLOCATIONS

| INTERNATIONAL | | | UNITED STATES | | | |
Region 1 MHz	Region 2 MHz	Region 3 MHz	Band MHz	Government Allocation	Non-Govt. Allocation	Remarks
149.9-150.05 LAND MOBILE-SATELLITE (Earth-to-space) 599B 609B RADIONAVIGATION-SATELLITE 608B 609 609A			149.9-150.05	RADIONAVIGATION SATELLITE MOBILE SATELLITE (Earth-to-space) 599B US319 US322 608B 609A	RADIONAVIGATION SATELLITE MOBILE SATELLITE (Earth-to-space) 599B US319 US322 608B 609A	
150.05-153 FIXED MOBILE except aero- nautical mobile RADIO ASTRONOMY 611 613 613A	150.05-156.7625 FIXED MOBILE 611 613 613A		150.05-150.8	FIXED MOBILE US216 G30	US216	

610

TABLES OF FREQUENCY ALLOCATIONS

INTERNATIONAL			UNITED STATES		
Region 1 MHz	Region 2 MHz	Region 3 MHz	Band MHz	Government Allocation	Non-Govt. Allocation	Remarks
			150.8– 156.2475	US216 613	LAND MOBILE US216 613 NG4 NG51 NG112 NG117 NG124 NG148	150.80–150.98 MHz Land Transportation 150.98–151.4825 MHz Public Safety 151.4825– 151.4975 MHz Industrial 151.4975–152.000 MHz Industrial Public Safety 152.00–152.255 MHz Domestic Public 152.255–152.465 MHz Land Transportation 152.465–152.495 MHz Industrial 152.495–152.855 MHz Domestic Public 152.885–153.7325 MHz Industrial 153.7325– 154.4825 MHz Industrial Public Safety 154.2 MHz Earth Tele- command 154.4825– 154.6375 MHz Industrial 154.6375– 156.2475 MHz Public Safety
153–154 FIXED MOBILE except aero- nautical mobile (R) Meteorological Aids						
154–156.7625 FIXED MOBILE except aero- nautical mobile (R) 613 613A			156.2475– 157.0375	US77 US106 US107 US266 613	MARITIME MOBILE US77 US106 US107 US266 613 NG117	

TABLES OF FREQUENCY ALLOCATIONS

INTERNATIONAL			Band MHz	UNITED STATES		Remarks
Region 1 MHz	Region 2 MHz	Region 3 MHz		Government Allocation	Non-Govt. Allocation	
156.7625–156.8375 MARITIME MOBILE (distress and calling) 501 613						
156.8375–174 FIXED MOBILE except aero-nautical mobile 6:3 613B 615	156.8375–174 FIXED MOBILE 613 616 617 618					
			157.0375–157.1875	MARITIME MOBILE US214 US266 613 G109	US214 US266 613	
			157.1875–157.45	US223 US266 613	MARITIME MOBILE LAND MOBILE US223 US266 613 NG111 NG154	
			157.45–161.575	US266 613	LAND MOBILE US266 613 NG6 NG28 NG70 NG111 NG112 NG124 NG148	157.45–157.725 MHz Land Transportation / 157.725–157.755 MHz Industrial / 157.755–158.115 MHz Domestic Public / 158.115–158.475 MHz Industrial / 158.715–159.480 MHz Public Safety / 159.480–161.575 MHz Land Transportation
			161.575–161.625	US77 613	MARITIME MOBILE US77 613 NG6 NG17	
			161.625–161.775	613	LAND MOBILE 613 NG6	Remote pickup broadcast

TABLES OF FREQUENCY ALLOCATIONS

	INTERNATIONAL			UNITED STATES		
Region 1 MHz	Region 2 MHz	Region 3 MHz	Band MHz	Government Allocation	Non-Govt. Allocation	Remarks
			161.775-162.0125	US266 613	MARITIME MOBILE LAND MOBILE US266 613 NG6 NG154	
			162.0125-173.2	FIXED MOBILE US8 US11 US13 US216 US223 US300 US312 613 G5	FIXED MOBILE US8 US11 US13 US216 US223 US300 US312 613	The Channeling Plan for assignments in this band is shown in Section 4.3.7 of the NTIA Manual.
			173.2-173.4	FIXED Land Mobile NG124	FIXED Land Mobile	Industrial Public safety
			173.4-174	FIXED MOBILE G5		The Channeling Plan of assignments in this band is shown in Section 4.3.7 of the NTIA Manual.
174-223 BROADCASTING 621 623 628 629	174-216 BROADCASTING Fixed Mobile 620	174-223 FIXED MOBILE BROADCASTING 619 624 625 626 630	174-216		BROADCASTING NG115 NG128 NG149	Television broadcasting
	216-220 FIXED MARITIME MOBILE Radiolocation 627 627A		216-220	MARITIME MOBILE Aeronautical Mobile Fixed Land Mobile Radiolocation US210 US229 US274 US317 627 G2	MARITIME MOBILE Aeronautical Mobile Fixed Land Mobile US210 US229 US274 US317 627 NG152	

TABLES OF FREQUENCY ALLOCATIONS

INTERNATIONAL			Band MHz	UNITED STATES		Remarks
Region 1 MHz	Region 2 MHz	Region 3 MHz		Government Allocation	Non-Govt. Allocation	
	220-225 AMATEUR FIXED MOBILE Radiolocation 627		220-222	LAND MOBILE Radiolocation 627 G2	LAND MOBILE 627	The Channeling Plan for Land Mobile assignments in this band is shown in Section 4.3.15 of the NTIA Manual.
			222-225	Radiolocation	AMATEUR	
223-230 BROADCASTING Fixed Mobile 622 628 629 631 632 635	225-235 FIXED MOBILE	223-230 FIXED MOBILE BROADCASTING AERONAUTICAL RADIONAVIGATION Radiolocation 636 637	225-235	FIXED MOBILE G27	627	The FAA provides air traffic control communications to the military services on selected frequencies in this band.
230-235 FIXED MOBILE 629 632 635 638 639		230-235 FIXED MOBILE AERONAUTICAL RADIONAVIGATION 637	235-267	FIXED MOBILE 501 592 642 G27 G100	501 592 642	The FAA provides air traffic control communications to the military services on selected frequencies in this band.
235-267 FIXED MOBILE 501 592 635 640 641 642						

TABLES OF FREQUENCY ALLOCATIONS

INTERNATIONAL			Band MHz	UNITED STATES		Remarks
Region 1 MHz	Region 2 MHz	Region 3 MHz		Government Allocation	Non-Govt. Allocation	
267-272 FIXED MOBILE Space Operation (space-to-Earth) 641 643			267-322	FIXED MOBILE		The FAA provides air traffic control communications to the military services on selected frequencies in this band.
272-273 SPACE OPERATION (space-to-Earth) FIXED MOBILE 641						
273-312 FIXED MOBILE 641						
312-315 FIXED MOBILE Mobile-Satellite (Earth-to-space) 641 641A						
315-322 FIXED MOBILE 641						
322-328.6 FIXED MOBILE RADIO ASTRONOMY 644			322-328.5	FIXED MOBILE 644 G27	644	The FAA provides air traffic control communications to the military services on selected frequencies in this band.
				G27 G100		
328.6-335.4 AERONAUTICAL RADIONAVIGATION 645 645A			328.6-335.4	AERONAUTICAL RADIONAVIGATION 645	AERONAUTICAL RADIONAVIGATION 645	

TABLES OF FREQUENCY ALLOCATIONS

INTERNATIONAL			Band MHz	UNITED STATES		Remarks
Region 1 MHz	Region 2 MHz	Region 3 MHz		Government Allocation	Non-Govt. Allocation	
335.4-387 FIXED MOBILE 641			335.4-399.9	FIXED MOBILE G27 G100		The FAA provides air traffic control communications to the military services on selected frequencies in this band.
387-390 FIXED MOBILE Mobile-Satellite (space-to-Earth) 641 641A						
390-399.9 FIXED MOBILE 641						
399.9-400.05 RADIONAVIGATION-SATELLITE 609 645B			399.9-400.05	399.9-400.05 RADIONAVIGATION-SATELLITE MOBILE-SATELLITE (Earth-to-space) US319 US326 645B	399.9-400.05 RADIONAVIGATION-SATELLITE MOBILE-SATELLITE (Earth-to-space) US319 US326 645B	
400.05-400.15 STANDARD FREQUENCY AND TIME SIGNAL-SATELLITE (400.1 MHz) 646 647			400.05-400.15	STANDARD FREQUENCY AND TIME SIGNAL-SATELLITE (400.1 MHz) 646	STANDARD FREQUENCY AND TIME SIGNAL-SATELLITE (400.1 MHz) 646	

TABLES OF FREQUENCY ALLOCATIONS

INTERNATIONAL			Band MHz	UNITED STATES		Remarks
Region 1 MHz	Region 2 MHz	Region 3 MHz		Government Allocation	Non-Govt. Allocation	
400.15-401 METEOROLOGICAL AIDS METEOROLOGICAL-SATELLITE (space-to-Earth) MOBILE-SATELLITE (space-to-Earth) 599B SPACE RESEARCH (space-to-Earth) 647A Space Operation (space-to-Earth) 647 647B			400.15-401	METEOROLOGICAL AIDS (radio-sonde) METEOROLOGICAL SATELLITE (space-to-Earth) SPACE RESEARCH (space-to-Earth) 647A MOBILE-SATELLITE (space-to-Earth) 599B US319 US320 US324 Space Operation (space-to-Earth) 647 647B US70	METEOROLOGICAL AIDS (radio-sonde) SPACE RESEARCH (space-to-Earth) 647A MOBILE-SATELLITE (space-to-Earth) 599B US319 US320 US324 Space Operation (space-to-Earth) 647 647B US70	SATELLITE COMMUNICATION (25)
401-402 METEOROLOGICAL AIDS SPACE OPERATION (space-to-Earth) Earth Exploration-Satellite (Earth-to-space) Fixed Meteorological-Satellite (Earth-to-space) Mobile except aeronautical mobile			401-402	METEOROLOGICAL AIDS (Radiosonde) SPACE OPERATION (space-to-Earth) Earth Exploration-Satellite (Earth-to-space) Meteorological-Satellite (Earth-to-space) US70	METEOROLOGICAL AIDS (Radiosonde) SPACE OPERATION (space-to-Earth) Earth Exploration-Satellite (Earth-to-space) Meteorological-Satellite (Earth-to-space) US70	

TABLES OF FREQUENCY ALLOCATIONS

INTERNATIONAL			Band MHz	UNITED STATES		Remarks
Region 1 MHz	Region 2 MHz	Region 3 MHz		Government Allocation	Non-Govt. Allocation	
402-403 METEOROLOGICAL AIDS Earth Exploration-Satellite (Earth-to-space) Fixed Meteorological-Satellite (Earth-to-space) Mobile except aeronautical mobile			402-403	METEOROLOGICAL AIDS (Radiosonde) Earth Exploration-Satellite (Earth-to-space) Meteorological-Satellite (Earth-to-space) US70	METEOROLOGICAL AIDS (Radiosonde) Earth Exploration-Satellite (Earth-to-space) Meteorological-Satellite (Earth-to-space) US70	
403-406 METEOROLOGICAL AIDS Fixed Mobile except aeronautical mobile 648			403-406	METEOROLOGICAL AIDS (Radiosonde) US70 G6	METEOROLOGICAL AIDS (Radiosonde) US70	
406-406.1 MOBILE-SATELLITE (Earth-to-space) 649 649A			406-406.1	MOBILE-SATELLITE (Earth-to-space) 649 649A	MOBILE-SATELLITE (Earth-to-space) 649 649A	Satellite Emergency Position Indicating Radiobeacon (EPIRB).
406.1-410 FIXED MOBILE except aeronautical mobile RADIO ASTRONOMY 648 650			406.1-410	FIXED MOBILE RADIO ASTRONOMY US13 US74 US117 G5 G6	RADIO ASTRONOMY US13 US74 US117	
410-420 FIXED MOBILE except aeronautical mobile Space Research (space-to-space) 651A			410-420	FIXED MOBILE Space Research (space-to-space) US13 G5 651A	US13	The Channeling Plan for assignments in these bands are shown in Section 4.3.9 of the NTIA Manual.
420-430 FIXED MOBILE except aeronautical mobile Radiolocation 651 652 653			420-450	RADIOLOCATION US7 US87 US217 US228 US230 664 668 G2 G8	Amateur US7 US87 US217 US228 US230 664 668 NG135	

TABLES OF FREQUENCY ALLOCATIONS

INTERNATIONAL			UNITED STATES			Remarks
Region 1 MHz	Region 2 MHz	Region 3 MHz	Band MHz	Government Allocation	Non-Govt. Allocation	
430-440 AMATEUR RADIOLOCATION	430-440 RADIOLOCATION Amateur					
653 654 655 656 657 658 659 661 662 663 664 665	653 658 659 660 660A 663 664					
440-450 FIXED MOBILE except aeronautical mobile Radiolocation						
651 652 653 666 667 668						
450-460 FIXED MOBILE			450-460	US87 668 669 670	LAND MOBILE US87 668 669 670 NG12 NG12 NG124 NG148	450-451 MHz Remote pickup broadcast
653 668 669 670						451-454 MHz Public Safety Industrial Land Transportation
						454-455 MHz Domestic Public
						455-456 MHz Remote pickup broadcast
						456-459 MHz Public Safety Industrial Land Transportation
						459-460 MHz Domestic Public

TABLES OF FREQUENCY ALLOCATIONS

INTERNATIONAL			Band MHz	UNITED STATES		Remarks
Region 1 MHz	Region 2 MHz	Region 3 MHz		Government Allocation	Non-Govt. Allocation	
470-470 FIXED MOBILE Meteorological-Satellite (space-to-Earth) 669 670 671 672	Meteorological-Satellite (space-to-Earth)		460-470	Meteorological-Satellite (space-to-Earth) US201 US209 US216 669 670 671	LAND MOBILE US201 US209 US216 669 670 671 NG124	460-462.5375 MHz Public Safety Industrial Land Transportation 462.5375- 462.7375 MHz Personal 462.7375- 467.5375 MHz Public Safety Industrial Land Transportation 467.5375- 467.7375 MHz Personal 467.7375-470 MHz Public Safety Industrial Land Transporta- tion
470-790 BROADCASTING 676 677A 683 684 685 686 686A 687 689 693 694	470-512 BROADCASTING Fixed Mobile 674 675	470-585 FIXED MOBILE BROADCASTING 673 677 679	470-512		BROADCASTING LAND MOBILE NG66 NG114 NG127 NG128 NG149	Broadcasting Public Safety Industrial Land Transporta- tion Domestic Public
	512-608 BROADCASTING 678	585-610 FIXED MOBILE BROADCASTING RADIONAVIGATION 688 689 690	512-608		BROADCASTING NG128 NG149	Television broadcasting

TABLES OF FREQUENCY ALLOCATIONS

INTERNATIONAL			UNITED STATES			
Region 1 MHz	Region 2 MHz	Region 3 MHz	Band MHz	Government Allocation	Non-Govt. Allocation	Remarks
	608-614 RADIO ASTRONOMY Mobile-Satellite except aeronautical mobile-satellite (Earth-to-space)	610-890 FIXED MOBILE BROADCASTING	608-614	RADIO ASTRONOMY US74 US246	RADIO ASTRONOMY US74 US246	
790-862 FIXED BROADCASTING 694 695 695A 696 697 700B 702	614-806 BROADCASTING Fixed Mobile 675 692 692A 693	677 688 689 690 691 693 701	614-806		BROADCASTING NG30 NG43 NG128 NG149	
862-890 FIXED MOBILE except aero-nautical mobile BROADCASTING 703 700B 704	806-890 FIXED MOBILE BROADCASTING 692A 700 700A		806-902	US116 US268 704A G2	LAND MOBILE US116 US268 704A NG30 NG43 NG63 NG151	806-821 MHz Conventional and Trunked Systems 821-824 MHz Public Safety 824-825 MHz Cellular 825-849 MHz Cellular 849-851 MHz Reserve 851-866 MHz Conventional and Trunked Systems 866-869 MHz Public Safety 869-870 MHz Cellular 890-894 MHz Cellular 894-896 MHz Reserve 896-901 MHz Private Land Mobile 901-902 MHz Reserve
890-942 FIXED MOBILE except aero-nautical mobile Radiolocation 700A 704A 705 704	890-902 FIXED MOBILE except aero-nautical mobile Radiolocation 700A 704A 705	890-942 FIXED MOBILE BROADCASTING Radiolocation 706				

TABLES OF FREQUENCY ALLOCATIONS

INTERNATIONAL			UNITED STATES			
Region 1 MHz	Region 2 MHz	Region 3 MHz	Band MHz	Government Allocation	Non-Govt. Allocation	Remarks
	902-928 FIXED Amateur Mobile except aeronautical mobile Radiolocation 705 707 707A		902-928	RADIOLOCATION US215 US218 US267 US275 707 G11 G59	US215 US218 US267 US275 707	ISM 915 ± 13 MHz
	928-942 FIXED MOBILE except aeronautical mobile Radiolocation 705		928-929	FIXED US116 US215 US268 G2	FIXED US116 US215 US268 NG120	
			929-932	US116 US215 US268 G2	LAND MOBILE US116 US215 US268 NG120	
			932-935	FIXED US215 US268 G2	FIXED US215 US268 NG120	The Channeling Plan for assignments in this band is shown in Section 4.3.14 of the NTIA Manual.
			935-940	US116 US215 US268 G2	LAND MOBILE US116 US215 US268 N120	
			940-941	US116 US268 G2	MOBILE US116 US268	
			941-944	FIXED US268 US301 US302 G2	FIXED US268 US301 US302 NG64 NG120	The Channeling Plan for assignments in this band is shown in Section 4.3.14 of the NTIA Manual.
942-960 FIXED MOBILE except aeronautical mobile BROADCASTING 703 704	942-960 FIXED MOBILE	942-960 FIXED MOBILE BROADCASTING 701	944-960	FIXED	FIXED NG64 NG120	

TABLES OF FREQUENCY ALLOCATIONS

INTERNATIONAL			Band MHz	UNITED STATES		Remarks
Region 1 MHz	Region 2 MHz	Region 3 MHz		Government Allocation	Non-Govt. Allocation	
AERONAUTICAL RADIONAVIGATION 709			960-1215	AERONAUTICAL RADIONAVIGATION US224 709	AERONAUTICAL RADIONAVIGATION US224 709	
RADIOLOCATION RADIONAVIGATION-SATELLITE (space-to-Earth) 710 711 712 712A 713			1215-1240	RADIOLOCATION RADIONAVIGATION-SATELLITE (space-to-Earth) 713 G56	713	
RADIOLOCATION RADIONAVIGATION-SATELLITE (space-to-Earth) 710 Amateur 711 712 712A 713 714			1240-1300	RADIOLOCATION 664 713 714 G56	Amateur 664 713 714	
RADIOLOCATION Amateur 664 711 712 712A 713 714						
AERONAUTICAL RADIONAVIGATION 717 Radiolocation 715 716 718			1300-1350	AERONAUTICAL RADIONAVIGATION Radiolocation 717 718 G2	AERONAUTICAL RADIONAVIGATION 717 718	
FIXED MOBILE RADIOLOCATION 718 719 720	RADIOLOCATION 714 718 720		1350-1400	FIXED MOBILE RADIOLOCATION US311 714 718 720 G2 G27 G114	US311 714 718 720	
EARTH EXPLORATION-SATELLITE (passive) RADIO ASTRONOMY SPACE RESEARCH (passive) 721 722			1400-1427	EARTH EXPLORATION-SATELLITE (Passive) RADIO ASTRONOMY SPACE RESEARCH (Passive) US74 US246 722	EARTH EXPLORATION-SATELLITE (Passive) RADIO ASTRONOMY SPACE RESEARCH (Passive) US74 US246 722	

TABLES OF FREQUENCY ALLOCATIONS

INTERNATIONAL			Band MHz	UNITED STATES		
Region 1 MHz	Region 2 MHz	Region 3 MHz		Government Allocation	Non-Govt. Allocation	Remarks
1427-1429 SPACE OPERATION (Earth-to-space) FIXED MOBILE except aeronautical mobile 722			1427-1429	FIXED MOBILE except aeronautical mobile SPACE OPERATION (Earth-to-space) 722 G30	SPACE OPERATION (Earth-to-space) Land Mobile (Telemetering and telecommand) Fixed (Telemetering) 722	
1429-1452 FIXED MOBILE except aeronautical mobile 722 723B	1429-1452 FIXED MOBILE 723 722		1429-1435	FIXED MOBILE 722 G30	Land Mobile (Telemetering and telecommand) Fixed (Telemetering) 722	
1452-1492 FIXED MOBILE except aeronautical mobile BROADCASTING 722A 722B BROADCASTING-SATELLITE 722A 722B 722 723B	1452-1492 FIXED MOBILE 723 BROADCASTING 722A 722B BROADCASTING-SATELLITE 722A 722B 722 722C	1452-1492 FIXED MOBILE 723 BROADCASTING 722A 722B BROADCASTING-SATELLITE 722A	1435-1525	MOBILE (Aeronautical telemetering) US78 722	MOBILE (Aeronautical telemetering) US78 722	
1492-1525 FIXED MOBILE except aeronautical mobile 722 723B	1492-1525 FIXED MOBILE 723 MOBILE-SATELLITE (space-to-Earth) 722 722C 723C	1492-1525 FIXED MOBILE 723 722				

TABLES OF FREQUENCY ALLOCATIONS

INTERNATIONAL			Band MHz	UNITED STATES		Remarks
Region 1 MHz	Region 2 MHz	Region 3 MHz		Government Allocation	Non-Govt. Allocation	
1525-1530 SPACE OPERATION (space-to-Earth) FIXED MARITIME MOBILE-SATELLITE (space-to-Earth) Earth Exploration-Satellite Land Mobile Satellite (space-to-Earth) 726B Mobile except aero-nautical mobile 724 722 723B 725 726A 726D	1525-1530 SPACE OPERATION (space-to-Earth) MOBILE-SATELLITE (space-to-Earth) Earth Exploration-Satellite Fixed Mobile 723 722 723A 726A 726D	1525-1530 SPACE OPERATION (space-to-Earth) FIXED MOBILE-SATELLITE (space-to-Earth) Earth Exploration-Satellite Mobile 723 724 722 726A 726D	1525-1530	MOBILE-SATELLITE (Space-to-Earth) Mobile (Aero-nautical tele-metry) 722 726A US78	MOBILE-SATELLITE (Space-to-Earth) Mobile (Aero-nautical tele-metry) 722 726A US78	
1530-1533 SPACE OPERATION (space-to-Earth) MARITIME MOBILE-SATELLITE (space-to-Earth) LAND MOBILE-SATELLITE (space-to-Earth) Earth Exploration-Satellite Fixed Mobile except aero-nautical mobile 722 723B 726A 726D	1530-1533 SPACE OPERATION (space-to-Earth) MARITIME MOBILE-SATELLITE (space-to-Earth) LAND MOBILE-SATELLITE (space-to-Earth) Earth Exploration-Satellite Fixed Mobile 723 722 726A 726C 726D		1530-1535	MARITIME MOBILE-SATELLITE (space-to-Earth) MOBILE-SATELLITE (space-to-Earth) Mobile (Aeronautical telemetering) US78 US272 US315 722 726A	MARITIME MOBILE-SATELLITE (space-to-Earth) MOBILE-SATELLITE (space-to-Earth) Mobile (Aeronautical telemetering) US78 US272 US315 722 726A	

TABLES OF FREQUENCY ALLOCATIONS

INTERNATIONAL			Band MHz	UNITED STATES		
Region 1 MHz	Region 2 MHz	Region 3 MHz		Government Allocation	Non-Govt. Allocation	Remarks
1533-1535 SPACE OPERATION (space-to-Earth) MARITIME MOBILE-SATELLITE (space-to-Earth) Earth Exploration-Satellite Fixed Mobile except aeronautical mobile Land Mobile-Satellite (space-to-Earth) 726B 722 723B 726A 726D	1533-1535 SPACE OPERATION (space-to-Earth) MARITIME MOBILE-SATELLITE (space-to-Earth) Earth Exploration-Satellite Fixed Mobile 723 Land Mobile-Satellite (space-to-Earth) 726B 722 726A 726C 726D					
1535-1544 MARITIME MOBILE-SATELLITE (space-to-Earth) Land Mobile-Satellite (space-to-Earth) 726B 722 726A 726C 726D 727			1535-1544	MARITIME MOBILE-SATELLITE (space-to-Earth) MOBILE-SATELLITE (space-to-Earth) US315 722 726A	MARITIME MOBILE-SATELLITE (space-to-Earth) MOBILE-SATELLITE (space-to-Earth) US315 722 726A	
1544-1545 MOBILE-SATELLITE (space-to-Earth) 722 726D 727 727A			1544-1545	MOBILE-SATELLITE (space-to-Earth) 722 727A	MOBILE-SATELLITE (space-to-Earth) 722 727A	
1545-1555 AERONAUTICAL MOBILE-SATELLITE (R) (space-to-Earth) 722 726A 726D 727 729 729A 730			1545-1549.5	AERONAUTICAL MOBILE-SATELLITE (R) (space-to-Earth) Mobile-Satellite (space-to-Earth) US308 US309 722 726A	AERONAUTICAL MOBILE-SATELLITE (R) (space-to-Earth) Mobile-Satellite (space-to-Earth) US308 US309 722 726A	

TABLES OF FREQUENCY ALLOCATIONS

INTERNATIONAL			Band MHz	UNITED STATES		Remarks
Region 1 MHz	Region 2 MHz	Region 3 MHz		Government Allocation	Non-Govt. Allocation	
	1555-1559 LAND MOBILE-SATELLITE (space-to-Earth) 722 726A 726D 727 730 730A 730B 730C		1549.5-1558.5	AERONAUTICAL MOBILE-SATELLITE (R) (space-to-Earth) MOBILE-SATELLITE (space-to-Earth) US308 US309 722 726A	AERONAUTICAL MOBILE-SATELLITE (R) (space-to-Earth) MOBILE-SATELLITE (space-to-Earth) US308 US309 722 726A	
			1558.5-1559	AERONAUTICAL MOBILE-SATELLITE (R) (space-to-Earth) US308 US309 722 726A	AERONAUTICAL MOBILE-SATELLITE (R) (space-to-Earth) US308 US309 722 726A	
	1559-1610 AERONAUTICAL RADIONAVIGATION RADIONAVIGATION-SATELLITE (space-to-Earth) 722 727 730 731		1559-1610	AERONAUTICAL RADIONAVIGATION RADIONAVIGATION-SATELLITE (space-to-Earth) G126 US208 US260 722	AERONAUTICAL RADIONAVIGATION RADIONAVIGATION-SATELLITE (space-to-Earth) US208 US260 722	
1610-1610.6 MOBILE-SATELLITE (Earth-to-space) AERONAUTICAL RADIONAVIGATION 722 727 730 731 731E 732 733 733A 733B 733E 733F	1610-1610.6 MOBILE-SATELLITE (Earth-to-space) AERONAUTICAL RADIONAVIGATION RADIODETERMINATION-SATELLITE (Earth-to-space) 722 731E 732 733 733A 733C 733D 733E	1610-1610.6 MOBILE-SATELLITE (Earth-to-space) AERONAUTICAL RADIONAVIGATION Radiodetermination-Satellite (Earth-to-space) 722 727 730 731E 732 733 733A 733B 733E	1610-1610.6	AERONAUTICAL RADIONAVIGATION RADIODETERMINATION SATELLITE (Earth-to-space) MOBILE-SATELLITE (Earth-to-space) US208 US260 US319 722 731E 732 733 733A 733E	AERONAUTICAL RADIONAVIGATION RADIODETERMINATION SATELLITE (Earth-to-space) MOBILE-SATELLITE (Earth-to-space) US208 US260 US319 722 731E 732 733 733A 733E	

TABLES OF FREQUENCY ALLOCATIONS

INTERNATIONAL			Band MHz	UNITED STATES		Remarks
Region 1 MHz	Region 2 MHz	Region 3 MHz		Government Allocation	Non-Govt. Allocation	
1610.6-1613.8 MOBILE-SATELLITE (Earth-to-space) RADIO ASTRONOMY AERONAUTICAL RADIONAVIGATION 722 727 730 731 731E 732 733 733A 733B 733E 733F 734	1610.6-1613.8 MOBILE-SATELLITE (Earth-to-space) RADIO ASTRONOMY AERONAUTICAL RADIONAVIGATION RADIODETERMINA-TION-SATELLITE (Earth-to-space) 722 731E 732 733 733A 733C 733D 733E 734	1610.6-1613.8 MOBILE-SATELLITE (Earth-to-space) RADIO ASTRONOMY AERONAUTICAL RADIONAVIGATION Radiodetermina-tion-Satellite (Earth-to-space) 722 727 730 731E 732 733 733A 733B 733E 734	1610.6-1613.8	AERONAUTICAL RADIONAVIGATION RADIODETERMINATION SATELLITE (Earth-to-space) MOBILE-SATELLITE (Earth-to-space) RADIO-ASTRONOMY US208 US260 US319 722 731E 732 733 733A 733E 734	AERONAUTICAL RADIONAVIGATION RADIODETERMINATION SATELLITE (Earth-to-space) MOBILE-SATELLITE (Earth-to-space) RADIO-ASTRONOMY US208 US260 US319 722 731E 732 733 733A 733E 734	
1613.8-1626.5 MOBILE-SATELLITE (Earth-to-space) AERONAUTICAL RADIONAVIGATION Mobile Satellite (space-to-Earth) 722 727 730 731 731E 731F 732 733 733A 733B 733E 733F	1613.8-1626.5 MOBILE-SATELLITE (Earth-to-space) AERONAUTICAL RADIONAVIGATION RADIODETERMINA-TION-SATELLITE (Earth-to-space) Mobile Satellite (space-to-Earth) 722 731E 731F 732 733 733A 733C 733D 733E	1613.8-1626.5 MOBILE-SATELLITE (Earth-to-space) AERONAUTICAL RADIONAVIGATION Radiodetermina-tion-Satellite (Earth-to-space) Mobile Satellite (space-to-Earth) 722 727 730 731E 731F 732 733 733A 733B 733E	1613.8-1626.5	AERONAUTICAL RADIONAVIGATION RADIODETERMINATION SATELLITE (Earth-to-space) MOBILE-SATELLITE (Earth-to-space) Mobile-Satellite (space-to-Earth) US208 US260 US319 722 731E 731F 732 733 733E	AERONAUTICAL RADIONAVIGATION RADIODETERMINATION SATELLITE (Earth-to-space) MOBILE-SATELLITE (Earth-to-space) Mobile-Satellite (space-to-Earth) US208 US260 US319 722 731E 731F 732 733 733E	
1626.5-1631.5 MARITIME MOBILE-SAT-ELLITE (Earth-to-space) Land Mobile-Satellite (Earth-to-Space) 726B 722 726A 726D 727 730	1626.5-1631.5 MOBILE-SATELLITE (Earth-to-space) 722 726A 726C 726D 727 730	1626.5-1631.5 MOBILE-SATELLITE (Earth-to-space)	1626.5-1645.5	MARITIME MOBILE-SATELLITE (Earth-to-space) MOBILE-SATELLITE (Earth-to-space) US315 722 726A	MARITIME MOBILE-SATELLITE (Earth-to-space) MOBILE-SATELLITE (Earth-to-space) US315 722 726A	

TABLES OF FREQUENCY ALLOCATIONS

INTERNATIONAL			UNITED STATES		Band MHz	Remarks
Region 1 MHz	Region 2 MHz	Region 3 MHz	Government Allocation	Non-Govt. Allocation		
1631.5-1634.5 MARITIME MOBILE-SATELLITE (Earth-to-space) LAND MOBILE-SATELLITE (Earth-to-space) 722 726A 726C 726D 727 730 734A						
1634.5-1645.5 MARITIME MOBILE-SATELLITE (Earth-to-space) Land Mobile-Satellite (Earth-to-space) 726B 722 726A 726C 726D 727 730						
1645.5-1646.5 MOBILE-SATELLITE (Earth-to-space) 722 726D 734B			MOBILE-SATELLITE (Earth-to-space) 722 734B	MOBILE-SATELLITE (Earth-to-space) 722 734B	1645.5-1646.5	
1646.5-1656.5 AERONAUTICAL MOBILE-SATELLITE (R) (Earth-to-space) 722 726A 726D 727 729A 730 735			AERONAUTICAL MOBILE-SATELLITE (R) (Earth-to-space) Mobile-Satellite (Earth-to-space) US308 US309 722 726A	AERONAUTICAL MOBILE-SATELLITE (R) (Earth-to-space) Mobile-Satellite (Earth-to-space) US308 US309 722 726A	1646.5-1651	
			AERONAUTICAL MOBILE-SATELLITE (R) (Earth-to-space) MOBILE-SATELLITE (Earth-to-space) US308 US309 722 726A	AERONAUTICAL MOBILE-SATELLITE (R) (Earth-to-space) MOBILE-SATELLITE (Earth-to-space) US308 US309 722 726A	1651-1660	
1656.5-1660 LAND MOBILE-SATELLITE (Earth-to-space) 722 726A 726D 727 730 730A 730B 730C 734A						
1660-1660.5 LAND MOBILE-SATELLITE (Earth-to-space) RADIO ASTRONOMY 722 726A 726D 730A 730B 730C 736			AERONAUTICAL MOBILE-SATELLITE (R) (Earth-to-space) RADIO ASTRONOMY US309 722 726A 736	AERONAUTICAL MOBILE-SATELLITE (R) (Earth-to-space) RADIO ASTRONOMY US309 722 726A 736	1660-1660.5	

TABLES OF FREQUENCY ALLOCATIONS

INTERNATIONAL			Band MHz	UNITED STATES		Remarks
Region 1 MHz	Region 2 MHz	Region 3 MHz		Government Allocation	Non-Govt. Allocation	
1660.5-1668.4 RADIO ASTRONOMY SPACE RESEARCH (passive) Fixed Mobile except aeronautical mobile 722 736 737 738 739			1660.5- 1668.4	RADIO ASTRONOMY SPACE RESEARCH (Passive) US74 US246 722	RADIO ASTRONOMY SPACE RESEARCH (Passive) US74 US246 722	
1668.4-1670 METEOROLOGICAL AIDS FIXED MOBILE except aeronautical mobile RADIO ASTRONOMY 722 736			1668.4-1670	METEOROLOGICAL AIDS (Radiosonde) RADIO ASTRONOMY US74 US99 722 736	METEOROLOGICAL AIDS (Radiosonde) RADIO ASTRONOMY US74 US99 722 736	
1670-1675 METEOROLOGICAL AIDS FIXED METEOROLOGICAL-SATELLITE (space-to-Earth) MOBILE 740A 722	1670-1690 METEOROLOGICAL AIDS FIXED METEOROLOGICAL- SATELLITE (space- to- Earth) MOBILE except aero- nautical mobile MOBILE-SATELLITE (Earth-to-space) 722 735A	1675-1690 FIXED METEOROLOGICAL AIDS METEOROLOGICAL- SATELLITE (space- to-Earth) MOBILE except aeronautical mobile 722	1670-1690	METEOROLOGICAL AIDS (Radiosonde) METEOROLOGICAL- SATELLITE (space- to-Earth) US211 722	METEOROLOGICAL AIDS (Radiosonde) METEOROLOGICAL- SATELLITE (space- to-Earth) US211 722	
1675-1690 METEOROLOGICAL AIDS FIXED METEOROLOGICAL- SATELLITE (space- to- Earth) MOBILE except aero- nautical mobile 722						

TABLES OF FREQUENCY ALLOCATIONS

INTERNATIONAL			Band MHz	UNITED STATES		Remarks
Region 1 MHz	Region 2 MHz	Region 3 MHz		Government Allocation	Non-Govt. Allocation	
1690-1700 METEOROLOGICAL AIDS METEOROLOGICAL-SATELLITE (space-to-Earth) Fixed Mobile except aero-nautical mobile 671 722 741	1690-1700 METEOROLOGICAL AIDS METEOROLOGICAL-SATELLITE (space-to-Earth) MOBILE-SATELLITE (Earth-to-space) 671 722 735A 740	1690-1700 METEOROLOGICAL AIDS METEOROLOGICAL SATELLITE (space-to-Earth) 671 722 740 742	1690-1700	METEOROLOGICAL AIDS (Radiosonde) METEOROLOGICAL-SATELLITE (space-to-Earth) 671 722	METEOROLOGICAL AIDS (Radiosonde) METEOROLOGICAL-SATELLITE (space-to-Earth) 671 722	
1700-1710 FIXED METEOROLOGICAL-SATELLITE (space-to-Earth) MOBILE except aero-nautical mobile 671 722	1700-1710 FIXED METEOROLOGICAL-SATELLITE (space-to-Earth) MOBILE except aeronautical mo-bile MOBILE-SATELLITE (Earth-to-space) 671 722 735A	1700-1710 FIXED METEOROLOGICAL-SATELLITE (space-to-Earth) MOBILE except aero-nautical mobile 671 722 743	1700-1713	FIXED METEOROLOGICAL-SATELLITE (space-to-Earth) 671 722 G118	METEOROLOGICAL-SATELLITE (space-to-Earth) Fixed 671 722	
1710-1930 FIXED MOBILE 740A 722 744 745 746 746A			1710-1850	FIXED MOBILE US256 722 G42	US256 722	
1930-1970 FIXED MOBILE 764A	1930-1970 FIXED MOBILE Mobile-Satellite (Earth-to-space) 746A	1930-1970 FIXED MOBILE 746A	1850-1990		FIXED MOBILE	
1970-1980 FIXED MOBILE 746A	1970-1980 FIXED MOBILE MOBILE-SATELLITE (Earth-to-space) 746A 746B 746C	1970-1980 FIXED MOBILE 746A		US331	US331	

TABLES OF FREQUENCY ALLOCATIONS

INTERNATIONAL			UNITED STATES			Remarks
Region 1 MHz	Region 2 MHz	Region 3 MHz	Band MHz	Government Allocation	Non-Govt. Allocation	
1980-2010 FIXED MOBILE MOBILE-SATELLITE (Earth-to-space) 746A 746B 746C			1990-2110	US90 US111 US219 US222	FIXED MOBILE US90 US111 US219 US222 NG23 NG118	
2010-2025 FIXED MOBILE 746A						
2025-2110 SPACE OPERATION (Earth-to-space)(space-to-space) EARTH EXPLORATION-SATELLITE (Earth-to-space)(space-to-space) FIXED MOBILE 747A SPACE RESEARCH (Earth-to-space)(space-to-space) 750A						
2110-2120 FIXED MOBILE SPACE RESEARCH (deep space) (Earth-to-space) 746A			2110-2150	US111 US252	FIXED MOBILE US111 US252 NG23 NG153	
2120-2160 FIXED MOBILE 746A	2120-2160 FIXED MOBILE Mobile-Satellite (space-to-Earth) 746A	2120-2160 FIXED MOBILE 746A				
			2150-2160		FIXED NG23	

TABLES OF FREQUENCY ALLOCATIONS

INTERNATIONAL			UNITED STATES			
Region 1 MHz	Region 2 MHz	Region 3 MHz	Government Allocation	Non-Govt. Allocation	Band MHz	Remarks
2160-2170 FIXED MOBILE 746A	2160-2170 FIXED MOBILE MOBILE-SATELLITE (space-to-Earth) 746A 746B 746C	2160-2170 FIXED MOBILE 746A		FIXED MOBILE NG23 NG153	2160-2200	
2170-2200 FIXED MOBILE MOBILE-SATELLITE (space-to-Earth) 746A 746B 746C						
2200-2290 SPACE OPERATION (space-to-Earth) (space-to-space) EARTH EXPLORATION-SATELLITE (space-to-Earth) FIXED MOBILE 747A SPACE RESEARCH (space-to-Earth) (space-to-space) 750A			FIXED (LOS* only) MOBILE (LOS only) including aeronautical telemetering, but excluding flight testing of manned aircraft) SPACE RESEARCH (space-to-Earth) (space-to-space) SPACE OPERATION (space-to-Earth) (space-to-space) EARTH EXPLORATION- SATELLITE (space-to-Earth) (space-to-space) G101 US303 750A	US303	2200-2290	* Line of sight.
2290-2300 FIXED MOBILE except aeronautical mobile SPACE RESEARCH (Deep Space) (space-to-Earth)			FIXED MOBILE except aeronautical mobile SPACE RESEARCH (space-to-Earth) (Deep Space only)	SPACE RESEARCH (space-to-Earth) (Deep Space only)	2290-2300	

TABLES OF FREQUENCY ALLOCATIONS

INTERNATIONAL			UNITED STATES		
Region 1 MHz	Region 2 MHz	Region 3 MHz	Band MHz	Government Allocation	Non-Govt. Allocation	Remarks
2300-2450 FIXED MOBILE Amateur Radiolocation 664 751A 752	2300-2450 FIXED MOBILE RADIOLOCATION Amateur 664 750B 751 751B 752		2300-2310		Amateur	
				G123	US253	
			2310-2360	Mobile Radiolocation Fixed 751B US276 US327 US328 G2 G120	BROADCASTING- SATELLITE Mobile 751B US276 US327 US328	
			2360-2390	MOBILE RADIOLOCATION Fixed US276 G2 G120	MOBILE US276	
			2390-2400	G122	AMATEUR	
			2400-2402	664 752 G123	Amateur 644 752	ISM 2450 ± 50 MHz
			2402-2417	664 752 G122	AMATEUR 664 752	ISM 2450 ± 50 MHz
			2417-2450	Radiolocation 664 752 G2 G124	Amateur 664 752	ISM 2450 ± 50 MHz
2450-2483.5 FIXED MOBILE Radiolocation 752 753	2450-2483.5 FIXED MOBILE RADIOLOCATION 751 752		2450-2483.5		FIXED MOBILE Radiolocation	ISM 2450 ± 50 MHz
				US41 752	US41 752	

TABLES OF FREQUENCY ALLOCATIONS

INTERNATIONAL			Band MHz	UNITED STATES		Remarks
Region 1 MHz	Region 2 MHz	Region 3 MHz		Government Allocation	Non-Govt. Allocation	
2483.5-2500 FIXED MOBILE MOBILE-SATELLITE (space-to-Earth) Radiolocation 733F 752 753 753A 753B 753C 753F	2483.5-2500 FIXED MOBILE MOBILE-SATELLITE (space-to-Earth) RADIODETERMIN-ATION-SATELLITE (space-to-Earth) 753A RADIOLOCATION 752 753D 753F	2483.5-2500 FIXED MOBILE MOBILE-SATELLITE (space-to-Earth) RADIOLOCATION Radiodetermina-tion-Satellite (space-to-Earth) 753A 752 753C 753F	2483.5-2500	RADIODETERMINA-TION - SATELLITE (space-to-Earth) 753A MOBILE-SATELLITE (space-to-Earth) US41 US319 752 753F	RADIODETERMINA-TION-SATELLITE (space-to-Earth) 753A MOBILE-SATELLITE (space-to-Earth) US41 US319 752 753F NG147	
2500-2520 FIXED 762 763 764 MOBILE except aero-nautical mobile MOBILE-SATELLITE (space-to-Earth) 754 754B 755A 756 759 760A	2500-2520 FIXED 762 764 FIXED-SATELLITE (space-to-Earth) 761 MOBILE except aeronautical mobile MOBILE-SATELLITE (space-to-Earth) 754 754A 755 755A 760A	2483.5-2500 FIXED MOBILE MOBILE-SATELLITE (space-to-Earth) RADIOLOCATION Radiodetermina-tion-Satellite (space-to-Earth) 753A 752 753C 753F	2500-2655	US205 US269 720	BROADCASTING-SATELLITE FIXED US205 US269 720 NG47 NG101 NG102	
2520-2655 FIXED 762 763 764 MOBILE except aero-nautical mobile BROADCASTING-SATELLITE 757 760 720 754 754B 756 757A 758 759	2520-2655 FIXED 762 764 FIXED-SATELLITE (space-to-Earth) 761 MOBILE except aero-nautical mobile BROADCASTING-SATELLITE 757 760 720 754 755	2520-2535 FIXED 762 764 FIXED-SATELLITE (space-to-Earth) 761 MOBILE except aero-nautical mobile BROADCASTING-SATELLITE 757 760 754				

TABLES OF FREQUENCY ALLOCATIONS

INTERNATIONAL				UNITED STATES		
Region 1 MHz	Region 2 MHz	Region 3 MHz	Band MHz	Government Allocation	Non-Govt. Allocation	Remarks
		2535-2655 FIXED 762 764 MOBILE except aeronautical mobile BROADCASTING-SATEL-LITE 757 760 720 757A				
2655-2670 FIXED 762 763 764 MOBILE except aeronautical mobile BROADCASTING-SATEL-LITE 757 760 Earth Exploration-Satellite (passive) Radio Astronomy Space Research (passive) 758 759 765 766	2655-2670 FIXED 762 764 FIXED-SATELLITE (Earth-to-space) 761 MOBILE except aeronautical mobile BROADCASTING-SATEL-LITE 757 760 Earth Exploration-Satellite (passive) Radio Astronomy Space Research (passive) 765 766	2655-2670 FIXED 762 764 FIXED-SATELLITE (Earth-to-space) 761 MOBILE except aeronautical mobile BROADCASTING-SATEL-LITE 757 760 Earth Exploration-Satellite (passive) Radio Astronomy Space Research (passive) 765 766	2655-2690	Earth Exploration-Satellite (Passive) Radio Astronomy Space Research (Passive) US205 US269	BROADCASTING-SATELLITE FIXED Earth Exploration-tion-Satellite (passive) Radio Astronomy Space Research (Passive) US205 US269 NG47 NG101 NG102	

TABLES OF FREQUENCY ALLOCATIONS

INTERNATIONAL			UNITED STATES			Remarks
Region 1 MHz	Region 2 MHz	Region 3 MHz	Band MHz	Government Allocation	Non-Govt. Allocation	
2670-2690 FIXED 762 763 764 MOBILE except aeronautical mobile MOBILE-SATELLITE (Earth-to-space) Earth Exploration-Satellite (passive) Space Research (passive) 764A 765 766	2670-2690 FIXED 762 764 FIXED-SATELLITE (Earth-to-space) 761 MOBILE except aeronautical mobile (space-to-Earth) MOBILE-SATELLITE (Earth-to-space) Earth Exploration-Satellite (passive) Radio Astronomy Space Research (passive) 764A 765 766	2670-2690 FIXED 762 764 FIXED-SATELLITE (Earth-to-space) 761 MOBILE except aeronautical mobile MOBILE-SATELLITE (Earth-to-space) Earth Exploration-Satellite (passive) Radio Astronomy Space Research (passive) 764A 765 766				
EARTH EXPLORATION-SATELLITE (passive) RADIO ASTRONOMY SPACE RESEARCH (passive) 767 768 769			2690-2700	EARTH EXPLORATION-SATELLITE (passive) RADIO ASTRONOMY SPACE RESEARCH (passive) US74 US246	EARTH EXPLORATION-SATELLITE (passive) RADIO ASTRONOMY SPACE RESEARCH (passive) US74 US246	
AERONAUTICAL RADIONAVIGATION 717 Radiolocation 770 771			2700-2900	AERONAUTICAL RADIONAVIGATION METEOROLOGICAL AIDS Radiolocation US18 717 770 G2 G15	US18 717 770	
RADIONAVIGATION 773 Radiolocation 772 775A			2900-3100	MARITIME RADIONAVIGATION Radiolocation US44 US316 775A G56	MARITIME RADIONAVIGATION Radiolocation US44 US316 775A	
RADIOLOCATION 713 777 778			3100-3300	RADIOLOCATION US110 713 778 G59	Radiolocation US110 713 778	See Part 7.18 of the NTIA Manual.

TABLES OF FREQUENCY ALLOCATIONS

INTERNATIONAL			Band MHz	UNITED STATES		Remarks
Region 1 MHz	Region 2 MHz	Region 3 MHz		Government Allocation	Non-Govt. Allocation	
3300-3400 RADIOLOCATION 778 779 780	3300-3400 RADIOLOCATION Amateur Fixed Mobile 778 780	3300-3400 RADIOLOCATION Amateur 778 779	3300-3500	RADIOLOCATION US108 664 778 G31	Amateur Radiolocation US108 664 778	
3400-3600 FIXED FIXED-SATELLITE (space-to-Earth) Mobile Radiolocation 781 785	3400-3500 FIXED FIXED-SATELLITE (space-to-Earth) Amateur Mobile Radiolocation 784 664 783	3400-3500 FIXED FIXED-SATELLITE (space-to-Earth) Amateur Mobile Radiolocation 784	3500-3600	AERONAUTICAL RADIONAVIGATION (Ground-based) RADIOLOCATION US110 G59 G110	Radiolocation US110	
3600-4200 FIXED FIXED-SATELLITE (space-to-Earth) Mobile	3500-3700 FIXED FIXED-SATELLITE (space-to-Earth) MOBILE except aeronautical mobile Radiolocation 784	786	3600-3700	AERONAUTICAL RADIONAVIGATION (Ground-based) RADIOLOCATION US110 US245 G59 G110	FIXED-SATELLITE (space-to-Earth) Radiolocation US110 US245	
	3700-4200 FIXED FIXED-SATELLITE (space-to-Earth) MOBILE except aeronautical mobile 787		3700-4200	US110 US245	FIXED FIXED-SATELLITE (space-to-Earth) NG41	
4200-4400 AERONAUTICAL RADIONAVIGATION 789 788 790 791			4200-4400	AERONAUTICAL RADIONAVIGATION US261 791	AERONAUTICAL RADIONAVIGATION US261 791	
4400-4500 FIXED MOBILE			4400-4500	FIXED MOBILE		

TABLES OF FREQUENCY ALLOCATIONS

| INTERNATIONAL | | | UNITED STATES | | | |
Region 1 MHz	Region 2 MHz	Region 3 MHz	Band MHz	Government Allocation	Non-Govt. Allocation	Remarks
4500-4800 FIXED FIXED-SATELLITE (space-to-Earth) 792A MOBILE			4500-4635	FIXED MOBILE US245	FIXED-SATELLITE (space-to-Earth) US245 792A	
			4635-4660	US245	FIXED-SATELLITE (Space-to-Earth) US245	
			4660-4685	G125	US245	
			4685-4800	G122	FIXED FIXED-SATELLITE (space-to-Earth) MOBILE US245 792A	
4800-4990 FIXED MOBILE 793 Radio Astronomy 720 778 794			4800-4990	FIXED MOBILE US203 US257 720 778	US203 US257 720 778	
4990-5000 FIXED MOBILE except aeronautical mobile RADIO ASTRONOMY Space Research (passive) 795			4990-5000	RADIO ASTRONOMY Space Research (Passive) US74 US246	RADIO ASTRONOMY Space Research (Passive) US74 US246	
5000-5250 AERONAUTICAL RADIONAVIGATION 733 796 797 797A 797B			5000-5250	AERONAUTICAL RADIONAVIGATION G126 US211 US260 US307 733 796 797 797A	AERONAUTICAL RADIONAVIGATION US211 US260 US307 733 796 797 797A	
5250-5255 RADIOLOCATION Space Research 713 798			5250-5350	RADIOLOCATION US110 713 G59	Radiolocation US110 713	See Part 7.18 of the NTIA Manual.

TABLES OF FREQUENCY ALLOCATIONS

INTERNATIONAL			Band MHz	UNITED STATES		Remarks
Region 1 MHz	Region 2 MHz	Region 3 MHz		Government Allocation	Non-Govt. Allocation	
5255-5350 RADIOLOCATION 713 798			5255-5350			
5350-5460 AERONAUTICAL RADIONAVIGATION 799 Radiolocation			5350-5460	AERONAUTICAL RADIONAVIGATION RADIOLOCATION US48 799 G56	AERONAUTICAL RADIONAVIGATION Radiolocation US48 799	
5460-5470 RADIONAVIGATION 799 Radiolocation			5460-5470	RADIONAVIGATION Radiolocation US49 US65 799 G56	RADIONAVIGATION Radiolocation US49 US65 799	See Part 7.18 of the NTIA Manual.
5470-5650 MARITIME RADIONAVIGATION Radiolocation 800 801 802			5470-5600	MARITIME RADIONAVIGATION Radiolocation US50 US65 G56	MARITIME RADIONAVIGATION Radiolocation US50 US65	
			5600-5650	MARITIME RADIONAVIGATION METEOROLOGICAL AIDS Radiolocation US51 US65 802 G56	MARITIME RADIONAVIGATION METEOROLOGICAL AIDS Radiolocation US51 US65 802	
5650-5725 RADIOLOCATION Amateur Space Research (deep space) 664 801 803 804 805			5650-5850	RADIOLOCATION 664 806 808 G2	Amateur 664 806 808	ISM 5800 ± 75 MHz
5725-5850 FIXED-SATELLITE (Earth-to-space) RADIOLOCATION Amateur 801 803 805 806 807 808		5725-5850 RADIOLOCATION Amateur 803 805 806 808				

TABLES OF FREQUENCY ALLOCATIONS

INTERNATIONAL			Band MHz	UNITED STATES		Remarks
Region 1 MHz	Region 2 MHz	Region 3 MHz		Government Allocation	Non-Govt. Allocation	
5850-5925 FIXED FIXED-SATELLITE (Earth-to-space) MOBILE 806	5850-5925 FIXED FIXED-SATELLITE (Earth-to-space) MOBILE Amateur Radiolocation 806	5850-5925 FIXED FIXED-SATELLITE (Earth-to-space) MOBILE Radiolocation 806	5850-5925	RADIOLOCATION US245 806 G2	FIXED-SATELLITE (Earth-to-space) Amateur US245 806	
5925-7075 FIXED FIXED-SATELLITE (Earth-to-space) MOBILE 791 809	5925-7075 FIXED-SATELLITE (Earth-to-space) 792A		5925-6425		FIXED FIXED-SATELLITE (Earth-to-space) NG41	
			6425-6525		FIXED-SATELLITE (Earth-to-space) MOBILE	
			6525-6875	791 809	791 809 NG122	
			6875-7075	809	FIXED FIXED-SATELLITE (Earth-to-space) MOBILE 809	
7075-7250 FIXED MOBILE 809 810 811			7075-7125	809	809 NG118	
			7125-7190	809	FIXED MOBILE 809 NG118	
			7190-7235	US252 809 G116	US252 809	
			7235-7250	FIXED SPACE RESEARCH (Earth-to-space) 809 FIXED 809	809 809	

TABLES OF FREQUENCY ALLOCATIONS

INTERNATIONAL Region 1 MHz	Region 2 MHz	Region 3 MHz	Band MHz	UNITED STATES Government Allocation	Non-Govt. Allocation	Remarks
7250-7300 FIXED FIXED-SATELLITE (space-to-Earth) MOBILE 812	FIXED-SATELLITE (space-to-Earth)		7250-7300	FIXED-SATELLITE (space-to-Earth) MOBILE-SATELLITE (space-to-Earth) Fixed G117		
7300-7450 FIXED FIXED-SATELLITE (space-to-Earth) MOBILE except aeronautical mobile 812	FIXED-SATELLITE (space-to-Earth) MOBILE except aeronautical mobile		7300-7450	FIXED FIXED-SATELLITE (space-to-Earth) Mobile-Satellite (space-to-Earth) G117		
7450-7550 FIXED FIXED-SATELLITE (space-to-Earth) METEOROLOGICAL-SATELLITE (space-to-Earth) MOBILE except aeronautical mobile	FIXED-SATELLITE (space-to-Earth) METEOROLOGICAL-SATELLITE (space-to-Earth)		7450-7550	FIXED FIXED-SATELLITE (space-to-Earth) METEOROLOGICAL-SATELLITE (space-to-Earth) Mobile-Satellite (space-to-Earth) G104 G117		
7550-7750 FIXED FIXED-SATELLITE (space-to-Earth) MOBILE except aeronautical mobile	FIXED-SATELLITE (space-to-Earth)		7550-7750	FIXED FIXED-SATELLITE (space-to-Earth) Mobile-Satellite (space-to-Earth) G117		
7750-7900 FIXED MOBILE except aeronautical mobile			7750-7900	FIXED		
7900-8025 FIXED FIXED-SATELLITE (Earth-to-space) MOBILE 812	FIXED-SATELLITE (Earth-to-space)		7900-8025	FIXED-SATELLITE (Earth-to-space) MOBILE-SATELLITE (Earth-to-space) Fixed G117		

TABLES OF FREQUENCY ALLOCATIONS

INTERNATIONAL			UNITED STATES			
Region 1 MHz	Region 2 MHz	Region 3 MHz	Band MHz	Government Allocation	Non-Govt. Allocation	Remarks
8025-8175 FIXED FIXED-SATELLITE (Earth-to-space) MOBILE Earth Exploration-Satellite (space-to-Earth) 813 815	8025-8175 EARTH EXPLORATION-SATELLITE (space-to-Earth) FIXED FIXED-SATELLITE (Earth-to-space) MOBILE 814	8025-8175 FIXED FIXED-SATELLITE (Earth-to-space) MOBILE Earth Exploration-Satellite (space-to-Earth) 813 815	8025-8175	EARTH EXPLORATION-SATELLITE (space-to-Earth) FIXED FIXED-SATELLITE (Earth-to-space) Mobile-Satellite (Earth-to-space) (No Airborne Transmission) US258 G117	US258	
8175-8215 FIXED FIXED-SATELLITE (Earth-to-space) METEOROLOGICAL-SATELLITE (Earth-to-space) MOBILE Earth Exploration-Satellite (space-to-Earth) 813 815	8175-8215 EARTH EXPLORATION-SATELLITE (space-to-Earth) FIXED FIXED-SATELLITE (Earth-to-space) METEOROLOGICAL-SATELLITE (Earth-to-space) MOBILE 814	8175-8215 FIXED FIXED-SATELLITE (Earth-to-space) METEOROLOGICAL-SATELLITE (Earth-to-space) MOBILE Earth Exploration-Satellite (space-to-Earth) 813 815	8175-8215	EARTH EXPLORATION-SATELLITE (space-to-Earth) FIXED FIXED-SATELLITE (Earth-to-space) METEOROLOGICAL-SATELLITE (Earth-to-space) Mobile-Satellite (Earth-to-space) (No Airborne Transmissions) US258 G104 G117	US258	
8215-8400 FIXED FIXED-SATELLITE (Earth-to-space) MOBILE Earth Exploration-Satellite (space-to-Earth) 813 815	8215-8400 EARTH EXPLORATION-SATELLITE (space-to-Earth) FIXED FIXED-SATELLITE (Earth-to-space) MOBILE 814	8215-8400 FIXED FIXED-SATELLITE (Earth-to-space) MOBILE Earth Exploration-Satellite (space-to-Earth) 813 815	8215-8400	EARTH EXPLORATION-SATELLITE (space-to-Earth) FIXED FIXED-SATELLITE (Earth-to-space) Mobile-Satellite (Earth-to-space) (No Airborne Transmissions) US258 G104 G117	US258	

TABLES OF FREQUENCY ALLOCATIONS

INTERNATIONAL Region 1 MHz	Region 2 MHz	Region 3 MHz	Band MHz	UNITED STATES Government Allocation	Non-Govt. Allocation	Remarks
8400-8500 FIXED MOBILE except aeronautical mobile SPACE RESEARCH (space-to-Earth) 816 817 818			8400-8450	FIXED SPACE RESEARCH (space-to-Earth) (Deep Space only)		
			8450-8500	FIXED SPACE RESEARCH (space-to-Earth)	SPACE RESEARCH (space-to-Earth)	
8500-8750 RADIOLOCATION 713 819 820			8500-9000	RADIOLOCATION US53 US110 713 G59	Radiolocation US53 US110 713	See Part 7.18 of the NTIA Manual.
8750-8850 RADIOLOCATION AERONAUTICAL RADIONAVIGATION 821 822						
8850-9000 RADIOLOCATION MARITIME RADIONAVIGATION 823 824						
9000-9200 AERONAUTICAL RADIONAVIGATION 717 Radiolocation 822			9000-9200	AERONAUTICAL RADIONAVIGATION Radiolocation US48 US54 717 G2 G19	AERONAUTICAL RADIONAVIGATION Radiolocation US48 US54 717	See Part 7.18 of the NTIA Manual.
9200-9300 RADIOLOCATION MARITIME RADIONAVIGATION 823 824 824A			9200-9300	MARITIME RADIONAVIGATION Radiolocation US110 823 824A G59	MARITIME RADIONAVIGATION Radiolocation US110 823 824A	See Part 7.18 of the NTIA Manual.
9300-9500 RADIONAVIGATION 825A Radiolocation 775A 824A 825			9300-9500	RADIONAVIGATION Meteorological Aids Radiolocation US51 US66 US67 US71 775A 824A 825A G56	RADIONAVIGATION Meteorological Aids Radiolocation US51 US66 US67 US71 775A 824A 825A	See Part 7.18 of the NTIA Manual.

TABLES OF FREQUENCY ALLOCATIONS

INTERNATIONAL			UNITED STATES		
Region 1 MHz	Region 2 MHz	Region 3 MHz	Band MHz	Government Allocation	Non-Govt. Allocation	Remarks
9500-9800 RADIOLOCATION RADIONAVIGATION			9500-10000	RADIOLOCATION US110 713 828	Radiolocation US110 713 828	See Part 7.18 of the NTIA Manual.
713						
9800-10000 RADIOLOCATION Fixed						
826 827 828						

TABLES OF FREQUENCY ALLOCATIONS

INTERNATIONAL			Band GHz	UNITED STATES		Remarks
Region 1 GHz	Region 2 GHz	Region 3 GHz		Government Allocation	Non-Govt. Allocation	
10-10.45 FIXED MOBILE RADIOLOCATION Amateur 828	10-10.45 RADIOLOCATION Amateur 828 829	10-10.45 FIXED MOBILE RADIOLOCATION Amateur 828	10-10.45	RADIOLOCATION US58 US108 828 G32	Amateur Radiolocation US58 US108 828 NG42	
10.45-10.5 RADIOLOCATION Amateur Amateur-Satellite 830			10.45-10.5	RADIOLOCATION US58 US108 G32	RADIOLOCATION Amateur Amateur-satellite US58 US108 NG42 NG134	
10.5-10.55 FIXED MOBILE Radiolocation	10.5-10.55 FIXED MOBILE RADIOLOCATION		10.5-10.55	RADIOLOCATION US59	RADIOLOCATION US59	
10.55-10.6 FIXED MOBILE except aeronautical mobile Radiolocation			10.55-10.6		FIXED	
10.6-10.68 EARTH EXPLORATION-SATELLITE (passive) FIXED MOBILE except aeronautical mobile RADIO ASTRONOMY SPACE RESEARCH (passive) Radiolocation 831 832			10.6-10.68	EARTH EXPLO- RATION- SATELLITE (Passive) SPACE RESEARCH (Passive) US265 US277	EARTH EXPLO- RATION- SATELLITE (Passive) FIXED SPACE RESEARCH (Passive) US265 US277	
10.68-10.7 EARTH EXPLORATION-SATELLITE (passive) RADIO ASTRONOMY SPACE RESEARCH (passive) 833 834			10.68-10.7	EARTH EXPLO- RATION- SATELLITE (Passive) RADIO ASTRONOMY SPACE RESEARCH (Passive) US74 US246	EARTH EXPLO- RATION- SATELLITE (Passive) RADIO ASTRONOMY SPACE RESEARCH (Passive) US74 US246	

TABLES OF FREQUENCY ALLOCATIONS

INTERNATIONAL			Band GHz	UNITED STATES		Remarks
Region 1 GHz	Region 2 GHz	Region 3 GHz		Government Allocation	Non-Govt. Allocation	
10.7-11.7 FIXED FIXED-SATELLITE (space-to-Earth) (Earth-to-space) 792A 835 MOBILE except aeronautical mobile	10.7-11.7 FIXED FIXED-SATELLITE (space-to-Earth) 792A MOBILE except aeronautical mobile		10.7-11.7	US211	FIXED FIXED-SATELLITE (space-to-Earth) 792A US211 NG41 NG104	
11.7-12.5 FIXED BROADCASTING BROADCASTING-SATELLITE Mobile except aeronautical mobile	11.7-12.1 FIXED 837 FIXED-SATELLITE (space-to-Earth) Mobile except aeronautical mobile 836 839	11.7-12.2 FIXED MOBILE except aeronautical mobile BROADCASTING BROADCASTING-SATELLITE	11.7-12.2	837 839	FIXED-SATELLITE (space-to-Earth) Mobile except aeronautical mobile 837 839 NG143 NG145	
838	12.1-12.2 FIXED-SATELLITE (space-to-Earth) 836 839 842	838				
	12.2-12.7 FIXED MOBILE except aeronautical mobile BROADCASTING BROADCASTING-SATELLITE 839 844 846	12.2-12.5 FIXED MOBILE except aeronautical mobile BROADCASTING 838 845	12.2-12.7	839 843 844	FIXED BROADCASTING-SATELLITE 839 843 844 NG139	
12.5-12.75 FIXED-SATELLITE (space-to-Earth) (Earth-to-space) 848 849 850	12.7-12.75 FIXED FIXED-SATELLITE (Earth-to-space) MOBILE except aero- nautical mobile	12.5-12.75 FIXED FIXED-SATELLITE (space-to-Earth) MOBILE except aero- nautical mobile BROADCASTING- SATELLITE 847	12.7-12.75		FIXED FIXED-SATELLITE (Earth-to-space) MOBILE NG53 NG118	

TABLES OF FREQUENCY ALLOCATIONS

INTERNATIONAL			Band GHz	UNITED STATES		Remarks
Region 1 GHz	Region 2 GHz	Region 3 GHz		Government Allocation	Non-Govt. Allocation	
12.75-13.25 FIXED FIXED-SATELLITE (Earth-to-space) 792A MOBILE Space Research (deep space) (space-to-Earth)			12.75-13.25	US251	FIXED FIXED-SATELLITE (Earth-to-space) MOBILE 792A US251 NG53 NG104 NG118	
13.25-13.4 AERONAUTICAL RADIONAVIGATION 851 852 853			13.25-13.4	AERONAUTICAL RADIONAVIGATION Space Research (Earth-to-space) 851	AERONAUTICAL RADIONAVIGATION Space Research (Earth-to-space) 851	
13.4-13.75 RADIOLOCATION Standard Frequency and Time Signal-Satellite (Earth-to-space) Space Research 713 853 854 855			13.4-14	RADIOLOCATION Space Research Standard Frequency and Time Signal Satellite (Earth-to-space) US110 713 G59	Radiolocation Space Research Standard Frequency and Time Signal-Satellite (Earth-to-space) US110 713	See Part 7.18 of the NTIA Manual.
13.75-14 FIXED-SATELLITE (Earth-to-space) RADIOLOCATION Standard Frequency and Time Signal-Satellite (Earth-to-space) 713 853 854 855 855A 855B						
14-14.25 FIXED-SATELLITE (Earth-to-space) 858 RADIONAVIGATION 856 Space Research 857 859			14-14.2	RADIONAVIGATION Space Research US287 US292	FIXED-SATELLITE (Earth-to-space) RADIONAVIGATION Space Research US287 US292	
14.25-14.3 FIXED-SATELLITE (Earth-to-space) 858 RADIONAVIGATION 856 Space Research 857 859 860 861			14.2-14.3	US287	FIXED-SATELLITE (Earth-to-space) US287	

TABLES OF FREQUENCY ALLOCATIONS

	INTERNATIONAL		UNITED STATES			
Region 1 GHz	Region 2 GHz	Region 3 GHz	Band GHz	Government Allocation	Non-Govt. Allocation	Remarks
14.3-14.4 FIXED FIXED-SATELLITE (Earth-to-space) 858 MOBILE except aeronautical mobile Radionavigation-Satellite 859	14.3-14.4 FIXED-SATELLITE (Earth-to-space) 858 Radionavigation-Satellite 859	14.3-14.4 FIXED FIXED-SATELLITE (Earth-to-space) 858 MOBILE except aeronautical mobile Radionavigation-Satellite 859	14.3-14.4	US287	FIXED-SATELLITE (Earth-to-space) US287	
14.4-14.47 FIXED FIXED-SATELLITE (Earth-to-space) 858 MOBILE except aeronautical mobile Space Research (space-to-Earth) 859			14.4-14.5	Fixed Mobile US203 US287 862	FIXED-SATELLITE (Earth-to-space) US203 US287 862	
14.47-14.5 FIXED FIXED-SATELLITE (Earth-to-space) 858 MOBILE except aeronautical mobile Radio Astronomy 859 862						
14.5-14.8 FIXED FIXED-SATELLITE (Earth-to-space) 863 MOBILE Space Research			14.5-14.7145	FIXED Mobile Space Research		
			14.7145-15.1365	MOBILE Fixed Space Research US310	US310	
14.8-15.35 FIXED MOBILE Space Research 720			15.1365-15.35	FIXED Mobile Space Research US211 720	US211 720	

TABLES OF FREQUENCY ALLOCATIONS

INTERNATIONAL			Band GHz	UNITED STATES		Remarks
Region 1 GHz	Region 2 GHz	Region 3 GHz		Government Allocation	Non-Govt. Allocation	
15.35-15.4 EARTH EXPLORATION-SATELLITE (passive) RADIO ASTRONOMY SPACE RESEARCH (passive) 864 865			15.35-15.4	EARTH EXPLO-RATION-SATELLITE (Passive) RADIO ASTRONOMY SPACE RESEARCH (Passive) US74 US246	EARTH EXPLO-RATION-SATELLITE (Passive) RADIO ASTRONOMY SPACE RESEARCH (Passive) US74 US246	
15.4-15.7 AERONAUTICAL RADIONAVIGATION 733 797			15.4-15.7	AERONAUTICAL RADIONAVIGATION US211 US260 733 797	AERONAUTICAL RADIONAVIGATION US211 US260 733 797	
15.7-16.6 RADIOLOCATION 866 867			15.7-16.6	RADIOLOCATION US110 G59	Radiolocation US110	See Part 7.18 and Section 8.2.46 of the NTIA Manual.
16.6-17.1 RADIOLOCATION Space Research (deep space) (Earth-to-space) 866 867			16.6-17.1	RADIOLOCATION Space Research (Deep Space) (Earth-to-space) US110 G59	Radiolocation US110	See Part 7.18 and Section 8.2.46 of the NTIA Manual.
17.1-17.2 RADIOLOCATION 866 867			17.1-17.2	RADIOLOCATION US110 G59	Radiolocation US110	See Part 7.18 and Section 8.2.46 of the NTIA Manual.
17.2-17.3 RADIOLOCATION Earth Exploration-Satellite (active) Space Research (active) 866 867			17.2-17.3	RADIOLOCATION Earth Explo-ration-Satellite (Active) Space Research (Active) US110 G59	Earth Explo-ration-Satellite (Active) Radiolocation Space Research (Active) US110	See Part 7.18 and Section 8.2.46 of the NTIA Manual.

TABLES OF FREQUENCY ALLOCATIONS

INTERNATIONAL			UNITED STATES			
Region 1 GHz	Region 2 GHz	Region 3 GHz	Band GHz	Government Allocation	Non-Govt. Allocation	Remarks
17.3–17.7 FIXED-SATELLITE (Earth-to-space) 869 Radiolocation 868	17.3–17.7 FIXED-SATELLITE (Earth-to-space) BROADCASTING-SATEL-LITE Radiolocation 868 868A 869A	17.3–17.7 FIXED-SATELLITE (Earth-to-space) 869 Radiolocation 868	17.3–17.7	Radiolocation US259 US271 G59	FIXED-SATELLITE (Earth-to-space) US259 US271 NG140	See Part 7.18 of the NTIA Manual.
17.7–18.1 FIXED FIXED-SATELLITE (space-to-Earth) (Earth-to-space) 869 MOBILE	17.7–17.8 FIXED FIXED-SATELLITE (space-to-Earth) (Earth-to-space) 869 BROADCASTING-SATEL LITE Mobile 869B 868A 869A	17.7–18.1 FIXED FIXED-SATELLITE (space-to-Earth) (Earth-to-space) 869 MOBILE	17.7–17.8	US271	FIXED FIXED-SATELLITE (space-to-Earth) (Earth-to-space) MOBILE US271 NG140 NG144	
	17.8–18.1 FIXED FIXED-SATELLITE (space-to-Earth) (Earth-to-space) 869 MOBILE		17.8–18.1		FIXED FIXED-SATELLITE (space-to-Earth) MOBILE US334 NG144	
18.1–18.4 FIXED FIXED-SATELLITE (space-to-Earth) (Earth-to-space) 870A MOBILE 870 870B			18.1–18.6	870 US334 G117	FIXED FIXED-SATELLITE (space-to-Earth) MOBILE 870 US334 NG144	
18.4–18.6 FIXED FIXED-SATELLITE (space-to-Earth) MOBILE						

TABLES OF FREQUENCY ALLOCATIONS

INTERNATIONAL			UNITED STATES			
Region 1 GHz	Region 2 GHz	Region 3 GHz	Band GHz	Government Allocation	Non-Govt Allocation	Remarks
18.6-18.8 FIXED FIXED-SATELLITE (space-to-Earth) 872 MOBILE except aeronautical mobile Earth Exploration-Satellite (passive) Space Research (passive) 871	18.6-18.8 EARTH EXPLORATION-SATELLITE (passive) FIXED FIXED-SATELLITE (space-to-Earth) 872 MOBILE except aeronautical mobile SPACE RESEARCH (passive) 871	18.6-18.8 FIXED FIXED-SATELLITE (space-to-Earth) 872 MOBILE except aeronautical mobile Earth Exploration-Satellite (passive) Space Research (passive) 871	18.6-18.8	EARTH EXPLORATION-SATELLITE (passive) SPACE RESEARCH (passive) US254 US255 US334 G117	FIXED FIXED-SATELLITE (space-to-Earth) EARTH EXPLORATION-SATELLITE (Passive) MOBILE except aeronautical mobile SPACE RESEARCH (Passive) US254 US255 US334 NG144	
18.8-19.7 FIXED FIXED-SATELLITE (space-to-Earth) MOBILE			18.8-19.7		FIXED FIXED-SATELLITE (space-to-Earth) MOBILE US334 NG144	
19.7-20.1 FIXED-SATELLITE (space-to-Earth) Mobile-Satellite (space-to-Earth) 873	19.7-20.1 FIXED-SATELLITE (space-to-Earth) MOBILE-SATELLITE (space-to-Earth) 873 873A 873B 873C 873D 873E	19.7-20.1 FIXED-SATELLITE (space-to-Earth) Mobile-Satellite (space-to-Earth) 873	19.7-20.1	US334 G117	FIXED-SATELLITE (space-to-earth) MOBILE-SATELLITE (space-to-Earth) 873A 873B 873C 873D 873E US334	
20.1-20.2 FIXED-SATELLITE (space-to-Earth) MOBILE-SATELLITE (space-to-Earth) 873 873A 873B 873C 873D			20.1-20.2	US334 G117	FIXED-SATELLITE (space-to-Earth) MOBILE-SATELLITE (space-to-Earth) 873A 873B 873C 873D US334	

TABLES OF FREQUENCY ALLOCATIONS

INTERNATIONAL				UNITED STATES		
Region 1 GHz	Region 2 GHz	Region 3 GHz	Band GHz	Government Allocation	Non-Govt. Allocation	Remarks
20.2-21.2 FIXED-SATELLITE (space-to-Earth) MOBILE-SATELLITE (space-to-Earth) Standard Frequency and Time Signal-Satellite (space-to-Earth) 873			20.2-21.2	FIXED-SATELLITE (space-to-Earth) MOBILE-SATELLITE (space-to-Earth) Standard Frequency and Time Signal-Satellite (space-to-Earth) G117	Standard Frequency and Time Signal-Satellite (space-to-Earth)	
21.2-21.4 EARTH EXPLORATION-SATELLITE (passive) FIXED MOBILE SPACE RESEARCH (passive)			21.2-21.4	EARTH EXPLO-RATION-SATELLITE (passive) FIXED MOBILE SPACE RESEARCH (passive) US263	EARTH EXPLO-RATION-SATELLITE (passive) FIXED MOBILE SPACE RESEARCH (passive) US263	
21.4-22 FIXED MOBILE BROADCASTING-SATELLITE 873F	21.4-22 FIXED MOBILE	21.4-22 FIXED MOBILE BROADCASTING-SATEL-LITE 873F 873G	21.4-22	FIXED MOBILE	FIXED MOBILE	
22-22.21 FIXED MOBILE except aeronautical mobile 874			22-22.21	FIXED MOBILE except aero-nautical mobile 874	FIXED MOBILE except aero-nautical mobile 874	

TABLES OF FREQUENCY ALLOCATIONS

INTERNATIONAL			Band GHz	UNITED STATES		Remarks
Region 1 GHz	Region 2 GHz	Region 3 GHz		Government Allocation	Non-Govt. Allocation	
22.21-22.5 EARTH EXPLORATION-SATELLITE (passive) FIXED MOBILE except aeronautical mobile RADIO ASTRONOMY SPACE RESEARCH (passive) 875 876			22.21-22.5	EARTH EXPLO- RATION- SATELLITE (passive) FIXED MOBILE except aero- nautical mobile RADIO ASTRONOMY SPACE RESEARCH (passive) US263 875	EARTH EXPLO- RATION- SATELLITE (passive) FIXED MOBILE except aero- nautical mobile RADIO ASTRONOMY SPACE RESEARCH (passive) US263 875	
22.5-22.55 FIXED MOBILE			22.5-22.55	FIXED MOBILE US211	FIXED MOBILE US211	
22.55-23 FIXED INTER-SATELLITE MOBILE 879			22.55-23	FIXED INTER-SATELLITE MOBILE US278 879	FIXED INTER-SATELLITE MOBILE US278 879	
23-23.55 FIXED INTER-SATELLITE MOBILE 879			23-23.55	FIXED INTER-SATELLITE MOBILE US278 879	FIXED INTER-SATELLITE MOBILE US278 879	
23.55-23.6 FIXED MOBILE			23.55-23.6	FIXED MOBILE	FIXED MOBILE	
23.6-24 EARTH EXPLORATION-SATELLITE (passive) RADIO ASTRONOMY SPACE RESEARCH (passive) 880			23.6-24	EARTH EXPLO- RATION- SATELLITE (passive) RADIO ASTRONOMY SPACE RESEARCH (passive) US74 US246	EARTH EXPLO- RATION- SATELLITE (passive) RADIO ASTRONOMY SPACE RESEARCH (passive) US74 US246	
24-24.05 AMATEUR AMATEUR-SATELLITE 881			24-24.05	US211 881	AMATEUR AMATEUR-SATELLITE US211 881	

TABLES OF FREQUENCY ALLOCATIONS

INTERNATIONAL			Band GHz	UNITED STATES		Remarks
Region 1 GHz	Region 2 GHz	Region 3 GHz		Government Allocation	Non-Govt. Allocation	
24.05-24.25 RADIOLOCATION Amateur Earth Exploration-Satellite (active) 881	Earth Exploration-Satellite (active)		24.05-24.25	RADIOLOCATION Earth Exploration-Satellite (active) US110 881 G59	Amateur Earth Exploration-Satellite (active) Radiolocation US110 881	ISM 24.125 ± 125 MHz
24.25-24.45 FIXED	24.25-24.45 RADIONAVIGATION	24.25-24.45 RADIONAVIGATION FIXED MOBILE	24.25-24.45	RADIONAVIGATION	RADIONAVIGATION	
24.45-24.65 FIXED INTER-SATELLITE 882E	24.45-24.65 INTER-SATELLITE RADIONAVIGATION	24.45-24.65 FIXED INTER-SATELLITE MOBILE RADIONAVIGATION 882E	24.45-24.65	INTER-SATELLITE RADIONAVIGATION 882E	INTER-SATELLITE RADIONAVIGATION 882E	
24.65-24.75 FIXED INTER-SATELLITE 882E	24.65-24.75 INTER-SATELLITE RADIOLOCATION-SATELLITE (Earth-to-space) 882G	24.65-24.75 FIXED INTER-SATELLITE MOBILE 882E 882F	24.65-24.75	INTER-SATELLITE RADIOLOCATION-SATELLITE (Earth-to-space)	INTER-SATELLITE RADIOLOCATION-SATELLITE (Earth-to-space)	
24.75-25.25 FIXED FIXED-SATELLITE (Earth-to-space) 882G	24.75-25.25 FIXED-SATELLITE (Earth-to-space) 882G	24.75-25.25 FIXED FIXED-SATELLITE (Earth-to-space) 882G MOBILE 882F	24.75-25.25	RADIONAVIGATION	RADIONAVIGATION	
25.25-25.5 FIXED INTER-SATELLITE 881A MOBILE Standard Frequency and Time Signal-Satellite (Earth-to-space)			25.25-25.5	FIXED INTER-SATELLITE MOBILE Standard Frequency and Time Signal-Satellite (Earth-to-space) 881A	Earth Exploration-Satellite (space-to-space) Standard Frequency and Time Signal-Satellite (Earth-to-space)	

TABLES OF FREQUENCY ALLOCATIONS

INTERNATIONAL			Band GHz	UNITED STATES		Remarks
Region 1 GHz	Region 2 GHz	Region 3 GHz		Government Allocation	Non-Govt. Allocation	
25.5-27 FIXED INTER-SATELLITE 881A MOBILE Earth Exploration-Satellite (space-to-Earth) Standard Frequency and Time Signal-Satellite (Earth-to-space)	INTER-SATELLITE 881A	Earth Exploration-Satellite (space-to-Earth) Standard Frequency and Time Signal-Satellite	25.5-27	FIXED INTER-SATELLITE MOBILE Earth Exploration-Satellite (space-to-Earth) Standard Frequency and Time Signal-Satellite (Earth-to-space) 881A	Earth Exploration-Satellite (space-to-space) Standard Frequency and Time Signal-Satellite (Earth-to-space)	
27-27.5 FIXED INTER-SATELLITE 881A MOBILE	27-27.5 FIXED FIXED-SATELLITE (Earth-to-space) INTER-SATELLITE 881A 881B MOBILE	FIXED-SATELLITE (Earth-to-space) INTER-SATELLITE 881A 881B MOBILE	27-27.5	FIXED INTER-SATELLITE MOBILE 881A	Earth Exploration-Satellite (space-to-space)	
27.5-28.5 FIXED FIXED-SATELLITE (Earth-to-space) MOBILE 882A 882B	FIXED-SATELLITE (Earth-to-space) 882D	FIXED-SATELLITE (Earth-to-space) 882D	27.5-29.5		FIXED FIXED-SATELLITE (Earth-to-space) MOBILE	
28.5-29.5 FIXED FIXED-SATELLITE (Earth-to-space) MOBILE Earth Exploration-Satellite (Earth-to-space) 882C 882B	FIXED-SATELLITE (Earth-to-space) 882D	FIXED-SATELLITE (Earth-to-space) 882D Earth Exploration-Satellite (Earth-to-space) 882C				

TABLES OF FREQUENCY ALLOCATIONS

INTERNATIONAL			Band GHz	UNITED STATES		Remarks
Region 1 GHz	Region 2 GHz	Region 3 GHz		Government Allocation	Non-Govt. Allocation	
29.5-29.9 FIXED-SATELLITE (Earth-to-space) 882D Earth Exploration-Satellite (Earth-to-space) 882C Mobile-Satellite (Earth-to-space) 882B 883 885 886	29.5-29.9 FIXED-SATELLITE (Earth-to-space) 882D MOBILE-SATELLITE (Earth-to-space) Earth Exploration-Satellite (Earth-to-space) 882C 873A 873B 873C 873E 882B 883	29.5-29.9 FIXED-SATELLITE (Earth-to-space) 882D MOBILE-SATELLITE (Earth-to-space) Earth Exploration-Satellite (Earth-to-space) 882C Mobile-Satellite (Earth-to-space) 882B 883	29.5-30	882	FIXED-SATELLITE (Earth-to-space) Mobile-Satellite (Earth-to-space) 882	
29.9-30 FIXED-SATELLITE (Earth-to-space) 882D MOBILE-SATELLITE (Earth-to-space) 882C Earth Exploration-Satellite (Earth-to-space) 882C 873A 873B 873C 882 882A 882B 883						
30-31 FIXED-SATELLITE (Earth-to-space) MOBILE-SATELLITE (Earth-to-space) Standard Frequency and Time Signal-Satellite (space-to-Earth) 883			30-31	FIXED-SATELLITE (Earth-to-space) MOBILE-SATELLITE (Earth-to-space) Standard Frequency and Time Signal-Satellite (space-to-Earth) G117	Standard Frequency and Time Signal-Satellite (space-to-Earth)	
31-31.3 FIXED MOBILE Standard Frequency and Time Signal-Satellite (space-to-Earth) Space Research 884 885 886			31-31.3	Standard Frequency and Time Signal-Satellite (space-to-Earth) US211 886	FIXED MOBILE Standard Frequency and Time Signal-Satellite (space-to-Earth) US211 886	

TABLES OF FREQUENCY ALLOCATIONS

INTERNATIONAL				UNITED STATES		
Region 1 GHz	Region 2 GHz	Region 3 GHz	Band GHz	Government Allocation	Non-Govt. Allocation	Remarks
31.3-31.8 EARTH EXPLORATION-SATELLITE (passive) RADIO ASTRONOMY SPACE RESEARCH (passive) Fixed Mobile except aeronautical mobile 888 889	31.3-31.5 EARTH EXPLORATION-SATELLITE (passive) RADIO ASTRONOMY SPACE RESEARCH (passive) 887		31.3-31.8	EARTH EXPLORATION-SATELLITE (passive) RADIO ASTRONOMY SPACE RESEARCH (passive) US74 US246	EARTH EXPLORATION-SATELLITE (passive) RADIO ASTRONOMY SPACE RESEARCH (passive) US74 US246	
	31.5-31.8 EARTH EXPLORATION-SATELLITE (passive) RADIO ASTRONOMY SPACE RESEARCH (passive) 888	31.5-31.8 EARTH EXPLORATION-SATELLITE (passive) RADIO ASTRONOMY SPACE RESEARCH (passive) Fixed Mobile except aeronautical mobile 888				
31.8-32 RADIONAVIGATION SPACE RESEARCH (deep space) (space-to-Earth) 892 893			31.8-32	RADIONAVIGATION US69 US211 US262	RADIONAVIGATION US69 US211 US262	
32-32.3 INTER-SATELLITE RADIONAVIGATION SPACE RESEARCH (deep space) (space-to-Earth) 892 893			32-33	INTER-SATELLITE RADIONAVIGATION US69 US262 US278 893	INTER-SATELLITE RADIONAVIGATION US69 US262 US278 893	
32.3-33 INTER-SATELLITE RADIONAVIGATION 892 893						
33-33.4 RADIONAVIGATION 892			33-33.4	RADIONAVIGATION US69	RADIONAVIGATION US69	
33.4-34.2 RADIOLOCATION 892 894			33.4-36	RADIOLOCATION US110 US252 897 G34	Radiolocation US110 US252 897	

TABLES OF FREQUENCY ALLOCATIONS

INTERNATIONAL			Band GHz	UNITED STATES		Remarks
Region 1 GHz	Region 2 GHz	Region 3 GHz		Government Allocation	Non-Govt. Allocation	
34.2-34.7 RADIOLOCATION SPACE RESEARCH (deep space) (Earth-to-space) 894						
34.7-35.2 RADIOLOCATION Space Research 896 894						
35.2-36 METEOROLOGICAL AIDS RADIOLOCATION 894 897						
36-37 EARTH EXPLORATION-SATELLITE (passive) FIXED MOBILE SPACE RESEARCH (passive) 898			36-37	EARTH EXPLORATION-SATELLITE (passive) FIXED MOBILE SPACE RESEARCH (passive) US263 898	EARTH EXPLORATION-SATELLITE (passive) FIXED MOBILE SPACE RESEARCH (passive) US263 898	
37-37.5 FIXED MOBILE SPACE RESEARCH (space-to-Earth)			37-38.6	FIXED MOBILE	FIXED MOBILE	
37.5-38 FIXED FIXED-SATELLITE (space-to-Earth) MOBILE SPACE RESEARCH (space-to-Earth) Earth Exploration-Satellite (space-to-Earth)						
38-39.5 FIXED FIXED-SATELLITE (space-to-Earth) MOBILE Earth exploration-Satellite (space-to-Earth)			38.6-39.5	US291	FIXED FIXED-SATELLITE (space-to-Earth) MOBILE US291	

TABLES OF FREQUENCY ALLOCATIONS

INTERNATIONAL			UNITED STATES			
Region 1 GHz	Region 2 GHz	Region 3 GHz	Band GHz	Government Allocation	Non-Govt. Allocation	Remarks
39.5-40 FIXED FIXED-SATELLITE (space-to-Earth) MOBILE MOBILE-SATELLITE (space-to-Earth) Earth Exploration-Satellite (space-to-Earth)			39.5-40	FIXED-SATELLITE (space-to-Earth) MOBILE-SATELLITE (space-to-Earth) US291 G117	FIXED FIXED-SATELLITE (space-to-Earth) MOBILE MOBILE-SATELLITE (space-to-Earth) US291	
40-40.5 EARTH EXPLORATION-SATELLITE (Earth-to-space) FIXED FIXED-SATELLITE (space-to-Earth) MOBILE MOBILE-SATELLITE (space-to-Earth) SPACE RESEARCH (Earth-to-space) Earth Exploration-Satellite (space-to-Earth) 900			40-40.5	FIXED-SATELLITE (space-to-Earth) MOBILE-SATELLITE (space-to-Earth) G117	FIXED-SATELLITE (space-to-Earth) MOBILE-SATELLITE (space-to-Earth) US291	
40.5-42.5 BROADCASTING-SATELLITE /BROADCASTING/ Fixed Mobile			40.5-42.5	US211	BROADCASTING-SATELLITE /BROADCASTING/ Fixed Mobile US211	
42.5-43.5 FIXED FIXED-SATELLITE (Earth-to-space) 901 MOBILE except aeronautical mobile RADIO ASTRONOMY 900			42.5-43.5	FIXED FIXED-SATELLITE (Earth-to-space) MOBILE except aero-nautical mobile RADIO ASTRONOMY 900	FIXED FIXED-SATELLITE (Earth-to-space) MOBILE except aero-nautical mobile RADIO ASTRONOMY 900	
43.5-47 MOBILE 902 MOBILE-SATELLITE RADIONAVIGATION RADIONAVIGATION-SATELLITE 903			43.5-45.5	FIXED-SATELLITE (Earth-to-space) MOBILE-SATELLITE (Earth-to-space) G117 903	FIXED-SATELLITE (Earth-to-space) MOBILE-SATELLITE (Earth-to-space) 900	
			45.5-47	MOBILE MOBILE-SATELLITE (Earth-to-space) RADIONAVIGATION-SATELLITE 903	MOBILE MOBILE-SATELLITE (Earth-to-space) RADIONAVIGATION-SATELLITE 903	

TABLES OF FREQUENCY ALLOCATIONS

INTERNATIONAL				UNITED STATES		
Region 1 GHz	Region 2 GHz	Region 3 GHz	Band GHz	Government Allocation	Non-Govt. Allocation	Remarks
47-47.2 AMATEUR AMATEUR-SATELLITE			47-47.2		AMATEUR AMATEUR-SATELLITE	
47.2-50.2 FIXED FIXED-SATELLITE (Earth-to-space) 901 MOBILE 905 904			47.2-50.2	FIXED FIXED-SATELLITE (Earth-to-space) MOBILE US264 US297 904	FIXED FIXED-SATELLITE (Earth-to-space) MOBILE US264 US297 904	
50.2-50.4 EARTH EXPLORATION-SATELLITE (passive) FIXED MOBILE SPACE RESEARCH (passive)			50.2-50.4	EARTH EXPLO- RATION- SATELLITE (passive) FIXED MOBILE SPACE RESEARCH (passive) US263	EARTH EXPLO- RATION- SATELLITE (passive) FIXED MOBILE SPACE RESEARCH (passive) US263	
50.4-51.4 FIXED FIXED-SATELLITE (Earth-to-space) MOBILE Mobile-Satellite (Earth-to-space)			50.4-51.4	FIXED FIXED-SATELLITE (Earth-to-space) MOBILE MOBILE-SATELLITE (Earth-to-space) G117	FIXED FIXED-SATELLITE (Earth-to-space) MOBILE MOBILE-SATELLITE (Earth-to-space)	
51.4-54.25 EARTH EXPLORATION-SATELLITE (passive) SPACE RESEARCH (passive) 906 907			51.4-54.25	EARTH EXPLO- RATION- SATELLITE (passive) RADIO ASTRONOMY SPACE RESEARCH (passive) US246	EARTH EXPLO- RATION- SATELLITE (passive) RADIO ASTRONOMY SPACE RESEARCH (passive) US246	

TABLES OF FREQUENCY ALLOCATIONS

| INTERNATIONAL | | | Band GHz | UNITED STATES | | Remarks |
Region 1 GHz	Region 2 GHz	Region 3 GHz		Government Allocation	Non-Govt. Allocation	
54.25-58.2 EARTH EXPLORATION-SATELLITE (passive) FIXED INTER-SATELLITE MOBILE 909 SPACE RESEARCH (passive) 908			54.25-58.2	EARTH EXPLORATION-SATELLITE (passive) FIXED INTER-SATELLITE MOBILE SPACE RESEARCH (passive) US263 909	EARTH EXPLORATION-SATELLITE (passive) FIXED INTER-SATELLITE MOBILE SPACE RESEARCH (passive) US263 909	
58.2-59 EARTH EXPLORATION-SATELLITE (passive) SPACE RESEARCH (passive) 906 907			58.2-59	EARTH EXPLORATION-SATELLITE (passive) RADIO ASTRONOMY SPACE RESEARCH (passive) US246	EARTH EXPLORATION-SATELLITE (passive) RADIO ASTRONOMY SPACE RESEARCH (passive) US246	
59-64 FIXED INTER-SATELLITE MOBILE 909 RADIOLOCATION 910 911			59-64	FIXED INTER-SATELLITE MOBILE RADIOLOCATION 909 910 911	FIXED INTER-SATELLITE MOBILE RADIOLOCATION 909 910 911	ISM 61.25 $\pm$ 250 MHz
64-65 EARTH EXPLORATION-SATELLITE (passive) SPACE RESEARCH (passive) 906 907			64-65	EARTH EXPLORATION-SATELLITE (passive) RADIO ASTRONOMY SPACE RESEARCH (passive) US246	EARTH EXPLORATION-SATELLITE (passive) RADIO ASTRONOMY SPACE RESEARCH (passive) US246	
65-66 EARTH EXPLORATION-SATELLITE SPACE RESEARCH Fixed Mobile			65-66	EARTH EXPLORATION-SATELLITE SPACE RESEARCH Fixed Mobile	EARTH EXPLORATION-SATELLITE SPACE RESEARCH Fixed Mobile	

TABLES OF FREQUENCY ALLOCATIONS

INTERNATIONAL			Band GHz	UNITED STATES		Remarks
Region 1 GHz	Region 2 GHz	Region 3 GHz		Government Allocation	Non-Govt. Allocation	
66-71 MOBILE 902 MOBILE-SATELLITE RADIONAVIGATION RADIONAVIGATION-SATELLITE 903			66-71	MOBILE MOBILE-SATELLITE RADIONAVIGATION RADIONAVIGATION-SATELLITE 903	MOBILE MOBILE-SATELLITE RADIONAVIGATION RADIONAVIGATION-SATELLITE 903	
71-74 FIXED FIXED-SATELLITE (Earth-to-space) MOBILE MOBILE-SATELLITE (Earth-to-space) 906			71-74	FIXED FIXED-SATELLITE (Earth-to-space) MOBILE MOBILE-SATELLITE (Earth-to-space) US270	FIXED FIXED-SATELLITE (Earth-to-space) MOBILE MOBILE-SATELLITE (Earth-to-space) US270	
74-75.5 FIXED FIXED-SATELLITE (Earth-to-space) MOBILE Space Research (space-to-Earth)			74-75.5	FIXED FIXED-SATELLITE (Earth-to-space) MOBILE US297	FIXED FIXED-SATELLITE (Earth-to-space) MOBILE US297	
75.5-76 AMATEUR AMATEUR-SATELLITE Space Research (space-to-Earth)			75.5-76		AMATEUR AMATEUR-SATELLITE	
76-81 RADIOLOCATION Amateur Amateur-Satellite Space Research (space-to-Earth) 912			76-77	RADIOLOCATION	RADIOLOCATION Amateur	
			77-81	912	RADIOLOCATION Amateur Amateur-Satellite 912	
81-84 FIXED FIXED-SATELLITE (space-to-Earth) MOBILE MOBILE-SATELLITE (space-to-Earth) Space Research (space-to-Earth)			81-84	FIXED FIXED-SATELLITE (space-to-Earth) MOBILE MOBILE-SATELLITE (space-to-Earth)	FIXED FIXED-SATELLITE (space-to-Earth) MOBILE MOBILE-SATELLITE (space-to-Earth)	

TABLES OF FREQUENCY ALLOCATIONS

INTERNATIONAL			Band GHz	UNITED STATES		Remarks
Region 1 GHz	Region 2 GHz	Region 3 GHz		Government Allocation	Non-Govt. Allocation	
FIXED MOBILE BROADCASTING BROADCASTING-SATELLITE 913			84-86	FIXED MOBILE US211 913	BROADCASTING BROADCASTING-SATELLITE FIXED MOBILE US211 913	
EARTH EXPLORATION-SATELLITE (passive) RADIO ASTRONOMY SPACE RESEARCH (passive) 907			86-92	EARTH EXPLO-RATION-SATELLITE (passive) RADIO ASTRONOMY SPACE RESEARCH (passive) US74 US246	EARTH EXPLO-RATION-SATELLITE (passive) RADIO ASTRONOMY SPACE RESEARCH (passive) US74 US246	
FIXED FIXED-SATELLITE (Earth-to-space) MOBILE RADIOLOCATION 914			92-95	FIXED FIXED-SATELLITE (Earth-to-space) MOBILE RADIOLOCATION 914	FIXED FIXED-SATELLITE (Earth-to-space) MOBILE RADIOLOCATION 914	
MOBILE 902 MOBILE-SATELLITE RADIONAVIGATION RADIONAVIGATION-SATELLITE Radiolocation 903 904			95-100	MOBILE MOBILE-SATELLITE RADIONAVIGATION RADIONAVIGATION-SATELLITE Radiolocation 902 903 904	MOBILE MOBILE-SATELLITE RADIONAVIGATION RADIONAVIGATION-SATELLITE Radiolocation 902 903 904	
EARTH EXPLORATION-SATELLITE (passive) FIXED MOBILE SPACE RESEARCH (passive) 722			100-102	EARTH EXPLO-RATION-SATELLITE (passive) SPACE RESEARCH (passive) US246 722	EARTH EXPLO-RATION-SATELLITE (passive) SPACE RESEARCH (passive) US246 722	
FIXED FIXED-SATELLITE (space-to-Earth) MOBILE 722			102-105	FIXED FIXED-SATELLITE (space-to-Earth) MOBILE US211 722	FIXED FIXED-SATELLITE (space-to-Earth) MOBILE US211 722	

TABLES OF FREQUENCY ALLOCATIONS

INTERNATIONAL			Band GHz	UNITED STATES		Remarks
Region 1 GHz	Region 2 GHz	Region 3 GHz		Government Allocation	Non-Govt. Allocation	
105-116 EARTH EXPLORATION-SATELLITE (passive) RADIO ASTRONOMY SPACE RESEARCH (passive) 722 907			105-116	EARTH EXPLO-RATION-SATELLITE (passive) RADIO ASTRONOMY SPACE RESEARCH (passive) US74 US246 722	EARTH EXPLO-RATION-SATELLITE (passive) RADIO ASTRONOMY SPACE RESEARCH (passive) US74 US246 722	
116-126 EARTH EXPLORATION-SATELLITE (passive) FIXED INTER-SATELLITE MOBILE 909 SPACE RESEARCH (passive) 722 915 916			116-126	EARTH EXPLO-RATION-SATELLITE (passive) FIXED INTER-SATELLITE MOBILE SPACE RESEARCH (passive) US211 US263 722 909 915 916	EARTH EXPLO-RATION-SATELLITE (passive) FIXED INTER-SATELLITE MOBILE SPACE RESEARCH (passive) US211 US263 722 909 915 916	ISM 122.5 ± 0.5 GHz
126-134 FIXED INTER-SATELLITE MOBILE 909 RADIOLOCATION 910			126-134	FIXED INTER-SATELLITE MOBILE RADIOLOCATION 909 910	FIXED INTER-SATELLITE MOBILE RADIOLOCATION 909 910	
134-142 MOBILE 902 MOBILE-SATELLITE RADIONAVIGATION RADIONAVIGATION-SATELLITE Radiolocation 903 917 918			134-142	MOBILE MOBILE-SATELLITE RADIONAVIGATION RADIONAVIGATION-SATELLITE Radiolocation 902 903 917 918	MOBILE MOBILE-SATELLITE RADIONAVIGATION RADIONAVIGATION-SATELLITE Radiolocation 902 903 917 918	
142-144 AMATEUR AMATEUR-SATELLITE			142-144		AMATEUR AMATEUR-SATELLITE	
144-149 RADIOLOCATION Amateur Amateur-Satellite 918			144-149	RADIOLOCATION 224 918	RADIOLOCATION Amateur Amateur-Satellite 918	

TABLES OF FREQUENCY ALLOCATIONS

INTERNATIONAL			UNITED STATES			
Region 1 GHz	Region 2 GHz	Region 3 GHz	Band GHz	Government Allocation	Non-Govt. Allocation	Remarks
149-150 FIXED FIXED-SATELLITE (space-to-Earth) MOBILE			149-150	FIXED FIXED-SATELLITE (space-to-Earth) MOBILE	FIXED FIXED-SATELLITE (space-to-Earth) MOBILE	
150-151 EARTH EXPLORATION-SATELLITE (passive) FIXED FIXED-SATELLITE (space-to-Earth) MOBILE SPACE RESEARCH (passive) 919			150-151	EARTH EXPLO- RATION- SATELLITE (passive) FIXED FIXED-SATELLITE (space-to-Earth) MOBILE SPACE RESEARCH (passive) US263 919	EARTH EXPLO- RATION- SATELLITE (passive) FIXED FIXED-SATELLITE (space-to-Earth) MOBILE SPACE RESEARCH (passive) US263 919	
151-156 FIXED FIXED-SATELLITE (space-to-Earth) MOBILE			151-164	FIXED FIXED-SATELLITE (space-to-Earth) MOBILE US211	FIXED FIXED-SATELLITE (space-to-Earth) MOBILE US211	
156-158 EARTH EXPLORATION-SATELLITE (passive) FIXED FIXED-SATELLITE (space-to-Earth) MOBILE						
158-164 FIXED FIXED-SATELLITE (space-to-Earth) MOBILE						
164-168 EARTH EXPLORATION-SATELLITE (passive) RADIO ASTRONOMY SPACE RESEARCH (passive)			164-168	EARTH EXPLO- RATION- SATELLITE (passive) RADIO ASTRONOMY SPACE RESEARCH (passive) US246	EARTH EXPLO- RATION- SATELLITE (passive) RADIO ASTRONOMY SPACE RESEARCH (passive) US246	
168-170 FIXED MOBILE			168-170	FIXED MOBILE	FIXED MOBILE	

TABLES OF FREQUENCY ALLOCATIONS

INTERNATIONAL			Band GHz	UNITED STATES		Remarks
Region 1 GHz	Region 2 GHz	Region 3 GHz		Government Allocation	Non-Govt. Allocation	
170-174.5 FIXED INTER-SATELLITE MOBILE 909 919			170-174.5	FIXED INTER-SATELLITE MOBILE 909 919	FIXED INTER-SATELLITE MOBILE 909 919	
174.5-176.5 EARTH EXPLORATION-SATELLITE (passive) FIXED INTER-SATELLITE MOBILE 909 SPACE RESEARCH (passive) 919			174.5-176.5	EARTH EXPLO- RATION- SATELLITE (passive) FIXED INTER-SATELLITE MOBILE SPACE RESEARCH (passive) US263 909 919	EARTH EXPLO- RATION- SATELLITE (passive) FIXED INTER-SATELLITE MOBILE SPACE RESEARCH (passive) US263 909 919	
176.5-182 FIXED INTER-SATELLITE MOBILE 909 919			176.5-182	FIXED INTER-SATELLITE MOBILE US211 909 919	FIXED INTER-SATELLITE MOBILE US211 909 919	
182-185 EARTH EXPLORATION-SATELLITE (passive) RADIO ASTRONOMY SPACE RESEARCH (passive) 920 921			182-185	EARTH EXPLO- RATION- SATELLITE (passive) RADIO ASTRONOMY SPACE RESEARCH (passive) US246	EARTH EXPLO- RATION- SATELLITE (passive) RADIO ASTRONOMY SPACE RESEARCH (passive) US246	
185-190 FIXED INTER-SATELLITE MOBILE 909 919			185-190	FIXED INTER-SATELLITE MOBILE US211 909 919	FIXED INTER-SATELLITE MOBILE US211 909 919	
190-200 MOBILE 902 MOBILE-SATELLITE RADIONAVIGATION RADIONAVIGATION-SATELLITE 722 903			190-200	MOBILE MOBILE-SATELLITE RADIONAVIGATION RADIONAVIGATION- SATELLITE 722 902 903	MOBILE MOBILE-SATELLITE RADIONAVIGATION RADIONAVIGATION- SATELLITE 722 902 903	

TABLES OF FREQUENCY ALLOCATIONS

INTERNATIONAL			Band GHz	UNITED STATES		Remarks
Region 1 GHz	Region 2 GHz	Region 3 GHz		Government Allocation	Non-Govt. Allocation	
200-202 EARTH EXPLORATION-SATELLITE (passive) FIXED MOBILE SPACE RESEARCH (passive) 722			200-202	EARTH EXPLO- RATION- SATELLITE (passive) FIXED MOBILE SPACE RESEARCH (passive) US263 722	EARTH EXPLO- RATION- SATELLITE (passive) FIXED MOBILE SPACE RESEARCH (passive) US263 722	
202-217 FIXED FIXED-SATELLITE (Earth-to-space) MOBILE 722			202-217	FIXED FIXED-SATELLITE (Earth-to-space) MOBILE 722	FIXED FIXED-SATELLITE (Earth-to-space) MOBILE 722	
217-231 EARTH EXPLORATION-SATELLITE (passive) RADIO ASTRONOMY SPACE RESEARCH (passive) 722 907			217-231	EARTH EXPLO- RATION- SATELLITE (passive) RADIO ASTRONOMY SPACE RESEARCH (passive) US74 US246 722	EARTH EXPLO- RATION- SATELLITE (passive) RADIO ASTRONOMY SPACE RESEARCH (passive) US74 US246 722	
231-235 FIXED FIXED-SATELLITE (space-to-Earth) MOBILE Radiolocation			231-235	FIXED FIXED-SATELLITE (space-to-Earth) MOBILE Radiolocation US211	FIXED FIXED-SATELLITE (space-to-Earth) MOBILE Radiolocation US211	
235-238 EARTH EXPLORATION-SATELLITE (passive) FIXED FIXED-SATELLITE (space-to-Earth) MOBILE SPACE RESEARCH (passive)			235-238	EARTH EXPLO- RATION- SATELLITE (passive) FIXED FIXED-SATELLITE (space-to-Earth) MOBILE SPACE RESEARCH (passive) US263	EARTH EXPLO- RATION- SATELLITE (passive) FIXED FIXED-SATELLITE (space-to-Earth) MOBILE SPACE RESEARCH (passive) US263	

TABLES OF FREQUENCY ALLOCATIONS

INTERNATIONAL			Band GHz	UNITED STATES		Remarks
Region 1 GHz	Region 2 GHz	Region 3 GHz		Government Allocation	Non-Govt. Allocation	
FIXED FIXED-SATELLITE (space-to-Earth) MOBILE Radiolocation			238-241	FIXED FIXED-SATELLITE (space-to-Earth) MOBILE Radiolocation	FIXED FIXED-SATELLITE (space-to-Earth) MOBILE Radiolocation	ISM 245 ± 1 GHz
RADIOLOCATION Amateur Amateur-Satellite 922			241-248	RADIOLOCATION 922	RADIOLOCATION Amateur Amateur-Satellite 922	
AMATEUR AMATEUR-SATELLITE			248-250		AMATEUR AMATEUR-SATELLITE	
EARTH EXPLORATION-SATELLITE (Passive) SPACE RESEARCH (Passive) 923			250-252	EARTH EXPLORATION-SATELLITE (Passive) SPACE RESEARCH (Passive) 923	EARTH EXPLORATION-SATELLITE (Passive) SPACE RESEARCH (Passive) 923	
MOBILE 902 MOBILE-SATELLITE RADIONAVIGATION RADIONAVIGATION-SATELLITE 903 923 924 925			252-265	MOBILE MOBILE-SATELLITE RADIONAVIGATION RADIONAVIGATION-SATELLITE US211 902 903 923 924	MOBILE MOBILE-SATELLITE RADIONAVIGATION RADIONAVIGATION-SATELLITE US211 902 903 923 924	
FIXED FIXED-SATELLITE (Earth-to-space) MOBILE RADIO ASTRONOMY 926			265-275	FIXED FIXED-SATELLITE (Earth-to-space) MOBILE RADIO ASTRONOMY 926	FIXED FIXED-SATELLITE (Earth-to-space) MOBILE RADIO ASTRONOMY 926	
(Not Allocated) 927			275-300	FIXED MOBILE 927	FIXED MOBILE 927	
			300-400	(Not allocated) 927	(Not allocated) 927	

Light, Vision, and Photometry

5.1 Introduction

Vision results from stimulation of the eye by light and consequent interaction through connecting nerves with the brain.[1] In physical terms, light constitutes a small section in the range of electromagnetic radiation, extending in wavelength from about 400 to 700 nanometers (nm) or billionths (10^{-9}) of a meter. (See Figure 5.1.)

Under ideal conditions, the human visual system can detect:

- Wavelength differences of 1 millimicron (10 Ä, 1 Angstrom unit = 10^{-8} cm)

- Intensity differences as little as 1 percent

- Forms subtending an angle at the eye of 1 arc-minute, and often smaller objects

Although the range of human vision is small compared with the total energy spectrum, human discrimination—the ability to detect differences in intensity or quality—is excellent.

5.2 Sources of Illumination

Light reaching an observer usually has been reflected from some object. The original source of such energy typically is radiation from molecules or atoms resulting from internal (atomic) changes. The exact type of emission is determined by:

- The ways in which the atoms or molecules are supplied with energy to replace what they radiate

---

1 Portions of this chapter were adapted from: Jerry C. Whitaker and K. B. Benson (eds.), *Standard Handbook of Video and Television Engineering*, 3rd ed., McGraw-Hill, New York, NY, 1999. Used with permission.

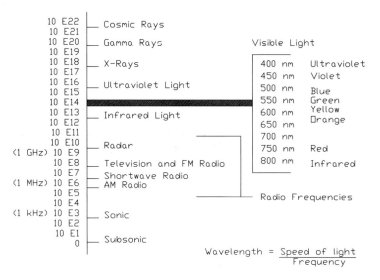

Figure 5.1 The electromagnetic spectrum.

• The physical state of the substance, whether solid, liquid, or gaseous

The most common source of radiant energy is the thermal excitation of atoms in the solid or gaseous state.

5.2.1 The Spectrum

When a beam of light traveling in air falls upon a glass surface at an angle, it is *refracted* or bent. The amount of refraction depends upon the wavelength, its variation with wavelength being known as *dispersion*. Similarly, when the beam, traveling in glass, emerges into air, it is refracted (with dispersion). A glass prism provides a refracting system of this type. Because different wavelengths are refracted by different amounts, an incident white beam is split up into several beams corresponding to the many wavelengths contained in the composite white beam. This is how the spectrum is obtained.

If a spectrum is allowed to fall upon a narrow slit arranged parallel to the edge of the prism, a narrow band of wavelengths passes through the slit. Obviously, the narrower the slit, the narrower the band of wavelengths or the "sharper" the spectral line. Also, more dispersion in the prism will cause a wider spectrum to be produced, and a narrower spectral line will be obtained for a given slit width.

It should be noted that purples are not included in the list of spectral colors. The purples belong to a special class of colors; they can be produced by mixing the light from two spectral lines, one in the red end of the spectrum, the other in the blue end.

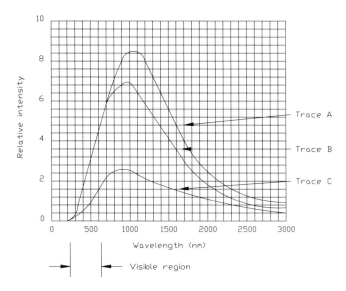

Figure 5.2 The typical radiating characteristics of tungsten: (trace *A*) the radiant flux from 1 cm² of a blackbody at 3000 K, (trace *B*) radiant flux from 1 cm² of tungsten at 3000 K, (trace *C*) radiant flux from 2.27 cm² of tungsten at 3000 K (equal to curve *A* in the visible region). (*After* [1].)

Purple (magenta is a more scientific name) is therefore referred to as a *nonspectral color.*

A plot of the power distribution of a source of light is indicative of the watts radiated at each wavelength per nanometer of wavelength. It is usual to refer to such a graph as an *energy distribution curve.*

Individual narrow bands of wavelengths of light are seen as strongly colored elements. Increasingly broader bandwidths retain the appearance of color, but with decreasing purity, as if white light had been added to them. A very broad band extending throughout the visible spectrum is perceived as white light. Many white light sources are of this type, such as the familiar tungsten-filament electric light bulb (see Figure 5.2). Daylight also has a broad band of radiation, as illustrated in Figure 5.3. The energy distributions shown in Figures 5.2 and 5.3 are quite different and, if the corresponding sets of radiation were seen side by side, would be different in appearance. Either one, particularly if seen alone, would represent a very acceptable white. A sensation of white light can also be induced by light sources that do not have a uniform energy distribution. Among these is fluorescent lighting, which exhibits sharp peaks of energy through the visible spectrum. Similarly, the light from a monochrome (black-and-white) video cathode ray tube is not uniform within the visible spectrum, generally exhibiting peaks in the yellow and blue regions of the spectrum; yet it appears as an acceptable white (see Figure 5.4).

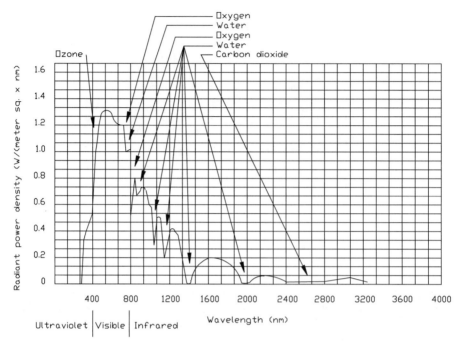

Figure 5.3 Spectral distribution of solar radiant power density at sea level, showing the ozone, oxygen, and carbon dioxide absorption bands. (*After* [1].)

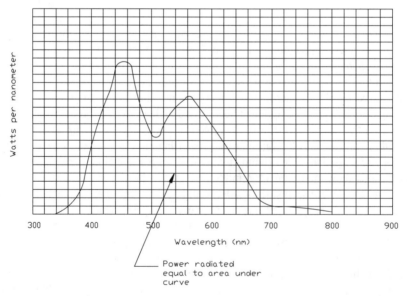

Figure 5.4 Power distribution of a monochrome video picture tube light source. (*After* [2].)

Table 5.1 Psychophysical and Psychological Characteristics of Color

Psychophysical Properties	Psychological Properties
Dominant wavelength	Hue
Excitation purity	Saturation
Luminance	Brightness
Luminous transmittance	Lightness
Luminous reflectance	Lightness

5.2.2 Monochrome and Color Vision

The color sensation associated with a light stimulus can be described in terms of three characteristics:

- Hue

- Saturation

- Brightness

The spectrum contains most of the principal hues: red, orange, yellow, green, blue, and violet. Additional hues are obtained from mixtures of red and blue light. These constitute the purple colors. Saturation pertains to the strength of the hue. Spectrum colors are highly saturated. White and grays have no hue and, therefore, have zero saturation. Pastel colors have low or intermediate saturation. Brightness pertains to the intensity of the stimulation. If a stimulus has high intensity, regardless of its hue, it is said to be "bright."

The psychophysical analogs of hue, saturation, and brightness are

- Dominant wavelength

- Excitation purity

- Luminance

This principle is illustrated in Table 5.1.

By using definitions and standard response functions, which have received international acceptance through the International Commission on Illumination, the dominant wavelength, purity, and luminance of any stimulus of known spectral energy distribution can be determined by simple computations. Although roughly analogous to their psychophysical counterparts, the psychological attributes of hue, saturation, and brightness pertain to observer responses to light stimuli and are not subject to calculation. These sensation characteristics—as applied to any given stimulus—depend in part on other visual stimuli in the field of view and upon the immediately preceding stimulations.

Color sensations arise directly from the action of light on the eye. They are normally associated, however, with objects in the field of view from which the light comes. The objects themselves are therefore said to have color. *Object colors* may be described in terms of their hues and saturations, such as with light stimuli. The intensity aspect is usually referred to in terms of lightness, rather than brightness. The psychophysical analogs of lightness are *luminous reflectance* for reflecting objects and *luminous transmittance* for transmitting objects.

At low levels of illumination, objects may differ from one another in their lightness appearances, but give rise to no sensation of hue or saturation. All objects appear as different shades of gray. Vision at low levels of illumination is called *scotopic vision*. This differs from *photopic vision*, which takes place at higher levels of illumination. Table 5.2 compares the luminosity values for photopic and scotopic vision.

Only the rods of the retina are involved in scotopic vision; cones play no part. Because the fovea centralis is free of rods, scotopic vision takes place outside the fovea. Visual acuity of scotopic vision is low compared with photopic vision.

At high levels of illumination, where cone vision predominates, all vision is color vision. Reproducing systems such as black-and-white photography and monochrome video cannot reproduce all three types of characteristics of colored objects. All images belong to the series of grays, differing only in relative brightness.

The relative brightness of the reproduced image of any object depends primarily upon the luminance of the object as seen by the photographic or video camera. Depending upon the camera pickup element or the film, the dominant wavelength and purity of the light may also be of consequence. Most films and video pickup elements currently in use exhibit sensitivity throughout the visible spectrum. Consequently, marked distortions in luminance as a function of dominant wavelength and purity are not encountered. However, their spectral sensitivities seldom conform exactly to that of the human observer. Some brightness distortions, therefore, do exist.

5.2.3 Luminosity Curve

A *luminosity curve* is a plot indicative of the relative brightnesses of spectrum colors of different wavelength or frequency. To a normal observer, the brightest part of a spectrum consisting of equal amounts of radiant flux per unit wavelength interval is at about 555 nm. Luminosity curves are, therefore, commonly normalized to have a value of *unity* at 555 nm. If, at some other wavelength, twice as much radiant flux as at 555 nm is required to obtain brightness equality with radiant flux at 555 nm, the luminosity at this wavelength is 0.5. The luminosity at any wavelength λ is, therefore, defined as the ratio P_{555}/P_λ, where P_λ denotes the amount of radiant flux at the wavelength λ, which is equal in brightness to a radiant flux of P_{555}.

The luminosity function that has been accepted as standard for photopic vision is given in Figure 5.5. Tabulated values at 10 nm intervals are given in Table 5.2. This function was agreed upon by the International Commission on Illumination (CIE) in 1924. It is based upon considerable experimental work that was conducted over a number of years. Chief reliance in arriving at this function was based on the step-by-step equality-of-brightness method. Flicker photometry provided additional data.

Table 5.2 Relative Luminosity Values for Photopic and Scotopic Vision

Wavelength, nm	Photopic Vision	Scotopic Vision
390	0.00012	0.0022
400	0.0004	0.0093
410	0.0012	0.0348
420	0.0040	0.0966
430	0.0116	0.1998
440	0.023	0.3281
450	0.038	0.4550
460	0.060	0.5670
470	0.091	0.6760
480	0.139	0.7930
490	0.208	0.9040
500	0.323	0.9820
510	0.503	0.9970
520	0.710	0.9350
530	0.862	0.8110
540	0.954	0.6500
550	0.995	0.4810
560	0.995	0.3288
570	0.952	0.2076
580	0.870	0.1212
590	0.757	0.0655
600	0.631	0.0332
610	0.503	0.0159
620	0.381	0.0074
630	0.265	0.0033
640	0.175	0.0015
650	0.107	0.0007
660	0.061	0.0003
670	0.032	0.0001
680	0.017	0.0001
690	0.0082	
700	0.0041	
710	0.0021	
720	0.00105	
730	0.00052	

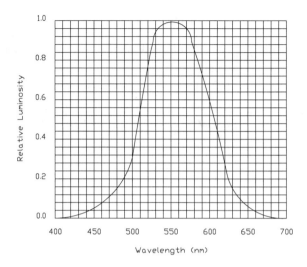

Figure 5.5 The photopic luminosity function. (*After* [2].)

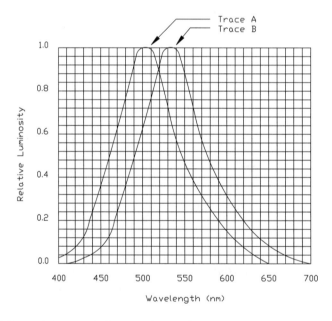

Figure 5.6 Scotopic luminosity function (trace *A*) as compared with photopic luminosity function (trace *B*). (*After* [2].)

In the scotopic range of intensities, the luminosity function is somewhat different from that of the photopic range. The two curves are compared in Figure 5.6. Values are listed in Table 5.2. While the two curves are similar in shape, there is a shift for the scotopic curve of about 40 nm to the shorter wavelengths.

5.2.4 Luminance

Brightness is a term used to describe one of the characteristics of appearance of a source of radiant flux or of an object from which radiant flux is being reflected or transmitted. Brightness specifications of two or more sources of radiant flux should be indicative of their actual relative appearances. These appearances will greatly depend upon the viewing conditions, including the state of adaptation of the observer's eye.

Luminance, as previously indicated, is a psychophysical analog of brightness. It is subject to physical determination, independent of particular viewing and adaptation conditions. Because it is an analog of brightness, however, it is defined to relate as closely as possible to brightness.

The best established measure of the relative brightnesses of different spectral stimuli is the luminosity function. In evaluating the luminance of a source of radiant flux consisting of many wavelengths of light, the amounts of radiant flux at the different wavelengths are weighted by the luminosity function. This converts radiant flux to luminous flux. As used in photometry, the term *luminance* applies only to extended sources of light, not to point sources. For a given amount (and quality) of radiant flux reaching the eye, brightness will vary inversely with the effective area of the source.

Luminance is described in terms of luminous flux per unit projected area of the source. The greater the concentration of flux in the angle of view of a source, the brighter it appears. Therefore, luminance is expressed in terms of amounts of flux per unit solid angle or *steradian*.

In considering the relative luminances of various objects of a scene to be captured and reproduced by a video system, it is convenient to normalize the luminance values so that the "white" in the region of principal illumination has a relative luminance value of 1.00. The relative luminance of any other object then becomes the ratio of its luminance to that of the white. This white is an object of highly diffusing surface with high and uniform reflectance throughout the visible spectrum. For purposes of computation it may be idealized to have 100 percent reflectance and perfect diffusion.

5.2.5 Luminance Discrimination

If an area of luminance B is viewed side by side with an equal area of luminance $B + \Delta B$, a value of ΔB may be established for which the brightnesses of the two areas are just noticeably different. The ratio of $\Delta B/B$ is known as *Weber's fraction*. The statement that this ratio is a constant, independent of B, is known as *Weber's law*.

Strictly speaking, the value of Weber's fraction is not independent of B. Furthermore, its value depends considerably on the viewer's state of adaptation. Values as determined for a dark-field surround are shown in Figure 5.7. It is seen that, at very low intensities, the value of $\Delta B/B$ is relatively large; that is, relatively large values of ΔB, as compared with B, are necessary for discrimination. A relatively constant value of

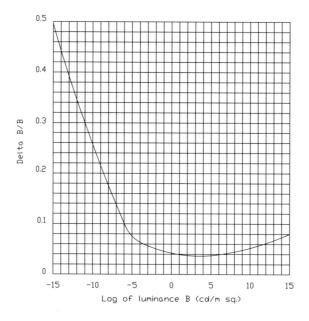

Figure 5.7 Weber's fraction $\Delta B/B$ as a function of luminance B for a dark-field surround. (*After* [3].)

roughly 0.02 is maintained through a brightness range of about 1 to 300 cd/m^2. The slight rise in the value of $\Delta B/B$ at high intensities as given in the graph may indicate lack of complete adaptation to the stimuli being compared.

The plot of $\Delta B/B$ as a function of B will change significantly if the comparisons between the two fields are made with something other than a dark surround. The greatest changes are for luminances below the adapting field. The loss of power of discrimination proceeds rapidly for luminances less by a factor of 10 than that of the adapting field. On the high-luminance side, adaptation is largely controlled by the comparison fields and is relatively independent of the adapting field.

Because of the luminance discrimination relationship expressed by Weber's law, it is convenient to express relative luminances of areas from either photographic or video images in logarithmic units. Because $\Delta(\log B)$ is approximately equal to $\Delta B/B$, equal small changes in $(\log B)$ correspond reasonably well with equal numbers of brightness discrimination steps.

5.2.6 Perception of Fine Detail

Detail is seen in an image because of brightness differences between small adjacent areas in a monochrome display or because of brightness, hue, or saturation differences in a color display. Visibility of detail in a picture is important because it determines the extent to which small or distant objects of a scene are visible, and because of its relationship to the "sharpness" appearance of the edges of objects.

"Picture definition" is probably the most acceptable term for describing the general characteristic of "crispness," "sharpness," or image-detail visibility in a picture. Picture definition depends upon characteristics of the eye, such as visual acuity, and upon a variety of characteristics of the picture-image medium, including its resolving power, luminance range, contrast, and image-edge gradients.

The extent to which a picture medium, such as a photographic or a video system, can reproduce fine detail is expressed in terms of *resolving power* or *resolution*. Resolution is a measure of the distance between two fine lines in the reproduced image that are visually distinct. The image is examined under the best possible conditions of viewing, including magnification.

Resolution in photography is usually expressed as the maximum number of lines (counting only the black ones or only the white ones) per millimeter that can be distinguished from one another. In addition to the photographic material itself, measured values of resolving power depend upon a number of factors. The most important ones typically are:

- Density differences between the black and the white lines of the test chart photographed

- Sharpness of focus of the test-chart image during exposure

- Contrast to which the photographic image is developed

- Composition of the developer

Resolution in a video system is expressed in terms of the maximum number of lines (counting both black and white) that are discernible when viewing a test chart. The value of horizontal (vertical lines) or vertical (horizontal lines) resolution is the number of lines equal to the dimension of the raster. Vertical resolution in a well-adjusted system equals the number of scanning lines, roughly 500 in conventional television. In normal broadcasting and reception practice, however, typical values of vertical resolution range from 350 to 400 lines.

5.2.7 Sharpness

The appearance evaluation of a picture image in terms of the edge characteristics of objects is called *sharpness*. The more clearly defined the line that separates dark areas from lighter ones, the greater the sharpness of the picture. Sharpness is, naturally, related to the transient curve in the image across an edge. The average gradient and the total density difference appear to be the most important characteristics. No physical measure has been devised, however, that predicts the sharpness (appearance) of an image in all cases.

Picture resolution and sharpness are to some extent interrelated, but they are by no means perfectly correlated. Pictures ranked according to resolution measures may be rated somewhat differently on the basis of sharpness. Both resolution and sharpness are related to the more general characteristic of picture definition. For pictures in which, under particular viewing conditions, effective resolution is limited by the visual acuity of the eye rather than by picture resolution, sharpness is probably a good indication of picture definition. If visual acuity is not the limiting factor, however, picture definition depends to an appreciable extent on both resolution and sharpness.

5.2.8 Response to Intermittent Excitation

The brightness sensation resulting from a single, short flash of light is a function of the duration of the flash and its intensity. For low-intensity flashes near the threshold of vision, stimuli of shorter duration than about 1/5 s are not seen at their full intensity. Their apparent intensities are nearly proportional to the action times of the stimuli.

With increasing intensity of the stimulus, the time necessary for the resulting sensation to reach its maximum becomes shorter. A stimulus of 5 mL reaches its maximum apparent intensity in about 1/10 s; a stimulus of 1000 mL reaches its maximum in less than 1/20 s. Also, for higher intensities, there is a brightness overshooting effect. For stimulus times longer than what is necessary for the maximum effect, the apparent brightness of the flash is decreased. A 1000 mL flash of 1/20 s will appear to be almost twice as bright as a flash of the same intensity that continues for 1/5 s. These effects are essentially the same for colors of equal luminances, independent of their chromatic characteristics.

Intermittent excitations at low frequencies are seen as successive individual light flashes. With increased frequency, the flashes appear to merge into one another, giving a coarse, pulsating *flicker effect*. Further increases in frequency result in finer and finer pulsations until, at a sufficiently high frequency, the flicker effect disappears.

The lowest frequency at which flicker is not seen is called the *critical fusion frequency* or simply the *critical frequency*. Over a wide range of stimuli luminances, the critical fusion frequency is linearly related to the logarithm of luminance. This relationship is called the *Ferry-Porter law*. Critical frequencies for several different wavelengths of light are plotted as functions of retinal illumination (*trolands*) in Figure 5.8. The second abscissa scale is plotted in terms of luminance, assuming a pupillary diameter of about 3 mm. At low luminances, critical frequencies differ for different wavelengths, being lowest for stimuli near the red end of the spectrum and highest for stimuli near the blue end. Above a retinal illumination of about 10 trolands (0.4 ft·L) the critical frequency is independent of wavelength. This is in the critical frequency range above approximately 18 Hz.

The critical fusion frequency increases approximately logarithmically with the increase in retinal area illuminated. It is higher for retinal areas outside the fovea than for those inside, although fatigue to flicker effects is rapid outside the fovea.

Intermittent stimulations sometimes result from rapid alternations between two color stimuli, rather than between one color stimulus and complete darkness. The critical frequency for such stimulations depends upon the relative luminance and chromatic characteristics of the alternating stimuli. The critical frequency is lower for chromatic differences than for luminance differences.

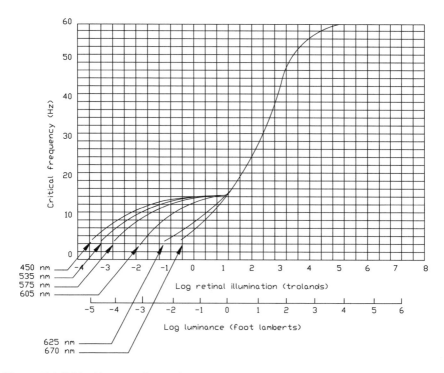

Figure 5.8 Critical frequencies as they relate to retinal illumination and luminance (1 ft·L ≅ cd/m²; 1 troland = retinal illuminance per square millimeter pupil area from the surface with luminance of 1 cd/m²). (*After* [4].)

5.3 References

1. *IES Lighting Handbook*, Illuminating Engineering Society of North America, New York, NY, 1981.
2. Fink, D. G., *Television Engineering*, 2nd ed., McGraw-Hill, New York, NY, 1952.
3. Hecht, S., "The Visual Discrimination of Intensity and the Weber-Fechner Law," *J. Gen Physiol.*, vol. 7, pg. 241, 1924.
4. Hecht, S., S. Shiaer, and E. L. Smith, "Intermittent Light Stimulation and the Duplicity Theory of Vision," Cold Spring Harbor Symposia on Quantitative Biology, vol. 3, pg. 241, 1935.

5.4 Bibliography

Benson, K. B., and J. C. Whitaker, *Television Engineering Handbook*, revised ed., McGraw-Hill, New York, NY, 1991.

Boynton, R. M., *Human Color Vision*, Holt, New York, NY, 1979.

Committee on Colorimetry, Optical Society of America, *The Science of Color*, New York, NY, 1953.

Davson, H., *Physiology of the Eye*, 4th ed., Academic, New York, NY, 1980.

Evans, R. M., W. T. Hanson, Jr., and W. L. Brewer, *Principles of Color Photography*, Wiley, New York, NY, 1953.

Kingslake, R. (ed.), *Applied Optics and Optical Engineering*, vol. 1, Academic, New York, NY, 1965.

Polysak, S. L., *The Retina*, University of Chicago Press, Chicago, IL, 1941.

Schade, O. H., "Electro-optical Characteristics of Television Systems," *RCA Review*, vol. 9, pp. 5–37, 245–286, 490–530, 653–686, 1948.

Wright, W. D., *Researches on Normal and Defective Colour Vision*, Mosby, St. Louis, MO, 1947.

Wright, W. D., *The Measurement of Colour*, 4th ed., Adam Hilger, London, England, 1969.

5.5 Tabular Data

Table 5.3 Typical Luminance Values (*After* [2].)

Illumination	Illuminance, ft-L
Sun at zenith	4.82×10^8
Perfectly reflecting, diffusing surface in sunlight	9.29×10^3
Moon, clear sky	2.23×10^3
Overcast sky	$9\text{-}20 \times 10^2$
Clear sky	$6\text{-}17.5 \times 10^2$
Motion-picture screen	10

Table 5.4 Conversion Factors for Illuminance Units (*After* [2].)

Parameter	Lux	Phot	Footcandle
Lux (meter-candle); lumens per square meter	1.00	1×10^{-4}	9.290×10^{-2}
Phot; lumens per square centimeter	1×10^4	1.00	9.290×10^2
Footcandle; lumens per square foot	1.06×10	1.076×10^{-3}	1.00
Multiply the quantity expressed in units of *X* by the conversion factor to obtain the quantity in units of *Y*.			

Table 5.5 Conversion Factors for Luminance and Retinal Illuminance Units (*After* [2].)

Multiply Quantity Expressed in Units of X by Conversion Factor to Obtain Quantity in Units of Y

X \ Y	Candelas per square centimeter	Candelas per square meter (nit)§	Candelas per square inch	Candelas per square foot	Lamberts	Millilamberts	Footlamberts	Trolands†‡
Candelas per square centimeter	1	1×10^4	6.452	9.290×10^2	3.142	3.142×10^3	2.919×10^3	7.854×10^3
Candelas per square meter (nit)§	1×10^{-4}	1	6.452×10^{-4}	9.290×10^{-2}	3.142×10^{-4}	3.142×10^{-1}	2.919×10^{-1}	7.854×10^{-1}
Candelas per square inch	1.550×10^{-1}	1.550×10^3	1	1.440×10^2	4.869×10^{-1}	4.869×10^2	4.524×10^2	1.217×10^3
Candelas per square foot	1.076×10^{-3}	1.076×10	6.944×10^{-3}	1	3.382×10^{-3}	3.382	3.142	8.454
Lamberts	3.183×10^{-1}	3.183×10^3	2.054	2.957×10^2	1	1×10^3	9.290×10^2	2.5×10^3
Millilamberts	3.183×10^{-4}	3.183	2.054×10^{-3}	2.957×10^{-1}	1×10^{-3}	1	9.290×10^{-1}	2.500
Footlamberts	3.426×10^{-4}	3.426	2.210×10^{-3}	3.183×10^{-1}	1.076×10^{-3}	1.076	1	2.691
Trolands‡	1.273×10^{-4}	1.273	8.213×10^{-4}	1.183×10^{-1}	4.000×10^{-4}	4.000×10^{-1}	3.716×10^{-1}	1

†In converting luminance to trolands it is necessary to multiply the ⎫ conversion factor by the square of the pupil
‡In converting trolands to luminance it is necessary to divide the ⎭ diameter in millimeters.

§As recommended at Session XII in 1951 of the International Commission on Illumination, one nit equals one candela per square meter.

Table 5.6 CIE Colorimetric Data (1931 Standard Observer)

Wave-length (nm)	Trichromatic Coefficients		Distribution Coefficients, Equal-Energy Stimulus			Energy Distributions for Standard Illuminants			
	r	g	$\bar{r}$	$\bar{g}$	$\bar{b}$	E_A	E_B	E_C	E_{D65}
380	0.0272	-0.0115	0.0000	0.0000	0.0012	9.80	22.40	33.00	49.98
390	0.0263	-0.0114	0.0001	0.0000	0.0036	12.09	31.30	47.40	54.65
400	0.0247	-0.0112	0.0003	0.0001	0.0121	14.71	41.30	63.30	82.75
410	0.0225	-0.0109	0.0008	-0.0004	0.0371	17.68	52.10	80.60	91.49
420	0.0181	-0.0094	0.0021	-0.0011	0.1154	20.99	63.20	98.10	93.43
430	0.0088	-0.0048	0.0022	-0.0012	0.2477	24.67	73.10	112.40	86.68
440	-0.0084	0.0048	-0.0026	0.0015	0.3123	28.70	80.80	121.50	104.86
450	-0.0390	0.0218	0.0121	0.0068	0.3167	33.09	85.40	124.00	117.01
460	0.0909	0.0517	-0.0261	0.0149	0.2982	37.81	88.30	123.10	117.81
470	-0.1821	0.1175	-0.0393	0.0254	0.2299	42.87	92.00	123.80	114.86
480	-0.3667	0.2906	-0.0494	0.0391	0.1449	48.24	95.20	123.90	115.92
490	-0.7150	0.6996	-0.0581	0.0569	0.0826	53.91	96.50	120.70	108.81
500	1.1685	1.3905	0.0717	0.0854	0.0478	59.86	94.20	112.10	109.35
510	-1.3371	1.9318	-0.0890	0.1286	0.0270	66.06	90.70	102.30	107.80
520	-0.9830	1.8534	-0.0926	0.1747	0.0122	72.50	89.50	96.90	104.79
530	-0.5159	1.4761	0.0710	0.2032	0.0055	79.13	92.20	98.00	107.69
540	0.1707	1.1628	0.0315	0.2147	0.0015	85.95	96.90	102.10	104.41
550	0.0974	0.9051	0.0228	0.2118	-0.0006	92.91	101.00	105.20	104.05
560	0.3164	0.6881	0.0906	0.1970	-0.0013	100.00	102.80	105.30	100.00
570	0.4973	0.5067	0.1677	0.1709	-0.0014	107.18	102.60	102.30	96.33
580	0.6449	0.3579	0.2543	0.1361	0.0011	114.44	101.00	97.80	95.79
590	0.7617	0.2402	0.3093	0.0975	-0.0008	121.73	99.20	93.20	88.69
600	0.8475	0.1537	0.3443	0.0625	-0.0005	129.04	98.00	89.70	90.01
610	0.9059	0.0494	0.3397	0.0356	0.0003	136.35	98.50	88.40	89.60
620	0.9425	0.0580	0.2971	0.0183	-0.0002	143.62	99.70	88.10	87.70
630	0.9649	0.0354	0.2268	0.0083	-0.0001	150.84	101.00	88.00	83.29
640	0.9797	0.0205	0.1597	0.0033	0.000	157.98	102.20	87.80	83.70
650	0.9888	0.0113	0.1017	0.0012	0.0000	165.03	103.90	88.20	80.03
660	0.9940	0.0061	0.0593	0.0004	0.0000	171.96	105.00	87.90	80.21
670	0.9966	0.0035	0.0315	0.0001	0.0000	178.77	104.90	86.30	82.28
680	0.9984	0.0016	0.0169	0.0000	0.0000	185.43	103.90	84.00	78.28
690	0.9996	0.0004	0.0082	0.0000	0.0000	191.93	101.60	80.20	69.72
700	1.0000	0.0000	0.0041	0.0000	0.0000	198.26	99.10	76.30	71.61
710	1.0000	0.0000	0.0021	0.0000	0.0000	204.41	96.20	72.40	74.15
720	1.0000	0.0000	0.0011	0.0000	0.0000	210.36	92.90	68.30	61.60
730	1.0000	0.0000	0.0005	0.0000	0.0000	216.12	89.40	64.40	69.89
740	1.0000	0.0000	0.0003	0.0000	0.0000	221.67	86.90	61.50	75.09
750	1.0000	0.0000	0.0001	0.0000	0.0000	227.00	85.20	59.20	63.59
760	1.0000	0.0000	0.0001	0.0000	0.0000	232.12	84.70	58.10	46.42
770	1.0000	0.0000	0.0000	0.0000	0.0000	237.01	85.40	58.20	66.81
780	1.0000	0.0000	0.0000	0.0000	0.0000	241.68	87.00	59.10	63.38

Table 5.6 CIE Colorimetric Data (continued)

Wave-length (nm)	Trichromatic Coefficients		Distribution Coefficients, Equal-Energy Stimulus			Distribution Coefficients Weighted by Illuminant C		
	x	y	$\bar{x}$	$\bar{y}$	$\bar{z}$	$E_c\bar{x}$	$E_c\bar{y}$	$E_c\bar{z}$
380	0.1741	0.0050	0.0014	0.0000	0.0065	0.0036	0.0000	0.0164
390	0.1738	0.0049	0.0042	0.0001	0.0201	0.0183	0.0004	0.0870
400	0.1733	0.0048	0.0143	0.0004	0.0679	0.0841	0.0021	0.3992
410	0.1726	0.0048	0.0435	0.0012	0.2074	0.3180	0.0087	1.5159
420	0.1714	0.0051	0.1344	0.0040	0.6456	1.2623	0.0378	6.0646
430	0.1689	0.0069	0.2839	0.0116	1.3856	2.9913	0.1225	14.6019
440	0.1644	0.0109	0.3483	0.0230	1.7471	3.9741	0.2613	19.9357
450	0.1566	0.0177	0.3362	0.0380	1.7721	3.9191	0.4432	20.6551
460	0.1440	0.0297	0.2908	0.0600	1.6692	3.3668	0.6920	19.3235
470	0.1241	0.0578	0.1954	0.0910	1.2876	2.2878	1.0605	15.0550
480	0.0913	0.1327	0.0956	0.1390	0.8130	1.1038	1.6129	9.4220
490	0.0454	0.2950	0.0320	0.2080	0.4652	0.3639	2.3591	5.2789
500	0.0082	0.5384	0.0049	0.3230	0.2720	0.0511	3.4077	2.8717
510	0.0139	0.7502	0.0093	0.5030	0.1582	0.0898	4.8412	1.5181
520	0.0743	0.8338	0.0633	0.7100	0.0782	0.5752	6.4491	0.7140
530	0.1547	0.8059	0.1655	0.8620	0.0422	1.5206	7.9357	0.3871
540	0.2296	0.7543	0.2904	0.9540	0.0203	2.7858	9.1470	0.1956
550	0.3016	0.6923	0.4334	0.9950	0.0087	4.2833	9.8343	0.0860
560	0.3731	0.6245	0.5945	0.9950	0.0039	5.8782	9.8387	0.0381
570	0.4441	0.5547	0.7621	0.9520	0.0021	7.3230	9.1476	0.0202
580	0.5125	0.4866	0.9163	0.8700	0.0017	8.4141	7.9897	0.0147
590	0.5752	0.4242	1.0263	0.7570	0.0011	8.9878	6.6283	0.0101
600	0.6270	0.3725	1.0622	0.6310	0.0008	8.9536	5.3157	0.0067
610	0.6658	0.3340	1.0026	0.5030	0.0003	8.3294	4.1788	0.0029
620	0.6915	0.3083	0.8544	0.3810	0.0002	7.0604	3.1485	0.0012
630	0.7079	0.2920	0.6424	0.2650	0.0000	5.3212	2.1948	0.0000
640	0,7190	0.2809	0.4479	0.1750	0.0000	3.6882	1.4411	0.0000
650	0.7260	0.2740	0.2835	0.1070	0.0000	2.3531	0.8876	0.0000
660	0.7300	0.2700	0.1649	0,0610	0.0000	1.3589	0.5028	0.0000
670	0.7320	0.2680	0.0874	0.0320	0.0000	0.7113	0.2606	0.0000
680	0.7334	0.2666	0.0468	0.0170	0.0000	0.3657	0.1329	0.0000
690	0.7344	0.2656	0.0227	0.0082	0.0000	0.1721	0.0621	0.0000
700	0.7347	0.2653	0.0114	0.0041	0.0000	0.0806	0.0290	0.0000
710	0.7347	0.2653	0.0058	0.0021	0.0000	0.0398	0.0143	0.0000
720	0.7347	0.2653	0.0029	0.0010	0.0000	0.0183	0.0064	0.0000
730	0.7347	0.2653	0.0014	0.0005	0.0000	0.0085	0.0030	0.0000
740	0.7347	0.2653	0.0007	0.0003	0.0000	0.0040	0.0017	0.0000
750	0.7347	0.2653	0.0003	0.0001	0.0000	0.0017	0.0006	0.0000
760	0.7347	0.2653	0.0002	0.0001	0.0000	0.0008	0.0003	0.0000
770	0.7347	0.2653	0.0001	0.0000	0.0000	0.0003	0.0000	0.0000

6

Circuit Fundamentals

6.1   Introduction

Electronic circuits are composed of elements such as resistors, capacitors, inductors, and voltage and current sources, all of which may be interconnected to permit the flow of electric currents. An *element* is the smallest component into which circuits can be subdivided. The points on a circuit element where they are connected in a circuit are called *terminals*.

Elements can have two or more terminals, as shown in Figure 6.1. The resistor, capacitor, inductor, and diode shown in the Figure 6.1*a* are two-terminal elements; the transistor in Figure 6.1*b* is a three-terminal element; and the transformer in Figure 6.1*c* is a four-terminal element.

Circuit elements and components also are classified as to their function in a circuit. An element is considered *passive* if it absorbs energy and *active* if it increases the level of energy in a signal. An element that receives energy from either a passive or active element is called a *load*. In addition, either passive or active elements, or components, can serve as loads.

The basic relationship of current and voltage in a two-terminal circuit where the voltage is constant and there is only one source of voltage is given in Ohm's law. This states that the voltage V between the terminals of a conductor varies in accordance with the current I. The ratio of voltage, current, and resistance R is expressed in Ohm's law as follows:

$$E = I \times R \tag{6.1}$$

Using Ohm's law, the calculation for power in watts can be developed from $P = E \times I$ as follows:

$$P = \frac{E^2}{R} \quad and \quad P = I^2 \times R \tag{6.2}$$

(a)

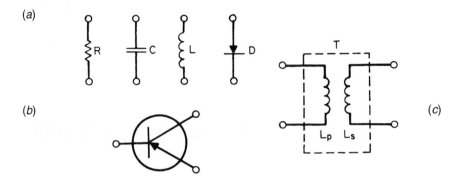

(b)

(c)

Figure 6.1 Schematic examples of circuit elements: (a) two-terminal element, (b) three-terminal element, (c) four-terminal element.

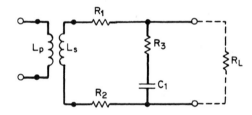

Figure 6.2 Circuit configuration composed of several elements and branches, and a closed loop (R_1, R, C_1, R_2, and L_s).

A circuit, consisting of a number of elements or components, usually amplifies or otherwise modifies a signal before delivering it to a load. The terminal to which a signal is applied is an *input port*, or *driving port*. The pair or group of terminals that delivers a signal to a load is the *output port*. An element or portion of a circuit between two terminals is a *branch*. The circuit shown in Figure 6.2 is made up of several elements and branches. R_1 is a branch, and R_1 and C_1 make up a two-element branch. The secondary of transformer T, a voltage source, and R_2 also constitute a branch. The point at which three or more branches join together is a *node*. A series connection of elements or branches, called a *path*, in which the end is connected back to the start is a *closed loop*.

6.2 Circuit Analysis

Relatively complex configurations of *linear circuit elements*, that is, where the signal gain or loss is constant over the signal amplitude range, can be analyzed by simplification into the equivalent circuits. After the restructuring of a circuit into an equivalent form, the current and voltage characteristics at various nodes can be calculated

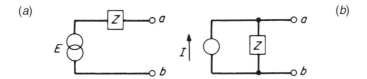

Figure 6.3 Equivalent circuits: (*a*) Thevenin's equivalent voltage source, (*b*) Norton's equivalent current source. (*After* [1].)

using network-analysis theorems, including Kirchoff's current and voltage laws, Thevenin's theorem, and Norton's theorem.

- **Kirchoff's current law (KCL).** The algebraic sum of the instantaneous currents entering a node (a common terminal of three or more branches) is zero. In other words, the currents from two branches entering a node add algebraically to the current leaving the node in a third branch.

- **Kirchoff's voltage law (KVL).** The algebraic sum of instantaneous voltages around a closed loop is zero.

- **Thevenin's theorem.** The behavior of a circuit at its terminals can be simulated by replacement with a voltage E from a dc source in series with an impedance Z (see Figure 6.3*a*).

- **Norton's theorem.** The behavior of a circuit at its terminals can be simulated by replacement with a dc source I in parallel with an impedance Z (see Figure 6.3*b*).

6.2.1 AC Circuits

Vectors are used commonly in ac circuit analysis to represent voltage or current values. Rather than using waveforms to show phase relationships, it is accepted practice to use vector representations (sometimes called *phasor diagrams*). To begin a vector diagram, a horizontal line is drawn, its left end being the *reference point*. Rotation in a counterclockwise direction from the reference point is considered to be positive. Vectors may be used to compare voltage drops across the components of a circuit containing resistance, inductance, and/or capacitance. Figure 6.4 shows the vector relationship in a series RLC circuit, and Figure 6.5 shows a parallel RLC circuit.

Power Relationship in AC Circuits

In a dc circuit, power is equal to the product of voltage and current. This formula also is true for purely resistive ac circuits. However, when a reactance—either inductive or capacitive—is present in an ac circuit, the dc power formula does not apply. The product of voltage and current is, instead, expressed in volt-amperes (VA) or kilovoltamperes (kVA). This product is known as the *apparent power*. When meters

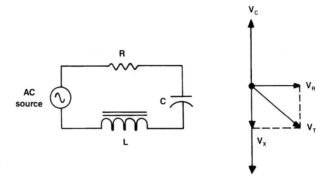

Figure 6.4 Voltage vectors in a series RLC circuit.

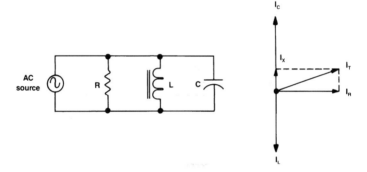

Figure 6.5 Current vectors in a parallel RLC circuit.

are used to measure power in an ac circuit, the apparent power is the voltage reading multiplied by the current reading. The *actual power* that is converted to another form of energy by the circuit is measured with a wattmeter, and is referred to as the *true power*. In ac power-system design and operation, it is desirable to know the ratio of true power converted in a given circuit to the apparent power of the circuit. This ratio is referred to as the *power factor*.

6.2.2 Complex Numbers

A complex number is represented by a *real part* and an *imaginary part*. For example, in $A = a + jb$, A is the complex number; a is the real part, sometimes written as Re(A); and b is the imaginary part of A, often written as Im(A). It is a convention to precede the imaginary component by the letter j (or i). This form of writing the real and imaginary components is called the *Cartesian form* and symbolizes the complex (or s) plane, wherein both the real and imaginary components can be indicated graphically

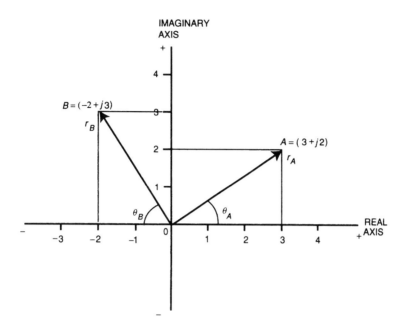

Figure 6.6 The s plane representing two complex numbers. (*From* [2]. *Used with permission.*)

[2]. To illustrate this, consider the same complex number A when represented graphically as shown in Figure 6.6. A second complex number B is also shown to illustrate the fact that the real and imaginary components can take on both positive and negative values. Figure 6.6 also shows an alternate form of representing complex numbers. When a complex number is represented by its magnitude and angle, for example, $A = r_A \angle \theta_A$, it is called the *polar representation*.

To see the relationship between the Cartesian and the polar forms, the following equations can be used:

$$r_A = \sqrt{a^2 + b^2} \tag{6.3}$$

$$\theta_A = \tan^{-1}\frac{b}{a} \tag{6.4}$$

Conceptually, a better perspective can be obtained by investigating the triangle shown in Figure 6.7, and considering the trigonometric relationships. From this figure, it can be seen that

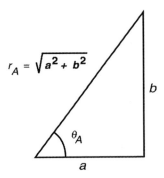

Figure 6.7 The relationship between Cartesian and polar forms. (*From* [2]. *Used with permission.*)

$$a = \text{Re}(A) = r_A \, \cos(\theta_A) \qquad\qquad (6.5)$$

$$b = \text{Im}(A) = r_A \, \sin(\theta_A) \qquad\qquad (6.6)$$

The well-known *Euler's identity* is a convenient conversion of the polar and Cartesian forms into an *exponential form*, given by

$$\exp(j\theta) = \cos\theta + j \sin\theta \qquad\qquad (6.7)$$

6.2.3 Phasors

The ac voltages and currents appearing in distribution systems can be represented by *phasors*, a concept useful in obtaining analytical solutions to one-phase and three-phase system design. A phasor is generally defined as a transform of sinusoidal functions from the time domain into the complex-number domain and given by the expression

$$\mathbf{V} = V \exp(j\theta) = P\{V \cos(\omega t + \theta)\} = V\angle\theta \qquad\qquad (6.8)$$

where $\mathbf{V}$ is the phasor, V is the magnitude of the phasor, and θ is the angle of the phasor. The convention used here is to use boldface symbols to symbolize phasor quantities. Graphically, in the time domain, the phasor $\mathbf{V}$ would be a simple sinusoidal wave shape as shown in Figure 6.8. The concept of a phasor leading or lagging another phasor becomes very apparent from the figure.

Phasor diagrams are also an effective medium for understanding the relationships between phasors. Figure 6.9 shows a phasor diagram for the phasors represented in Figure 6.8. In this diagram, the convention of positive angles being read counterclockwise is used. The other alternative is certainly possible as well. It is quite apparent that a purely capacitive load could result in the phasors shown in Figures 6.8 and 6.9.

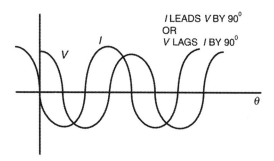

Figure 6.8 Waveforms representing leading and lagging phasors. (*From* [2]. *Used with permission.*)

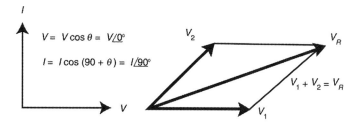

Figure 6.9 Phasor diagram showing phasor representation and phasor operation. (*From* [2]. *Used with permission.*)

6.2.4 Per Unit System

In the *per unit system*, basic quantities such as voltage and current are represented as certain percentages of base quantities. When so expressed, these per unit quantities do not need units, thereby making numerical analysis in power systems somewhat easier to handle. Four quantities encompass all variables required to solve a power system problem. These quantities are

- Voltage
- Current
- Power
- Impedance

Out of these, only two base quantities, corresponding to voltage (V_b) and power (S_b), are required to be defined. The other base quantities can be derived from these two. Consider the following. Let

V_b = voltage base, kV
S_b = power base, MVA
I_b = current base, A
Z_b = impedance base, Q

Then,

$$Z_b = \frac{V_b^2}{S_b} \ \Omega \tag{6.9}$$

$$I_b = \frac{V_b 10^3}{Z_b} \ A \tag{6.10}$$

6.2.5 Principles of Resonance

All RF systems rely on the principles of resonance for operation. Three basic systems exist:

- Series resonance circuits

- Parallel resonance circuits

- Cavity resonators

Series Resonant Circuits

When a constant voltage of varying frequency is applied to a circuit consisting of an inductance, capacitance, and resistance (all in series), the current that flows depends upon frequency in the manner shown in Figure 6.10. At low frequencies, the capacitive reactance of the circuit is large and the inductive reactance is small, so that most of the voltage drop is across the capacitor, while the current is small and leads the applied voltage by nearly 90°. At high frequencies, the inductive reactance is large and the capacitive reactance is low, resulting in a small current that lags nearly 90° behind the applied voltage; most of the voltage drop is across the inductance. Between these two extremes is the *resonant frequency*, at which the capacitive and inductive reactances are equal and, consequently, neutralize each other, leaving only the resistance of the circuit to oppose the flow of current. The current at this resonant frequency is, accordingly, equal to the applied voltage divided by the circuit resistance, and it is very large if the resistance is low.

The characteristics of a series resonant circuit depend primarily upon the ratio of inductive reactance ωL to circuit resistance R, known as the circuit Q:

$$Q = \frac{\omega L}{R} \tag{6.11}$$

The circuit Q also may be defined by:

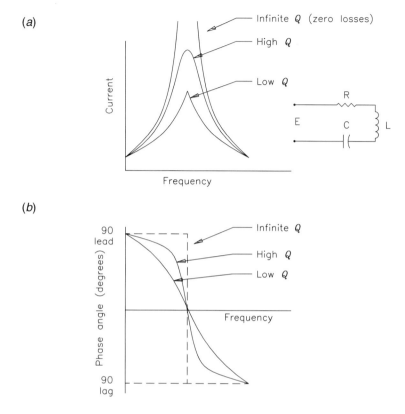

Figure 6.10 Characteristics of a series resonant circuit as a function of frequency for a constant applied voltage and different circuit Qs: (a) magnitude, (b) phase angle.

$$Q = 2\pi \left(\frac{E_s}{E_d} \right) \qquad\qquad (6.12)$$

Where:
E_s = energy stored in the circuit
E_d = energy dissipated in the circuit during one cycle

Most of the loss in a resonant circuit is the result of coil resistance; the losses in a properly constructed capacitor are usually small in comparison with those of the coil.

The general effect of different circuit resistances (different values of Q) is shown in Figure 6.10. As illustrated, when the frequency differs appreciably from the resonant frequency, the actual current is practically independent of circuit resistance and is nearly the current that would be obtained with no losses. On the other hand, the current at the resonant frequency is determined solely by the resistance. The effect of increasing the resistance of a series circuit is, accordingly, to flatten the resonance curve by re-

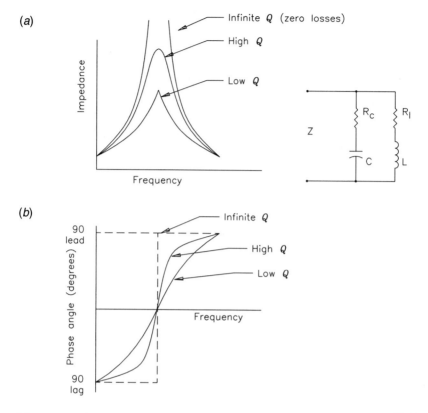

Figure 6.11 Characteristics of a parallel resonant circuit as a function of frequency for different circuit Qs: (a) magnitude, (b) phase angle.

ducing the current at resonance. This broadens the top of the curve, giving a more uniform current over a band of frequencies near the resonant point. This broadening is achieved, however, by reducing the selectivity of the tuned circuit.

Parallel Resonant Circuits

A parallel circuit consisting of an inductance branch in parallel with a capacitance branch offers an impedance of the character shown in Figure 6.11. At low frequencies, the inductive branch draws a large lagging current while the leading current of the capacitive branch is small, resulting in a large lagging line current and a low lagging circuit impedance. At high frequencies, the inductance has a high reactance compared with the capacitance, resulting in a large leading line current and a corresponding low circuit impedance that is leading in phase. Between these two extremes is a frequency at which the lagging current taken by the inductive branch and the leading current entering the capacitive branch are equal. Being 180° out of phase, they neutralize, leaving only a small resultant in-phase current flowing in the line; the impedance of the parallel circuit is, therefore, high.

The effect of circuit resistance on the impedance of the parallel circuit is similar to the influence that resistance has on the current flowing in a series resonant circuit, as is evident when Figures 6.10 and 6.11 are compared. Increasing the resistance of a parallel circuit lowers and flattens the peak of the impedance curve without appreciably altering the sides, which are relatively independent of the circuit resistance.

The resonant frequency F_0 of a parallel circuit can be taken as the same frequency at which the same circuit is in series resonance:

$$F_0 = \frac{1}{2\pi \sqrt{LC}} \tag{6.13}$$

Where:
L = inductance in the circuit
C = capacitance in the circuit

When the circuit Q is large, the frequencies corresponding to the maximum impedance of the circuit and to unity power factor of this impedance coincide, for all practical purposes, with the resonant frequency defined in this way. When the circuit Q is low, however, this rule does not necessarily apply.

6.3   Passive/Active Circuit Components

A voltage applied to a passive component results in the flow of current and the dissipation or storage of energy. Typical passive components are resistors, coils or inductors, and capacitors. For an example, the flow of current in a resistor results in radiation of heat; from a light bulb, the radiation of light as well as heat.

On the other hand, an active component either (1) increases the level of electric energy or (2) provides available electric energy as a voltage. As an example of (1), an amplifier produces an increase in energy as a higher voltage or power level, while for (2), batteries and generators serve as energy sources.

Active components can generate more alternating signal power into an output load resistance than the power absorbed at the input at the same frequency. Active components are the major building blocks in system assemblies such as amplifiers and oscillators.

6.4   References

1. Fink, Donald G., and Don Christiansen (eds.), *Electronic Engineers' Handbook*, McGraw-Hill, New York, NY, 1982.
2. Chowdhury, Badrul, "Power Distribution and Control," in *The Electronics Handbook*, Jerry C. Whitaker (ed.), pp. 1003, CRC Press, Boca Raton, FL, 1996.

6.5 Bibliography

Benson, K. Blair, and Jerry C. Whitaker, *Television and Audio Handbook for Technicians and Engineers*, McGraw-Hill, New York, NY, 1990.

Benson, K. Blair, *Audio Engineering Handbook*, McGraw-Hill, New York, NY, 1988.

Whitaker, Jerry C., and K. Blair Benson (eds.), *Standard Handbook of Video and Television Engineering*, McGraw-Hill, New York, NY, 2000.

Whitaker, Jerry C., *Television Engineers' Field Manual*, McGraw-Hill, New York, NY, 2000.

Resistors and Resistive Materials

7.1 Introduction

Resistors are components that have a nearly 0° phase shift between voltage and current over a wide range of frequencies with the average value of resistance independent of the instantaneous value of voltage or current. Preferred values of ratings are given ANSI standards or corresponding ISO or MIL standards. Resistors are typically identified by their construction and by the resistance materials used. Fixed resistors have two or more terminals and are not adjustable. Variable resistors permit adjustment of resistance or voltage division by a control handle or with a tool.

7.2 Resistor Types

There are a wide variety of resistor types, each suited to a particular application or group of applications. Low-wattage fixed resistors are usually identified by color-coding on the body of the device, as illustrated in Figure 7.1. The major types of resistors are identified in the following sections.

7.2.1 Wire-Wound Resistor

The resistance element of most wire-wound resistors is resistance wire or ribbon wound as a single-layer helix over a ceramic or fiberglass core, which causes these resistors to have a residual series inductance that affects phase shift at high frequencies, particularly in large-size devices. Wire-wound resistors have low noise and are stable with temperature, with temperature coefficients normally between ±5 and 200 ppm/°C. Resistance values between 0.1 and 100,000 W with accuracies between 0.001 and 20 percent are available with power dissipation ratings between 1 and 250 W at 70°C. The resistance element is usually covered with a vitreous enamel, which can be molded in plastic. Special construction includes such items as enclosure in an aluminum casing for heatsink mounting or a special winding to reduce inductance.

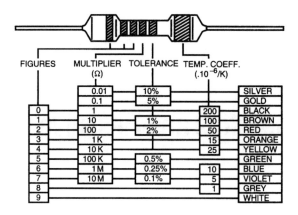

Figure 7.1 Color code for fixed resistors in accordance with IEC publication 62. (*From* [1]. *Used with permission.*)

Resistor connections are made by self-leads or to terminals for other wires or printed circuit boards.

7.2.2 Metal Film Resistor

Metal film, or *cermet*, resistors have characteristics similar to wire-wound resistors except at much lower inductance. They are available as axial lead components in 1/8, 1/4, or 1/2 W ratings, in chip resistor form for high-density assemblies, or as resistor networks containing multiple resistors in one package suitable for printed circuit insertion, as well as in tubular form similar to high-power wire-wound resistors. Metal film resistors are essentially printed circuits using a thin layer of resistance alloy on a flat or tubular ceramic or other suitable insulating substrate. The shape and thickness of the conductor pattern determine the resistance value for each metal alloy used. Resistance is trimmed by cutting into part of the conductor pattern with an abrasive or a laser. Tin oxide is also used as a resistance material.

7.2.3 Carbon Film Resistor

Carbon film resistors are similar in construction and characteristics to axial lead metal film resistors. Because the carbon film is a granular material, random noise may be developed because of variations in the voltage drop between granules. This noise can be of sufficient level to affect the performance of circuits providing high grain when operating at low signal levels.

7.2.4 Carbon Composition Resistor

Carbon composition resistors contain a cylinder of carbon-based resistive material molded into a cylinder of high-temperature plastic, which also anchors the external leads. These resistors can have noise problems similar to carbon film resistors, but their use in electronic equipment for the last 50 years has demonstrated their outstanding reliability, unmatched by other components. These resistors are commonly available at values from 2.7 Ω with tolerances of 5, 10, and 20 percent in 1/8-, 1/4-, 1/2-, 1-, and 2-W sizes.

7.2.5 Control and Limiting Resistors

Resistors with a large negative temperature coefficient, *thermistors*, are often used to measure temperature, limit inrush current into motors or power supplies, or to compensate bias circuits. Resistors with a large positive temperature coefficient are used in circuits that have to match the coefficient of copper wire. Special resistors also include those that have a low resistance when cold and become a nearly open circuit when a critical temperature or current is exceeded to protect transformers or other devices.

7.2.6 Resistor Networks

A number of metal film or similar resistors are often packaged in a single module suitable for printed circuit mounting. These devices see applications in digital circuits, as well as in fixed attenuators or padding networks.

7.2.7 Adjustable Resistors

Cylindrical wire-wound power resistors can be made adjustable with a metal clamp in contact with one or more turns not covered with enamel along an axial stripe. Potentiometers are resistors with a movable arm that makes contact with a resistance element, which is connected to at least two other terminals at its ends. The resistance element can be circular or linear in shape, and often two or more sections are mechanically coupled or ganged for simultaneous control of two separate circuits. Resistance materials include all those described previously.

Trimmer potentiometers are similar in nature to conventional potentiometers except that adjustment requires a tool.

Most potentiometers have a *linear taper*, which means that resistance changes linearly with control motion when measured between the movable arm and the "low," or counterclockwise, terminal. Gain controls, however, often have a *logarithmic taper* so that attenuation changes linearly in decibels (a logarithmic ratio). The resistance element of a potentiometer may also contain taps that permit the connection of other components as required in a specialized circuit.

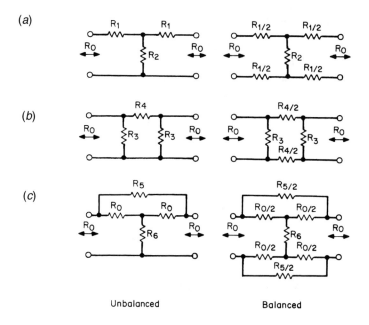

Figure 7.2 Unbalanced and balanced fixed attenuator networks for equal source and load resistance: (*a*) *T* configuration, (*b*) π configuration, (*c*) bridged-*T* configuration.

7.2.8 Attenuators

Variable attenuators are adjustable resistor networks that show a calibrated increase in attenuation for each switched step. For measurement of audio, video, and RF equipment, these steps may be decades of 0.1, 1, and 10 dB. Circuits for unbalanced and balanced fixed attenuators are shown in Figure 7.2. Fixed attenuator networks can be cascaded and switched to provide step adjustment of attenuation inserted in a constant-impedance network.

Audio attenuators generally are designed for a circuit impedance of 150 Ω, although other impedances can be used for specific applications. Video attenuators are generally designed to operate with unbalanced 75-Ω grounded-shield coaxial cable. RF attenuators are designed for use with 75- or 50-Ω coaxial cable.

7.3   References

1. Whitaker, Jerry C. (ed.), *The Electronics Handbook*, CRC Press, Boca Raton, FL, 1996.

7.4 Bibliography

Benson, K. Blair, and Jerry C. Whitaker, *Television and Audio Handbook for Technicians and Engineers*, McGraw-Hill, New York, NY, 1990.

Benson, K. Blair, *Audio Engineering Handbook*, McGraw-Hill, New York, NY, 1988.

Whitaker, Jerry C., and K. Blair Benson (eds.), *Standard Handbook of Video and Television Engineering*, McGraw-Hill, New York, NY, 2000.

Whitaker, Jerry C., *Television Engineers' Field Manual*, McGraw-Hill, New York, NY, 2000.

7.5 Tabular Data

Table 7.1 Resistivity of Selected Ceramics (*From* [1]. *Used with permission.*)

Ceramic	Resistivity, $\Omega \cdot cm$
Borides	
Chromium diboride (CrB_2)	21×10^{-6}
Hafnium diboride (HfB_2)	$10–12 \times 10^{-6}$ at room temp.
Tantalum diboride (TaB_2)	68×10^{-6}
Titanium diboride (TiB_2) (polycrystalline)	
85% dense	$26.5–28.4 \times 10^{-6}$ at room temp.
85% dense	9.0×10^{-6} at room temp.
100% dense, extrapolated values	$8.7–14.1 \times 10^{-6}$ at room temp.
	3.7×10^{-6} at liquid air temp.
Titanium diboride (TiB_2) (monocrystalline)	
Crystal length 5 cm, 39 deg. and 59 deg. orientation with respect to growth axis	$6.6 \pm 0.2 \times 10^{-6}$ at room temp.
Crystal length 1.5 cm, 16.5 deg. and 90 deg. orientation with respect to growth axis	$6.7 \pm 0.2 \times 10^{-6}$ at room temp.
Zirconium diboride (ZrB_2)	9.2×10^{-6} at 20°C
	1.8×10^{-6} at liquid air temp.
Carbides: boron carbide (B_4C)	$0.3–0.8$

Table 7.2 Electrical Resistivity of Various Substances in 10^{-8} $\Omega \cdot$ m (*From* [1]. *Used with permission.*)

T/K	Aluminum	Barium	Beryllium	Calcium	Cesium	Chromium	Copper
1	0.000100	0.081	0.0332	0.045	0.0026		0.00200
10	0.000193	0.189	0.0332	0.047	0.243		0.00202
20	0.000755	0.94	0.0336	0.060	0.86		0.00280
40	0.0181	2.91	0.0367	0.175	1.99		0.0239
60	0.0959	4.86	0.067	0.40	3.07		0.0971
80	0.245	6.83	0.075	0.65	4.16		0.215
100	0.442	8.85	0.133	0.91	5.28	1.6	0.348
150	1.006	14.3	0.510	1.56	8.43	4.5	0.699
200	1.587	20.2	1.29	2.19	12.2	7.7	1.046
273	2.417	30.2	3.02	3.11	18.7	11.8	1.543
293	2.650	33.2	3.56	3.36	20.5	12.5	1.678
298	2.709	34.0	3.70	3.42	20.8	12.6	1.712
300	2.733	34.3	3.76	3.45	21.0	12.7	1.725
400	3.87	51.4	6.76	4.7		15.8	2.402
500	4.99	72.4	9.9	6.0		20.1	3.090
600	6.13	98.2	13.2	7.3		24.7	3.792
700	7.35	130	16.5	8.7		29.5	4.514
800	8.70	168	20.0	10.0		34.6	5.262
900	10.18	216	23.7	11.4		39.9	6.041

Table 7.2 Electrical Resistivity of Various Substances in 10^{-8} $\Omega \bullet$ m (Continued)

T/K	Gold	Hafnium	Iron	Lead	Lithium	Magnesium	Manganese
1	0.0220	1.00	0.0225		0.007	0.0062	7.02
10	0.0226	1.00	0.0238		0.008	0.0069	18.9
20	0.035	1.11	0.0287		0.012	0.0123	54
40	0.141	2.52	0.0758		0.074	0.074	116
60	0.308	4.53	0.271		0.345	0.261	131
80	0.481	6.75	0.693	4.9	1.00	0.557	132
100	0.650	9.12	1.28	6.4	1.73	0.91	132
150	1.061	15.0	3.15	9.9	3.72	1.84	136
200	1.462	21.0	5.20	13.6	5.71	2.75	139
273	2.051	30.4	8.57	19.2	8.53	4.05	143
293	2.214	33.1	9.61	20.8	9.28	4.39	144
298	2.255	33.7	9.87	21.1	9.47	4.48	144
300	2.271	34.0	9.98	21.3	9.55	4.51	144
400	3.107	48.1	16.1	29.6	13.4	6.19	147
500	3.97	63.1	23.7	38.3		7.86	149
600	4.87	78.5	32.9			9.52	151
700	5.82		44.0			11.2	152
800	6.81		57.1			12.8	
900	7.86					14.4	

Table 7.2 Electrical Resistivity of Various Substances in 10^{-8} $\Omega \bullet$ m (Continued)

T/K	Molybdenum	Nickel	Palladium	Platinum	Potassium	Rubidium	Silver
1	0.00070	0.0032	0.0200	0.002	0.0008	0.0131	0.00100
10	0.00089	0.0057	0.0242	0.0154	0.0160	0.109	0.00115
20	0.00261	0.0140	0.0563	0.0484	0.117	0.444	0.0042
40	0.0457	0.068	0.334	0.409	0.480	1.21	0.0539
60	0.206	0.242	0.938	1.107	0.90	1.94	0.162
80	0.482	0.545	1.75	1.922	1.34	2.65	0.289
100	0.858	0.96	2.62	2.755	1.79	3.36	0.418
150	1.99	2.21	4.80	4.76	2.99	5.27	0.726
200	3.13	3.67	6.88	6.77	4.26	7.49	1.029
273	4.85	6.16	9.78	9.6	6.49	11.5	1.467
293	5.34	6.93	10.54	10.5	7.20	12.8	1.587
298	5.47	7.12	10.73	10.7	7.39	13.1	1.617
300	5.52	7.20	10.80	10.8	7.47	13.3	1.629
400	8.02	11.8	14.48	14.6			2.241
500	10.6	17.7	17.94	18.3			2.87
600	13.1	25.5	21.2	21.9			3.53
700	15.8	32.1	24.2	25.4			4.21
800	18.4	35.5	27.1	28.7			4.91
900	21.2	38.6	29.4	32.0			5.64

Table 7.2 Electrical Resistivity of Various Substances in 10^{-8} $\Omega \bullet$ m (Continued)

T/K	Sodium	Strontium	Tantalum	Tungsten	Vanadium	Zinc	Zirconium
1	0.0009	0.80	0.10	0.000016		0.0100	0.250
10	0.0015	0.80	0.102	0.000137	0.0145	0.0112	0.253
20	0.016	0.92	0.146	0.00196	0.039	0.0387	0.357
40	0.172	1.70	0.751	0.0544	0.304	0.306	1.44
60	0.447	2.68	1.65	0.266	1.11	0.715	3.75
80	0.80	3.64	2.62	0.606	2.41	1.15	6.64
100	1.16	4.58	3.64	1.02	4.01	1.60	9.79
150	2.03	6.84	6.19	2.09	8.2	2.71	17.8
200	2.89	9.04	8.66	3.18	12.4	3.83	26.3
273	4.33	12.3	12.2	4.82	18.1	5.46	38.8
293	4.77	13.2	13.1	5.28	19.7	5.90	42.1
298	4.88	13.4	13.4	5.39	20.1	6.01	42.9
300	4.93	13.5	13.5	5.44	20.2	6.06	43.3
400		17.8	18.2	7.83	28.0	8.37	60.3
500		22.2	22.9	10.3	34.8	10.82	76.5
600		26.7	27.4	13.0	41.1	13.49	91.5
700		31.2	31.8	15.7	47.2		104.2
800		35.6	35.9	18.6	53.1		114.9
900			40.1	21.5	58.7		123.1

Table 7.3 Electrical Resistivity of Various Metallic Elements at (approximately) Room Temperature (*From* [1]. *Used with permission.*)

Element	T/K	Electrical Resistivity $10^{-8}\ \Omega\cdot m$	Element	T/K	Electrical Resistivity $10^{-8}\ \Omega\cdot m$
Antimony	273	39	Polonium	273	40
Bismuth	273	107	Praseodymium	290–300	70.0
Cadmium	273	6.8	Promethium	290–300	75
Cerium	290–300	82.8	Protactinium	273	17.7
Cobalt	273	5.6	Rhenium	273	17.2
Dysprosium	290–300	92.6	Rhodium	273	4.3
Erbium	290–300	86.0	Ruthenium	273	7.1
Europium	290–300	90.0	Samarium	290–300	94.0
Gadolinium	290–300	131	Scandium	290–300	56.2
Gallium	273	13.6	Terbium	290–300	115
Holmium	290–300	81.4	Thallium	273	15
Indium	273	8.0	Thorium	273	14.7
Iridium	273	4.7	Thulium	290–300	67.6
Lanthanum	290–300	61.5	Tin	273	11.5
Lutetium	290–300	58.2	Titanium	273	39
Mercury	273	94.1	Uranium	273	28
Neodymium	290–300	64.3	Ytterbium	290–300	25.0
Niobium	273	15.2	Yttrium	290–300	59.6
Osmium	273	8.1			

Table 7.4 Electrical Resistivity of Selected Alloys in Units of $10^{-8}\ \Omega \cdot$ m (From [1]. Used with permission.)

	273 K	293 K	300 K	350 K	400 K
Alloy—Aluminum-Copper					
Wt % Al					
99[a]	2.51	2.74	2.82	3.38	3.95
95[a]	2.88	3.10	3.18	3.75	4.33
90[b]	3.36	3.59	3.67	4.25	4.86
85[b]	3.87	4.10	4.19	4.79	5.42
80[b]	4.33	4.58	4.67	5.31	5.99
70[b]	5.03	5.31	5.41	6.16	6.94
60[b]	5.56	5.88	5.99	6.77	7.63
50[b]	6.22	6.55	6.67	7.55	8.52
40[c]	7.57	7.96	8.10	9.12	10.2
30[c]	11.2	11.8	12.0	13.5	15.2
25[f]	16.3[aa]	17.2	17.6	19.8	22.2
15[h]	—	12.3	—	—	—
19[g]	1.8[aa]	11.0	11.1	11.7	12.3
5[e]	9.43	9.61	9.68	10.2	10.7
1[b]	4.46	4.60	4.65	5.00	5.37

	273 K	293 K	300 K	350 K	400 K
Alloy—Copper-Nickel					
Wt % Cu					
99[c]	2.71	2.85	2.91	3.27	3.62
95[c]	7.60	7.71	7.82	8.22	8.62
90[c]	13.69	13.89	13.96	14.40	14.81
85[c]	19.63	19.83	19.90	2032	20.70
80[c]	25.46	25.66	25.72	26.12[aa]	26.44[aa]
70[i]	36.67	36.72	36.76	36.85	36.89
60[i]	45.43	45.38	45.35	45.20	45.01
50[i]	50.19	50.05	50.01	49.73	49.50
40[c]	47.42	47.73	47.82	48.28	48.49
30[i]	40.19	41.79	42.34	44.51	45.40
25[c]	33.46	35.11	35.69	39.67[aa]	42.81[aa]
15[c]	22.00	23.35	23.85	27.60	31.38
10[c]	16.65	17.82	18.26	21.51	25.19
5[c]	11.49	12.50	12.90	15.69	18.78
1[c]	7.23	8.08	8.37	10.63[aa]	13.18[aa]

Table 7.4 Electrical Resistivity of Selected Alloys in Units of $10^{-8}\ \Omega \cdot$ m (Continued)

Alloy—Aluminum-Magnesium					Wt % Cu	Alloy—Copper-Palladium				
2.96	3.18	3.26	3.82	4.39	99c	2.10	2.23	2.27	2.59	2.92
5.05	5.28	5.36	5.93	6.51	95c	4.21	4.35	4.40	4.74	5.08
7.52	7.76	7.85	8.43	9.02	90c	6.89	7.03	7.08	7.41	7.74
—	—	—	—	—	85	9.48	9.61	9.66	10.01	10.36
—	—	—	—	—	80	11.99	12.12	12.16	12.51[aa]	12.87
—	—	—	—	—	70	16.87	17.01	17.06	17.41	17.78
—	—	—	—	—	60	21.73	21.87	21.92	22.30	22.69
—	—	—	—	—	50	27.62	27.79	27.86	28.25	28.64
—	—	—	—	—	40	35.31	35.51	35.57	36.03	36.47
—	—	—	—	—	30	46.50	46.66	46.71	47.11	47.47
—	—	—	—	—	25	46.25	46.45	46.52	46.99[aa]	47.43[aa]
—	—	—	—	—	15	36.52	36.99	37.16	38.28	39.35
17.1	17.4	17.6	18.4	19.2	10b	28.90	29.51	29.73	31.19[aa]	32.56[aa]
13.1	13.4	13.5	14.3	15.2	5b	20.00	20.75	21.02	22.84[aa]	24.54[aa]
5.92	6.25	6.37	7.20	8.03	1a	11.90	12.67	12.93[aa]	14.82[aa]	16.68[aa]

Table 7.4 Electrical Resistivity of Selected Alloys in Units of 10^{-8} $\Omega \bullet$ m (Continued)

Alloy—Copper-Gold

Wt % Cu					
99[c]	1.73	1.86[aa]	1.91[aa]	2.24[aa]	2.58[aa]
95[c]	2.41	2.54[aa]	2.59[aa]	2.92[aa]	3.26[aa]
90[c]	3.29	4.42[aa]	3.46[aa]	3.79[aa]	4.12[aa]
85[c]	4.20	4.33	4.38[aa]	4.71[aa]	5.05[aa]
80[c]	5.15	5.28	5.32	5.65	5.99
70[c]	7.12	7.25	7.30	7.64	7.99
60[c]	9.18	9.13	9.36	9.70	10.05
50[c]	11.07	11.20	11.25	11.60	11.94
40[c]	12.70	12.85	12.90[aa]	13.27[aa]	13.65[aa]
30[c]	13.77	13.93	13.99[aa]	14.38[aa]	14.78[aa]
25[c]	13.93	14.09	14.14	14.54	14.94
15[c]	12.75	12.91	12.96[aa]	13.36[aa]	13.77
10[c]	10.70	10.86	10.91	11.31	11.72
5[c]	7.25	7.41[aa]	7.46	7.87	8.28
1[c]	3.40	3.57	3.62	4.03	4.45

Alloy—Copper-Zinc

Wt % Cu					
99[b]	1.84	1.97	2.02	2.36	2.71
95[b]	2.78	2.92	2.97	3.33	3.69
90[b]	3.66	3.81	3.86	4.25	4.63
85[b]	4.37	4.54	4.60	5.02	5.44
80[b]	5.01	5.19	5.26	5.71	6.17
70[b]	5.87	6.08	6.15	6.67	7.19
60	—	—	—	—	—
50	—	—	—	—	—
40	—	—	—	—	—
30	—	—	—	—	—
25	—	—	—	—	—
15	—	—	—	—	—
10	—	—	—	—	—
5	—	—	—	—	—
1	—	—	—	—	—

Table 7.4 Electrical Resistivity of Selected Alloys in Units of 10^{-8} $\Omega \cdot$ m (Continued)

Alloy—Gold-Palladium

Wt % Au					
99[c]	2.69	2.86	2.91	3.32	3.73
95[c]	5.21	5.35	5.41	5.79	6.17
90[i]	8.01	8.17	8.22	8.56	8.93
85[b]	10.50[aa]	10.66	10.72[aa]	11.100[aa]	11.48[aa]
80[b]	12.75	12.93	12.99	13.45	13.93
70[c]	18.23	18.46	18.54	19.10	19.67
60[b]	26.70	26.94	27.02	27.63[aa]	28.23[aa]
50[b]	27.23	27.63	27.76	28.64[aa]	29.42[aa]
40[a]	24.65	25.23	25.42	26.74	27.95
30[b]	20.82	21.49	21.72	23.35	24.92
25[b]	18.86	19.53	19.77	21.51	23.19
15[a]	15.08	15.77	16.01	17.80	19.61
10[a]	13.25	13.95	14.20[aa]	16.00[aa]	17.81[aa]
5[a]	11.49[aa]	12.21	12.46[aa]	14.26[aa]	16.07[aa]
1[a]	10.07	10.85[aa]	11.12[aa]	12.99[aa]	14.80[aa]

Alloy—Gold-Silver

Wt % Au					
99[b]	2.58	2.75	2.80[aa]	3.22[aa]	3.63[aa]
95[a]	4.58	4.74	4.79	5.19	5.59
90[i]	6.57	6.73	6.78	7.19	7.58
85[j]	8.14	8.30	8.36[aa]	8.75	9.15
80[j]	9.34	9.50	9.55	9.94	10.33
70[j]	10.70	10.86	10.91	11.29	11.68[aa]
60[j]	10.92	11.07	11.12	11.50	11.87
50[j]	10.23	10.37	10.42	10.78	11.14
40[j]	8.92	9.06	9.11	9.46[aa]	9.81
30[a]	7.34	7.47	7.52	7.85	8.19
25[a]	6.46	6.59	6.63	6.96	7.30[aa]
15[a]	4.55	4.67	4.72	5.03	5.34
10[a]	3.54	3.66	3.71	4.00	4.31
5[i]	2.52	2.64[aa]	2.68[aa]	2.96[aa]	3.25[aa]
1[b]	1.69	1.80	1.84[aa]	2.12[aa]	2.42[aa]

Table 7.4 Electrical Resistivity of Selected Alloys in Units of $10^{-8}\ \Omega \cdot m$ (Continued)

Wt % Fe	Alloy—Iron-Nickel				
99[a]	10.9	12.0	12.4	—	18.7
95[c]	18.7	19.9	20.2	—	26.8
90[c]	24.2	25.5	25.9	—	33.2
85[c]	27.8	29.2	29.7	—	37.3
80[c]	30.1	31.6	32.2	—	40.0
70[b]	32.3	33.9	34.4	—	42.4
60[c]	53.8	57.1	58.2	—	73.9
50[d]	28.4	30.6	31.4	—	43.7
40[d]	19.6	21.6	22.5	—	34.0
30[c]	15.3	17.1	17.7	—	27.4
25[b]	14.3	15.9	16.4	—	25.1
15[c]	12.6	13.8	14.2	—	21.1
10[c]	11.4	12.5	12.9	—	18.9
5[c]	9.66	10.6	10.9	—	16.1[aa]
1[b]	7.17	7.94	8.12	—	12.8

[a] Uncertainty in resistivity is ±2%.
[b] Uncertainty in resistivity is ±3%.
[c] Uncertainty in resistivity is ±5%.
[d] Uncertainty in resistivity is ±7% below 300 K and ±5% at 300 and 400 K.
[e] Uncertainty in resistivity is ±7%.
[f] Uncertainty in resistivity is ±8%.
[g] Uncertainty in resistivity is ±10%.
[h] Uncertainty in resistivity is ±12%.
[i] Uncertainty in resistivity is ±4%.
[j] Uncertainty in resistivity is ±1%.
[k] Uncertainty in resistivity is ±3% up to 300 K and ± 4% above 300 K.
[m] Uncertainty in resistivity is ±2% up to 300 K and ± 4% above 300 K.
[a] Crystal usually a mixture of α-hcp and fcc lattice.
[aa] In temperature range where no experimental data are available.

Table 7.5 Resistivity of Semiconducting Minerals (*From* [1]. *Used with permission.*)

Mineral	$\rho, \Omega \cdot m$	Mineral	$\rho, \Omega \cdot m$
Diamond (C)	2.7	Gersdorffite, NiAsS	1 to 160 $\times 10^{-6}$
Sulfides		Glaucodote, (Co, Fe)AsS	5 to 100 $\times 10^{-6}$
Argentite, Ag_2S	1.5 to 2.0 $\times 10^{-3}$	Antimonide	
Bismuthinite, Bi_2S_3	3 to 570	Dyscrasite, Ag_3Sb	0.12 to 1.2 $\times 10^{-6}$
Bornite, $Fe_2S_3 \cdot nCu_2S$	1.6 to 6000 $\times 10^{-6}$	Arsenides	
Chalcocite, Cu_2S	80 to 100 $\times 10^{-6}$	Allemonite, $SbAs_2$	70 to 60,000
Chalcopyrite, $Fe_2S_3 \cdot Cu_2S$	150 to 9000 $\times 10^{-6}$	Lollingite, $FeAs_2$	2 to 270 $\times 10^{-6}$
Covellite, CuS	0.30 to 83 $\times 10^{-6}$	Nicollite, NiAs	0.1 to 2 $\times 10^{-6}$
Galena, PbS	6.8 $\times 10^{-6}$ to 9.0 $\times 10^{-2}$	Skutterudite, $CoAs_3$	1 to 400 $\times 10^{-6}$
Haverite, MnS_2	10 to 20	Smaltite, $CoAs_2$	1 to 12 $\times 10^{-6}$
Marcasite, FeS_2	1 to 150 $\times 10^{-3}$	Tellurides	
Metacinnabarite, $4HgS$	2 $\times 10^{-6}$ to 1 $\times 10^{-3}$	Altaite, PbTe	20 to 200 $\times 10^{-6}$
Millerite, NiS	2 to 4 $\times 10^{-7}$	Calavarite, $AuTe_2$	6 to 12 $\times 10^{-6}$
Molybdenite, MoS_2	0.12 to 7.5	Coloradoite, HgTe	4 to 100 $\times 10^{-6}$
Pentlandite, $(Fe, Ni)_9S_8$	1 to 11 $\times 10^{-6}$	Hessite, Ag_2Te	4 to 100 $\times 10^{-6}$
Pyrrhotite, Fe_7S_8	2 to 160 $\times 10^{-6}$	Nagyagite, $Pb_6Au(S, Te)_{14}$	20 to 80 $\times 10^{-6}$
Pyrite, FeS_2	1.2 to 600 $\times 10^{-3}$	Sylvanite, $AgAuTe_4$	4 to 20 $\times 10^{-6}$
Sphalerite, ZnS	2.7 $\times 10^{-3}$ to 1.2 $\times 10^4$	Oxides	
Antimony-sulfur compounds		Braunite, Mn_2O_3	0.16 to 1.0
Berthierite, $FeSb_2S_4$	0.0083 to 2.0	Cassiterite, SnO_2	4.5 $\times 10^{-4}$ to 10,000
Boulangerite, $Pb_5Sb_4S_{11}$	2 $\times 10^3$ to 4 $\times 10^4$	Cuprite, Cu_2O	10 to 50
Cylindrite, $Pb_3Sn_4Sb_2S_{14}$	2.5 to 60	Hollandite, $(Ba, Na, K)Mn_8O_{16}$	2 to 100 $\times 10^{-3}$
Franckeite, $Pb_5Sn_3Sb_2S_{14}$	1.2 to 4	Ilmenite, $FeTiO_3$	0.001 to 4
Hauchecornite, $Ni_9(Bi, Sb)_2S_8$	1 to 83 $\times 10^{-6}$	Magnetite, Fe_3O_4	52 $\times 10^{-6}$
Jamesonite, $Pb_4FeSb_6S_{14}$	0.020 to 0.15	Manganite, $MnO \cdot OH$	0.018 to 0.5
Tetrahedrite, Cu_3SbS_3	0.30 to 30,000	Melaconite, CuO	6000
Arsenic-sulfur compounds		Psilomelane, $KMnO \cdot MnO_2 \cdot nH_2$	0.04 to 6000
Arsenopyrite, FeAsS	20 to 300 $\times 10^{-6}$	Pyrolusite, MnO_2	0.007 to 30
Cobaltite, CoAsS	6.5 to 130 $\times 10^{-3}$	Rutile, TiO_2	29 to 910
Enargite, Cu_3AsS_4	0.2 to 40 $\times 10^{-3}$	Uraninite, UO	1.5 to 200

Source: Carmichael, R.S., ed., 1982. *Handbook of Physical Properties of Rocks*, Vol. I., CRC Press, Boca Raton, FL.

Capacitance and Capacitors

8.1   Introduction

A system of two conducting bodies (which are frequently identified as *plates*) located in an electromagnetic field and having equal charges of opposite signs $+Q$ and $-Q$ can be called a *capacitor* [1]. The capacitance C of this system is equal to the ratio of the charge Q (absolute value) to the voltage V (again, absolute value) between the bodies; that is,

$$C = \frac{Q}{V} \qquad\qquad (8.1)$$

Capacitance C depends on the size and shape of the bodies and their mutual location. It is proportional to the dielectric *permittivity* ε of the media where the bodies are located. The capacitance is measured in *farads* (F) if the charge is measured in coulombs (C) and the voltage in volts (V). One farad is a very big unit; practical capacitors have capacitances that are measured in micro- (μF, or 10^{-6}F), nano- (nF, or 10^{-9}F), and picofarads (pF, or 10^{-12}F).

The calculation of capacitance requires knowledge of the electrostatic field between the bodies. The following two theorems [2] are important in these calculations.

The integral of the flux density D over a closed surface is equal to the charge Q enclosed by the surface (the Gauss theorem), that is,

$$\oint D\, ds = Q \qquad\qquad (8.2)$$

This result is valid for linear and nonlinear dielectrics. For a linear and isotropic media $D = \varepsilon E$, where E is the electric field. The magnitude E of the field is measured in volt per meter, the magnitude D of the flux in coulomb per square meter, and the dielectric permittivity has the dimension of farad per meter. The dielectric permittivity is usually represented as $\varepsilon = \varepsilon_0 K_d$ where ε_0 is the permittivity of air ($\varepsilon_0 = 8.86 \times 10^{-12}$ F/m) and K_d is the dielectric constant.

The electric field is defined by an electric potential ϕ. The directional derivative of the potential taken with the minus sign is equal to the component of the electric field in this direction. The voltage V_{AB} between the points A and B, having the potentials ϕ_A and ϕ_B, respectively (the potential is also measured in volts), is equal to

$$V_{AB} = \int_A^B E dl = \phi_A - \phi_B \tag{8.3}$$

This result is the second basic relationship. The left hand side of equation 8.3 is a *line integral*. At each point of the line AB there exist two vectors: E defined by the field and dl that defines the direction of the line at this point.

8.2   Practical Capacitors

A wide variety of capacitors are in common usage. Capacitors are passive components in which current leads voltage by nearly 90° over a wide range of frequencies. Capacitors are rated by capacitance, voltage, materials, and construction.

A capacitor may have two voltage ratings:

- *Working voltage*—the normal operating voltage that should not be exceeded during operation

- Test or *forming voltage*—which stresses the capacitor and should occur only rarely in equipment operation

Good engineering practice dictates that components be used at only a fraction of their maximum ratings. The primary characteristics of common capacitors are given in Table 8.1. Some common construction practices are illustrated in Figure 8.1.

8.2.1   Polarized/Nonpolarized Capacitors

Polarized capacitors can be used in only those applications where a positive sum of all dc and peak-ac voltages is applied to the positive capacitor terminal with respect to its negative terminal. These capacitors include all tantalum and most aluminum electrolytic capacitors. These devices are commonly used in power supplies or other electronic equipment where these restrictions can be met.

Nonpolarized capacitors are used in circuits where there is no direct voltage bias across the capacitor. They are also the capacitor of choice for most applications requiring capacity tolerances of 10 percent or less.

Table 8.1 Parameters and Characteristics of Discrete Capacitors (*From* [1]. *Used with permission.*)

Capacitor Type	Range	Rated Voltage, V_R	TC ppm/°C	Tolerance, ±%	Insulation Resistance, MΩμF	Dissipation Factor, %	Dielectric Absorption, %	Temperature Range, °C	Comments, Applications	Cost
Polycarbonate	100 pF–30 μF	50–800	±50	10	5×10^5	0.2	0.1	−55/+125	High quality, small, low TC	High
Polyester/Mylar	1000 pF–50 μF	50–600	+400	10	10^5	0.75	0.3	−55/+125	Good, popular	Medium
Polypropylene	100 pF–50 μF	100–800	−200	10	10^5	0.2	0.1	−55/+105	High quality low absorption	High
Polystyrene	10 pF–2.7 μF	100–600	−100	10	10^6	0.05	0.04	−55/+85	High quality, large, low TC, signal filters	Medium
Polysulfone	1000 pF–1 μF		+80	5	10^5	0.3	0.2	−55/+150	High temperature	High
Parylene	5000 pF–1 μF		±100	10	10^5	0.1	0.1	−55/+125	High temperature	High
Kapton	1000 pF–1 μF		+100	10	10^5	0.3	0.3	−55/+220	High temperature	High
Teflon	1000 pF–2 μF	50–200	−200	10	5×10^6	0.04	0.04	−70/+250	High temperature lowest absorption	High
Mica	5 pF–0.01 μF	100–600	−50	5	2.5×10^4	0.001	0.75	−55/+125	Good at RF, low TC	High
Glass	5 pF–1000 pF	100–600	+140	5	10^6	0.001		−55/+125	Excellent long-term stability	High
Porcelain	100 pF–0.1 μF	50–400	+120	5	5×10^5	0.10	4.2	−55/+125	Good long-term stability	High
Ceramic (NPO)	100 pF–1 μF	50–400	±30	10	5×10^3	0.02	0.75	−55/+125	Active filters, low TC	Medium
Ceramic	10 pF–1 μF	50–30,000						−55/+125	Small, very popular selectable TC	Low
Paper	0.01 μF–10 μF	200–1600	±800	10	5×10^3	1.0	2.5	−55/+125	Motor capacitors	Low
Aluminum	0.1 μF–1.6 F	3–600	+2500	−10/+100	100	10	8.0	−40/+85	Power supply filters short life	High
Tantalum (Foil)	0.1 μF–1000 μF	6–100	+800	−10/+100	20	4.0	8.5	−55/+85	High capacitance small size, low inductance	High
Thin-film	10 pF–200 pF	6–30	+100	10	10^6	0.01		−55/+125		High
Oil	0.1 μF–20 μF	200–10,000				0.5			High voltage filters, large, long life	
Vacuum	1 pF–1000 pF	2,000–3,600							Transmitters	

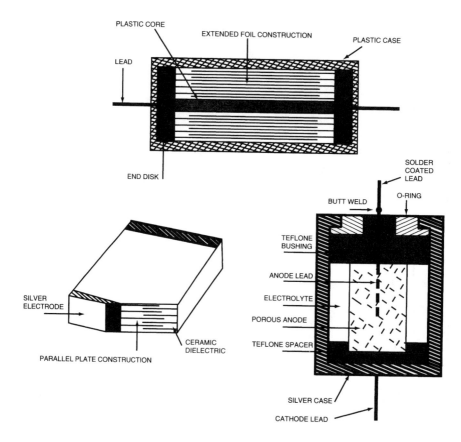

PLASTIC CORE
EXTENDED FOIL CONSTRUCTION
PLASTIC CASE
LEAD
END DISK

SILVER
ELECTRODE
PARALLEL PLATE CONSTRUCTION
CERAMIC
DIELECTRIC

SOLDER
COATED
LEAD
BUTT WELD
O-RING
TEFLONE
BUSHING
ANODE LEAD
ELECTROLYTE
POROUS ANODE
TEFLONE SPACER
SILVER CASE
CATHODE LEAD

Figure 8.1 Construction of discrete capacitors. (*From* [1]. *Used with permission.*)

8.2.2 Operating Losses

Losses in capacitors occur because an actual capacitor has various resistances. These losses are usually measured as the *dissipation factor* at a frequency of 120 Hz. Leakage resistance in parallel with the capacitor defines the time constant of discharge of a capacitor. This time constant can vary between a small fraction of a second to many hours depending on capacitor construction, materials, and other electrical leakage paths, including surface contamination.

The *equivalent series resistance* of a capacitor is largely the resistance of the conductors of the capacitor plates and the resistance of the physical and chemical system of the capacitor. When an alternating current is applied to the capacitor, the losses in the equivalent series resistance are the major causes of heat developed in the device. The same resistance also determines the maximum attenuation of a filter or bypass capacitor and the loss in a coupling capacitor connected to a load.

The *dielectric absorption* of a capacitor is the residual fraction of charge remaining in a capacitor after discharge. The residual voltage appearing at the capacitor terminals after discharge is of little concern in most applications but can seriously affect the performance of *analog-to-digital* (A/D) converters that must perform precision measurements of voltage stored in a sampling capacitor.

The *self-inductance* of a capacitor determines the high-frequency impedance of the device and its ability to bypass high-frequency currents. The self-inductance is determined largely by capacitor construction and tends to be highest in common metal foil devices.

8.2.3 Film Capacitors

Plastic is a preferred dielectrical material for capacitors because it can be manufactured with minimal imperfections in thin films. A metal-foil capacitor is constructed by winding layers of metal, plastic, metal, and plastic into a cylinder and then making a connection to the two layers of metal. A *metallized foil capacitor* uses two layers, each of which has a very thin layer of metal evaporated on one surface, thereby obtaining a higher capacity per volume in exchange for a higher equivalent series resistance. Metallized foil capacitors are self-repairing in the sense that the energy stored in the capacitor is often sufficient to burn away the metal layer surrounding the void in the plastic film.

Depending on the dielectric material and construction, capacitance tolerances between 1 and 20 percent are common, as are voltage ratings from 50 to 400 V. Construction types include axial leaded capacitors with a plastic outer wrap, metal-encased units, and capacitors in a plastic box suitable for printed circuit board insertion.

Polystyrene has the lowest dielectric absorption of 0.02 percent, a temperature coefficient of -20 to -100 ppm/°C, a temperature range to 85°C, and extremely low leakage. Capacitors between 0.001 and 2 μF can be obtained with tolerances from 0.1 to 10 percent.

Polycarbonate has an upper temperature limit of 100°C, with capacitance changes of about 2 percent up to this temperature. Polypropylene has an upper temperature limit of 85°C. These capacitors are particularly well suited for applications where high in-rush currents occur, such as switching power supplies. Polyester is the lowest-cost material with an upper temperature limit of 125°C. Teflon and other high-temperature materials are used in aerospace and other critical applications.

8.2.4 Foil Capacitors

Mica capacitors are made of multiple layers of silvered mica packaged in epoxy or other plastic. Available in tolerances of 1 to 20 percent in values from 10 to 10,000 pF, mica capacitors exhibit temperature coefficients as low as 100 ppm. Voltage ratings between 100 and 600 V are common. Mica capacitors are used mostly in high-frequency filter circuits where low loss and high stability are required.

8.2.5 Electrolytic Capacitors

Aluminum foil electrolytic capacitors can be made nonpolar through use of two cathode foils instead of anode and cathode foils in construction. With care in manufacturing, these capacitors can be produced with tolerance as tight as 10 percent at voltage ratings of 25 to 100 V peak. Typical values range from 1 to 1000 µF.

8.2.6 Ceramic Capacitors

Barium titanate and other ceramics have a high dielectric constant and a high breakdown voltage. The exact formulation determines capacitor size, temperature range, and variation of capacitance over that range (and consequently capacitor application). An alphanumeric code defines these factors, a few of which are given here.

- Ratings of Y5V capacitors range from 1000 pF to 6.8 µF at 25 to 100 V and typically vary +22 to –82 percent in capacitance from –30 to + 85°C.

- Ratings of Z5U capacitors range to 1.5 µF and vary +22 to –56 percent in capacitance from +10 to +85°C. These capacitors are quite small in size and are used typically as bypass capacitors.

- X7R capacitors range from 470 pF to 1 µF and vary 15 percent in capacitance from –55 to + 125°C.

Nonpolarized (NPO) rated capacitors range from 10 to 47,000 pF with a temperature coefficient of 0 to +30 ppm over a temperature range of –55 to +125°C.

Ceramic capacitors come in various shapes, the most common being the radial-lead disk. Multilayer monolithic construction results in small size, which exists both in radial-lead styles and as chip capacitors for direct surface mounting on a printed circuit board.

8.2.7 Polarized-Capacitor Construction

Polarized capacitors have a negative terminal—the cathode—and a positive terminal—the anode—and a liquid or gel between the two layers of conductors. The actual dielectric is a thin oxide film on the cathode, which has been chemically roughened for maximum surface area. The oxide is formed with a forming voltage, higher than the normal operating voltage, applied to the capacitor during manufacture. The direct current flowing through the capacitor forms the oxide and also heats the capacitor.

Whenever an electrolytic capacitor is not used for a long period of time, some of the oxide film is degraded. It is reformed when voltage is applied again with a leakage current that decreases with time. Applying an excessive voltage to the capacitor causes a severe increase in leakage current, which can cause the electrolyte to boil. The resulting steam may escape by way of the rubber seal or may otherwise damage the capacitor. Application of a reverse voltage in excess of about 1.5 V will cause forming to begin on the unetched anode electrode. This can happen when pulse voltages superimposed on a dc voltage cause a momentary voltage reversal.

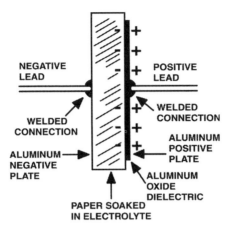

Figure 8.2 The basic construction of an aluminum electrolytic capacitor.

8.2.8 Aluminum Electrolytic Capacitors

Aluminum electrolytic capacitors use very pure aluminum foil as electrodes, which are wound into a cylinder with an interlayer paper or other porous material that contains the electrolyte. (See Figure 8.2.) Aluminum ribbon staked to the foil at the minimum inductance location is brought through the insulator to the anode terminal, while the cathode foil is similarly connected to the aluminum case and cathode terminal.

Electrolytic capacitors typically have voltage ratings from 6.3 to 450 V and rated capacitances from 0.47 μF to several hundreds of microfarads at the maximum voltage to several farads at 6.3 V. Capacitance tolerance may range from ±20 to +80/–20 percent. The operating temperature range is often rated from –25 to +85°C or wider. Leakage current of an electrolytic capacitor may be rated as low as 0.002 times the capacity times the voltage rating to more than 10 times as much.

8.2.9 Tantalum Electrolytic Capacitors

Tantalum electrolytic capacitors are the capacitors of choice for applications requiring small size, 0.33- to 100-μF range at 10 to 20 percent tolerance, low equivalent series resistance, and low leakage current. These devices are well suited where the less costly aluminum electrolytic capacitors have performance issues. Tantalum capacitors are packaged in hermetically sealed metal tubes or with axial leads in epoxy plastic, as illustrated in Figure 8.3.

8.2.10 Capacitor Failure Modes

Mechanical failures relate to poor bonding of the leads to the outside world, contamination during manufacture, and shock-induced short-circuiting of the aluminum foil

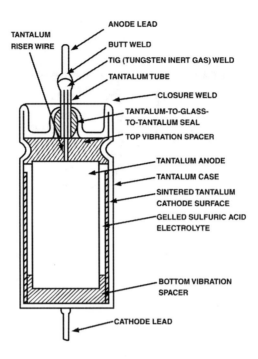

Figure 8.3 Basic construction of a tantalum capacitor.

plates. Typical failure modes include short-circuits caused by foil impurities, manufacturing defects (such as burrs on the foil edges or tab connections), breaks or tears in the foil, and breaks or tears in the separator paper.

Short-circuits are the most frequent failure mode during the useful life period of an electrolytic capacitor. Such failures are the result of random breakdown of the dielectric oxide film under normal stress. Proper capacitor design and processing will minimize such failures. Short-circuits also can be caused by excessive stress, where voltage, temperature, or ripple conditions exceed specified maximum levels.

Open circuits, although infrequent during normal life, can be caused by failure of the internal connections joining the capacitor terminals to the aluminum foil. Mechanical connections can develop an oxide film at the contact interface, increasing contact resistance and eventually producing an open circuit. Defective weld connections also can cause open circuits. Excessive mechanical stress will accelerate weld-related failures.

Temperature Cycling

Like semiconductor components, capacitors are subject to failures induced by thermal cycling. Experience has shown that thermal stress is a major contributor to failure in aluminum electrolytic capacitors. Dimensional changes between plastic and metal

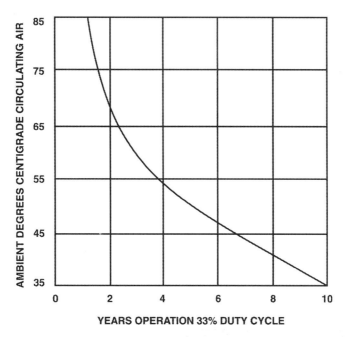

Figure 8.4 Life expectancy of an electrolytic capacitor as a function of operating temperature.

materials can result in microscopic ruptures at termination joints, possible electrode oxidation, and unstable device termination (changing series resistance). The highest-quality capacitor will fail if its voltage and/or current ratings are exceeded. Appreciable heat rise (20°C during a 2-hour period of applied sinusoidal voltage) is considered abnormal and may be a sign of incorrect application of the component or impending failure of the device.

Figure 8.4 illustrates the effects of high ambient temperature on capacitor life. Note that operation at 33 percent duty cycle is rated at 10 years when the ambient temperature is 35°C, but the life expectancy drops to just 4 years when the same device is operated at 55°C. A common rule of thumb is this: In the range of +75°C through the full-rated temperature, stress and failure rates double for each 10°C increase in operating temperature. Conversely, the failure rate is reduced by half for every 10°C decrease in operating temperature.

Electrolyte Failures

Failure of the electrolyte can be the result of application of a reverse bias to the component, or of a drying of the electrolyte itself. Electrolyte vapor transmission through the end seals occurs on a continuous basis throughout the useful life of the capacitor. This loss has no appreciable effect on reliability during the useful life period of the

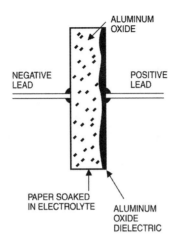

Figure 8.5 Failure mechanism of a leaky aluminum electrolytic capacitor. As the device ages, the aluminum oxide dissolves into the electrolyte, causing the capacitor to become leaky at high voltages.

product cycle. When the electrolyte loss approaches 40 percent of the initial electrolyte content of the capacitor, however, the electrical parameters deteriorate and the capacitor is considered to be worn out.

As a capacitor dries out, three failure modes may be experienced: leakage, a downward change in value, or *dielectric absorption*. Any one of these can cause a system to operate out of tolerance or fail altogether.

The most severe failure mode for an electrolytic is increased leakage, illustrated in Figure 8.5. Leakage can cause loading of the power supply, or upset the dc bias of an amplifier. Loading of a supply line often causes additional current to flow through the capacitor, possibly resulting in dangerous overheating and catastrophic failure.

A change of device operating value has a less devastating effect on system performance. An aluminum electrolytic capacitor has a typical tolerance range of about ±20 percent. A capacitor suffering from drying of the electrolyte can experience a drastic drop in value (to just 50 percent of its rated value, or less). The reason for this phenomenon is that after the electrolyte has dried to an appreciable extent, the charge on the negative foil plate has no way of coming in contact with the aluminum oxide dielectric. This failure mode is illustrated in Figure 8.6. Remember, it is the aluminum oxide layer on the positive plate that gives the electrolytic capacitor its large rating. The dried-out paper spacer, in effect, becomes a second dielectric, which significantly reduces the capacitance of the device.

Capacitor Life Span

The life expectancy of a capacitor—operating in an ideal circuit and environment—will vary greatly, depending upon the grade of device selected. Typical operating life,

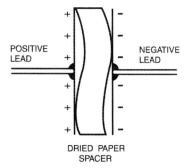

POSITIVE LEAD

NEGATIVE LEAD

DRIED PAPER SPACER

Figure 8.6 Failure mechanism of an electrolytic capacitor exhibiting a loss of capacitance. After the electrolyte dries, the plates can no longer come in contact with the aluminum oxide. The result is a decrease in capacitor value.

according to capacitor manufacturer data sheets, range from a low of 3 to 5 years for inexpensive electrolytic devices, to a high of greater than 10 years for computer-grade products. Catastrophic failures aside, expected life is a function of the rate of electrolyte loss by means of vapor transmission through the end seals, and the operating or storage temperature. Properly matching the capacitor to the application is a key component in extending the life of an electrolytic capacitor. The primary operating parameters include:

Rated voltage—the sum of the dc voltage and peak ac voltage that can be applied continuously to the capacitor. Derating of the applied voltage will decrease the failure rate of the device.

Ripple current—the rms value of the maximum allowable ac current, specified by product type at 120 Hz and +85°C (unless otherwise noted). The ripple current may be increased when the component is operated at higher frequencies or lower ambient temperatures.

Reverse voltage—the maximum voltage that can be applied to an electrolytic without damage. Electrolytic capacitors are polarized, and must be used accordingly.

8.3   References

1. Filanovsky, I. M., "Capacitance and Capacitors," in *The Electronics Handbook*, Jerry C. Whitaker (ed.), CRC Press, Boca Raton, FL, 1996.
2. Stuart, R. D., *Electromagnetic Field Theory*, Addison-Wesley, Reading, MA, 1965.

8.4   Bibliography

Benson, K. Blair, and Jerry C. Whitaker, *Television and Audio Handbook for Technicians and Engineers*, McGraw-Hill, New York, NY, 1990.
Benson, K. Blair, *Audio Engineering Handbook*, McGraw-Hill, New York, NY, 1988.

Whitaker, Jerry C., and K. Blair Benson (eds.), *Standard Handbook of Video and Television Engineering*, McGraw-Hill, New York, NY, 2000.

Whitaker, Jerry C., *Television Engineers' Field Manual*, McGraw-Hill, New York, NY, 2000.

Inductors and Magnetic Properties

9.1 Introduction

The elemental magnetic particle is the spinning electron. In magnetic materials, such as iron, cobalt, and nickel, the electrons in the third shell of the atom are the source of magnetic properties. If the spins are arranged to be parallel, the atom and its associated domains or clusters of the material will exhibit a magnetic field. The magnetic field of a magnetized bar has lines of magnetic force that extend between the ends, one called the north pole and the other the south pole, as shown in Figure 9.1a. The lines of force of a magnetic field are called *magnetic flux lines*.

9.1.1 Electromagnetism

A current flowing in a conductor produces a magnetic field surrounding the wire as shown in Figure 9.2a. In a coil or solenoid, the direction of the magnetic field relative to the electron flow (− to +) is shown in Figure 9.2b. The attraction and repulsion between two iron-core electromagnetic solenoids driven by direct currents is similar to that of two permanent magnets.

The process of magnetizing and demagnetizing an iron-core solenoid using a current being applied to a surrounding coil can be shown graphically as a plot of the magnetizing field strength and the resultant magnetization of the material, called a *hysteresis loop* (Figure 9.3). It will be found that the point where the field is reduced to zero, a small amount of magnetization, called *remnance*, remains.

9.1.2 Magnetic Shielding

In effect, the shielding of components and circuits from magnetic fields is accomplished by the introduction of a magnetic short circuit in the path between the field source and the area to be protected. The flux from a field can be redirected to flow in a partition or shield of magnetic material, rather than in the normal distribution pattern

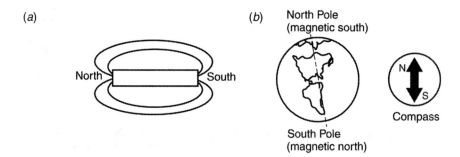

Figure 9.1 The properties of magnetism: (*a*) lines of force surrounding a bar magnet, (*b*) relation of compass poles to the earth's magnetic field.

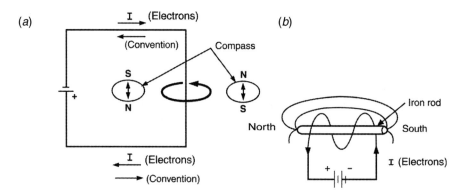

Figure 9.2 Magnetic field surrounding a current-carrying conductor: (*a*) Compass at right indicates the polarity and direction of a magnetic field circling a conductor carrying direct current. *I* indicates the direction of electron flow. Note: The convention for flow of electricity is from + to –, the reverse of the actual flow. (*b*) Direction of magnetic field for a coil or solenoid.

between north and south poles. The effectiveness of shielding depends primarily upon the thickness of the shield, the material, and the strength of the interfering field.

Some alloys are more effective than iron. However, many are less effective at high flux levels. Two or more layers of shielding, insulated to prevent circulating currents from magnetization of the shielding, are used in low-level audio, video, and data applications.

9.2   Inductors and Transformers

Inductors are passive components in which voltage leads current by nearly 90° over a wide range of frequencies. Inductors are usually coils of wire wound in the form of a cylinder. The current through each turn of wire creates a magnetic field that passes through every turn of wire in the coil. When the current changes, a voltage is induced

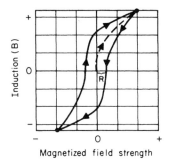

Figure 9.3 Graph of the magnetic hysteresis loop resulting from magnetization and de-magnetization of iron. The dashed line is a plot of the induction from the initial magnetization. The solid line shows a reversal of the field and a return to the initial magnetization value. *R* is the remaining magnetization (*remnance*) when the field is reduced to zero.

in the wire and every other wire in the changing magnetic field. The voltage induced in the same wire that carries the changing current is determined by the inductance of the coil, and the voltage induced in the other wire is determined by the *mutual inductance* between the two coils. A transformer has at least two coils of wire closely coupled by the common magnetic core, which contains most of the magnetic field within the transformer.

Inductors and transformers vary widely in size, weighing less than 1 g or more than 1 ton, and have specifications ranging nearly as wide.

9.2.1 Losses in Inductors and Transformers

Inductors have resistive losses because of the resistance of the copper wire used to wind the coil. An additional loss occurs because the changing magnetic field causes *eddy currents* to flow in every conductive material in the magnetic field. Using thin magnetic laminations or powdered magnetic material reduces these currents.

Losses in inductors are measured by the *Q*, or quality, factor of the coil at a test frequency. Losses in transformers are sometimes given as a specific insertion loss in decibels. Losses in power transformers are given as core loss in watts when there is no load connected and as a regulation in percent, measured as the relative voltage drop for each secondary winding when a rated load is connected.

Transformer loss heats the transformer and raises its temperature. For this reason, transformers are rated in watts or volt-amperes and with a temperature code designating the maximum hotspot temperature allowable for continued safe long-term operation. For example, class A denotes 105°C safe operating temperature. The volt-ampere rating of a power transformer must be always larger than the dc power output from the rectifier circuit connected because volt-amperes, the product of the rms currents and rms voltages in the transformer, are larger by a factor of about 1.6 than the product of the dc voltages and currents.

Inductors also have capacitance between the wires of the coil, which causes the coil to have a self-resonance between the winding capacitance and the self-inductance of the coil. Circuits are normally designed so that this resonance is outside of the frequency range of interest. Transformers are similarly limited. They also have capacitance to the other winding(s), which causes *stray coupling*. An electrostatic shield between windings reduces this problem.

9.2.2 Air-Core Inductors

Air-core inductors are used primarily in radio frequency applications because of the need for values of inductance in the microhenry or lower range. The usual construction is a multilayer coil made self-supporting with adhesive-covered wire. An inner diameter of 2 times coil length and an outer diameter 2 times as large yields maximum Q, which is also proportional to coil weight.

9.2.3 Ferromagnetic Cores

Ferromagnetic materials have a permeability much higher than air or vacuum and cause a proportionally higher inductance of a coil that has all its magnetic flux in this material. Ferromagnetic materials in audio and power transformers or inductors usually are made of silicon steel laminations stamped in the forms of letters E or I (Figure 9.4). At higher frequencies, powdered ferric oxide is used. The continued magnetization and remagnetization of silicon steel and similar materials in opposite directions does not follow the same path in both directions but encloses an area in the magnetization curve and causes a hysteresis loss at each pass, or twice per ac cycle.

All ferromagnetic materials show the same behavior; only the numbers for permeability, core loss, saturation flux density, and other characteristics are different. The properties of some common magnetic materials and alloys are given in Table 9.1.

9.2.4 Shielding

Transformers and coils radiate magnetic fields that can induce voltages in other nearby circuits. Similarly, coils and transformers can develop voltages in their windings when subjected to magnetic fields from another transformer, motor, or power circuit. Steel mounting frames or chassis conduct these fields, offering less reluctance than air.

The simplest way to reduce the stray magnetic field from a power transformer is to wrap a copper strip as wide as the coil of wire around the transformer enclosing all three legs of the core. Shielding occurs by having a short circuit turn in the stray magnetic field outside of the core.

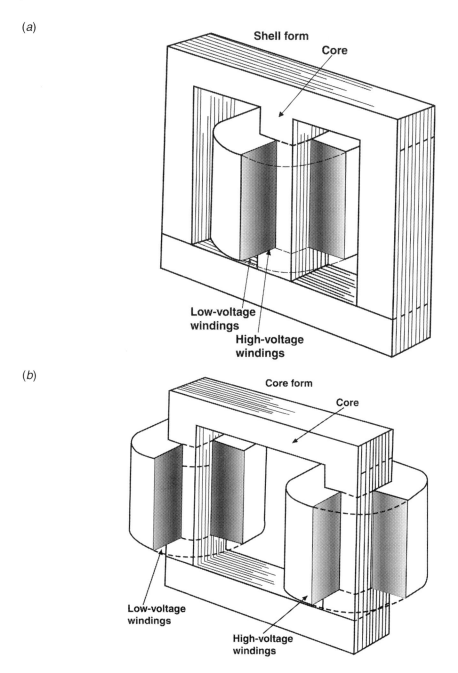

Figure 9.4 Physical construction of a power transformer: (a) E-shaped device with the low- and high-voltage windings stacked as shown, (b) construction using a box core with physical separation between the low- and high-voltage windings.

Table 9.1 Properties of Magnetic Materials and Magnetic Alloys (*From* [1]. *Used with permission.*)

Material (Composition)	Initial Relative Permeability, μ_i/μ_0	Maximum Relative Permeability, μ_{max}/μ_0	Coercive Force H_e, A/m (Oe)	Residual Field B_r, Wb/m^2(G)	Saturation Field B_s, Wb/m^2(G)	Electrical Resistivity ρ ×10^{-8} Ω·m	Uses
			Soft				
Commercial iron (0.2 imp.)	250	9,000	≈80 (1)	0.77 (7,700)	2.15 (21,500)	10	Relays
Purified iron (0.05 imp.)	10,000	200,000	4 (0.05)	—	2.15 (21,500)	10	
Silicon-iron (4 Si)	1,500	7,000	20 (0.25)	0.5 (5,000)	1.95 (19,500)	60	Transformers
Silicon-iron (3 Si)	7,500	55,000	8 (0.1)	0.95 (9,500)	2 (20,000)	50	Transformers
Silicon-iron (3 Si)	—	116,000	4.8 (0.06)	1.22 (12,200)	2 (20,100)	50	Transformers
Mu metal (5 Cu, 2 Cr, 77 Ni)	20,000	100,000	4 (0.05)	0.23 (2,300)	0.65 (6,500)	62	Transformers
78 Permalloy (78.5 Ni)	8,000	100,000	4(0.05)	0.6 (6,000)	1.08 (10,800)	16	Sensitive relays
Supermalloy (79 Ni, 5 Mo)	100,000	1,000,000	0.16 (0.002)	0.5 (5,000)	0.79 (7,900)	60	Transformers
Permendur (50 Cs)	800	5,000	160 (2)	1.4 (14,000)	2.45 (24,500)	7	Electromagnets
Mn-Zn ferrite	1,500	2,500	16 (0.2)	—	0.34 (3,400)	20 × 10^6	Core material
Ni-Zn ferrite	2,500	5,000	8 (0.1)	—	0.32 (3,200)	10^{11}	for coils

Source: After Plonus, M.A. 1978. Applied Electromagnetics. McGraw-Hill, New York.

9.3   References

1. Whitaker, Jerry C. (ed.), *The Electronics Handbook*, CRC Press, Boca Raton, FL, 1996.

9.4   Bibliography

Benson, K. Blair, and Jerry C. Whitaker, *Television and Audio Handbook for Technicians and Engineers*, McGraw-Hill, New York, NY, 1990.
Benson, K. Blair, *Audio Engineering Handbook*, McGraw-Hill, New York, NY, 1988.
Whitaker, Jerry C., *Television Engineers' Field Manual*, McGraw-Hill, New York, NY, 2000.

9.5   Tabular Data

Table 9.2 Magnetic Properties of Transformer Steels (*From* [1]. *Used with permission.*)

Ordinary Transformer Steel			High Silicon Transformer Steels		
B (Gauss)	H (Oersted)	Permeability $= B/H$	B	H	Permeability
2,000	0.60	3,340	2,000	0.50	4,000
4,000	0.87	4,600	4,000	0.70	5,720
6,000	1.10	5,450	6,000	0.90	6,670
8,000	1.48	5,400	8,000	1.28	6,250
10,000	2.28	4,380	10,000	1.99	5,020
12,000	3.85	3,120	12,000	3.60	3,340
14,000	10.9	1,280	14,000	9.80	1,430
16,000	43.0	372	16,000	47.4	338
18,000	149	121	18,000	165	109

Table 9.3 Characteristics of High-Permeability Materials (*From* [1]. *Used with permission.*)

Material	Form	Approximate % Composition					Typical Heat Treatment, °C	Permeability at $B = 20$, G	Maximum Permeability	Saturation flux density B, G	Hysteresis loss[a], W_h, ergs/cm^2	Coercive force[a] H_a O	Resistivity $\mu \cdot \Omega$cm	Density, g/cm^3
		Fe	Ni	Co	Mo	Other								
Cold rolled steel	Sheet	98.5	–	–	–	–	950 Anneal	180	2,000	21,000	–	1.8	10	7.88
Iron	Sheet	99.91	–	–	–	–	950 Anneal	200	5,000	21,500	5,000	1.0	10	7.88
Purified iron	Sheet	99.95	–	–	–	–	1480 H_2 + 880	5,000	180,000	21,500	300	0.05	10	7.88
4% Silicon-iron	Sheet	96	–	–	–	4 Si	800 Anneal	500	7,000	19,700	3,500	0.5	60	7.65
Grain oriented[b]	Sheet	97	–	–	–	3 Si	800 Anneal	1,500	30,000	20,000	–	0.15	47	7.67
45 Permalloy	Sheet	54.7	45	–	–	0.3 Mn	1050 Anneal	2,500	25,000	16,000	1,200	0.3	45	8.17
45 Permalloy[c]	Sheet	54.7	45	–	–	0.3 Mn	1200 H_2 Anneal	4,000	50,000	16,000	–	0.07	45	8.17
Hipernik	Sheet	50	50	–	–	–	1200 H_2 Anneal	4,500	70,000	16,000	220	0.05	50	8.25
Monimax	Sheet	–	–	–	–	–	1125 H_2 Anneal	2,000	35,000	15,000	–	0.1	80	8.27
Sinimax	Sheet	–	–	–	–	–	1125 H_2 Anneal	3,000	35,000	11,000	–	–	90	–
78 Permalloy	Sheet	21.2	78.5	–	–	0.3 Mn	1050 + 600 Q^d	8,000	100,000	10,700	200	0.05	16	8.60
4–79 Permalloy	Sheet	16.7	79	–	4	0.3 Mn	1100 + Q	20,000	100,000	8,700	200	0.05	55	8.72
Mu metal	Sheet	18	75	–	–	2 Cr, 5 Cu	1175 H_2	20,000	100,000	6,500	–	0.05	62	8.58
Supermalloy	Sheet	15.7	79	–	5	0.3 Mn	1300 H_2+ Q	100,000	800,000	8,000	–	0.002	60	8.77
Permendur	Sheet	49.7	–	50	–	0.3 Mn	800 Anneal	800	5,000	24,500	12,000	2.0	7	8.3
2 V Permendur	Sheet	49	–	49	–	2 V	800 Anneal	800	4,500	24,000	6,000	2.0	26	8.2
Hiperco	Sheet	64	–	34	–	Cr	850 Anneal	650	10,000	24,200	–	1.0	25	8.0
2–81 Permalloy	Insulated powder	17	81	–	2	–	650 Anneal	125	130	8,000	–	<1.0	10^6	7.8
Carbonyl iron	Insulated powder	99.9	–	–	–	–	–	55	132	–	–	–	–	7.86
Ferroxcube III	Sintered powder	$MnFe_2O_4 + ZnFe_2O_4$					–	1,000	1,500	2,500	–	0.1	10^8	5.0

[a] At saturation.
[b] Properties in direction of rolling.
[c] Similar properties for Nicaloi, 4750 alloy, Carpenter 49, Armco 48.
[d] Q, quench or controlled cooling.

Table 9.4 Characteristics of Permanent Magnet Alloys (*From* [1]. *Used with permission.*)

Material	% composition (remainder Fe)	Heat treatment[a] (temperature, °C)	Magnetizing force H_{max}, O	Coercive force H_c, O	Residual induction B_r, G	Energy product BH_{max} $\times 10^{-6}$	Method of fabrication[b]	Mechanical properties[c]	Weight lb/In.³
Carbon steel	1 Mn, 0.9 C	Q 800	300	50	10,000	0.20	HR, M, P	H, S	0.280
Tungsten steel	5 W, 0.3 Mn, 0.7 C	Q 850	300	70	10,300	0.32	HR, M, P	H, S	0.292
Chromium steel	3.5 Cr, 0.9 C, 0.3 Mn	Q 830	300	65	9,700	0.30	HR, M, P	H, S	0.280
17% Cobalt steel	17 Co, 0.75 C, 2.5 Cr, 8 W	–	1,000	150	9,500	0.65	HR, M, P	H, S	–
36% Cobalt steel	36 Co, 0.7 C, 4 Cr, 5 W	Q 950	1,000	240	9,500	0.97	HR, M, P	H, S	0.296
Remalloy or Comol	17 Mo, 12 Co	Q 1200, B 700	1,000	250	10,500	1.1	HR, M, P	H	0.295
Alnico I	12 Al, 20 Ni, 5 Co	A 1200, B 700	2,000	440	7,200	1.4	C, G	H, B	0.249
Alnico II	10 Al, 17 Ni, 2.5 Co, 6 Cu	A 1200, B 600	2,000	550	7,200	1.6	C, G	H, B	0.256
Alnico II (sintered)	10 Al, 17 Ni, 2.5 Co, 6 Cu	A 1300	2,000	520	6,900	1.4	Sn, G	H	0.249
Alnico IV	12 Al, 28 Ni, 5 Co	Q 1200, B 650	3,000	700	5,500	1.3	Sn, C, G	H	0.253
Alnico V	8 Al, 14 Ni, 24 Co, 3 Cu	AF 1300, B 600	2,000	550	12,500	4.5	C, G	H, B	0.264
Alnico VI	8 Al, 15 Ni, 24 Co, 3 Cu, 1 Ti	–	3,000	750	10,000	3.5	C, G	H, B	0.268
Alnico XII	6 Al, 18 Ni, 35 Co, 8 Ti	–	3,000	950	5,800	1.5	C, G	H, B	0.26
Vicalloy I	52 Co, 10 V	B 600	1,000	300	8,800	1.0	C, CR, M, P	D	0.295
Vicalloy II (wire)	52 Co, 14 V	CW + B 600	2,000	510	10,000	3.5	C, CR, M, P	D	0.292
Cunife (wire)	60 Cu, 20 Ni	CW + B 600	2,400	550	5,400	1.5	C, CR, M, P	D, M	0.311
Cunico	50 Cu, 21 Ni, 29 Co	–	3,200	660	3,400	0.80	C, CR, M, P	D, M	0.300
Vectolite	$30Fe_2O_3, 44Fe_3O_4, 26C_2O_3$	–	3,000	1,000	1,600	0.60	Sn, G	W	0.113
Silmanal	86.8Ag, 8.8Mn, 4.4Al	–	20,000	6,000[d]	550	0.075	C, CR, M, P	D, M	0.325
Platinum-cobalt	77 Pt, 23 Co	Q 1200, B 650	15,000	3,600	5,900	6.5	C, CR, M	D	–
Hyflux	Fine powder	–	2,000	390	6,600	0.97	–	–	0.176

[a]Q, quenched in oil or water; A, air cooled; B, baked; F, cooled in magnetic field; CW, cold worked.

[b]HR, hot rolled or forged; CR, cold rolled or drawn; M, machined; G, must be ground; P, punched; C, cast; Sn, sintered.

[c]H, hard; B, brittle; S, strong; D, strong; M, ductile; M, malleable; W, weak.

[d]Value given is intrinsic H_c.

Table 9.5 Properties of Antiferromagnetic Compounds (*From* [1]. *Used with permission.*)

Compound	Crystal Symmetry	θ_N[a] K	θ_P[b] K	$(P_A)_{\text{eff}}$[c] μ_B	P_A[d] μ_B
$CoCl_2$	Rhombohedral	25	−38.1	5.18	3.1 ± 0.6
CoF_2	Tetragonal	38	50	5.15	3.0
CoO	Tetragonal	291	330	5.1	3.8
Cr	Cubic	475			
Cr_2O_3	Rhombohedral	307	485	3.73	3.0
CrSb	Hexagonal	723	550	4.92	2.7
$CuBr_2$	Monoclinic	189	246	1.9	
$CuCl_2 \cdot 2H_2O$	Orthorhombic	4.3	4–5	1.9	
$CuCl_2$	Monoclinic	~70	109	2.08	
$FeCl_2$	Hexagonal	24	−48	5.38	4.4 ± 0.7
FeF_2	Tetragonal	79–90	117	5.56	4.64
FeO	Rhombohedral	198	507	7.06	3.32
$\alpha\text{-}Fe_2O_3$	Rhombohedral	953	2940	6.4	5.0
α-Mn	Cubic	95			
$MnBr_2 \cdot 4H_2O$	Monoclinic	2.1	$\begin{Bmatrix} 2.5 \\ 1.3 \end{Bmatrix}$	5.93	
$MnCl_2 \cdot 4H_2O$	Monoclinic	1.66	1.8	5.94	
MnF_2	Tetragonal	72–75	113.2	5.71	5
MnO	Rhombohedral	122	610	5.95	5.0
β-MnS	Cubic	160	982	5.82	5.0
MnSe	Cubic	~173	361	5.67	
MnTe	Hexagonal	310–323	690	6.07	5.0
$NiCl_2$	Hexagonal	50	−68	3.32	
NiF_2	Tetragonal	78.5–83	115.6	3.5	2.0
NiO	Rhombohedral	533–650	~2000	4.6	2.0
$TiCl_3$		100			
V_2O_3		170			

[a] θ_N = Néel temperature, determined from susceptibility maxima or from the disappearance of magnetic scattering.
[b] θ_P = a constant in the Curie-Weiss law written in the form $\chi_A = C_A/(T + \theta_P)$, which is valid for antiferromagnetic material for $T > \theta_N$.
[c] $(P_A)_{\text{eff}}$ = effective moment per atom, derived from the atomic Curie constant $C_A = (P_A)^2_{\text{eff}}(N^2/3R)$ and expressed in units of the Bohr magneton, $\mu_B = 0.9273 \times 10^{-20}$ erg G^{-1}.
[d] P_A = magnetic moment per atom, obtained from neutron diffraction measurements in the ordered state.

Table 9.6 Saturation Constants for Magnetic Substances (*From* [1]. *Used with permission.*)

Substance	Field Intensity	Induced Magnetization (For Saturation)	Substance	Field Intensity	Induced Magnetization (For Saturation)
Cobalt	9,000	1,300	Nickel, hard	8,000	400
Iron, wrought	2,000	1,700	annealed	7,000	515
cast	4,000	1,200	Vicker's steel	15,000	1,600
Manganese steel	7,000	200			

Table 9.7 Saturation Constants and Curie Points of Ferromagnetic Elements (*From*[1]. *Used with permission.*)

Element	$\sigma_s{}^a$ (20°C)	$M_s{}^b$ (20°C)	σ_s (0 K)	$n_B{}^c$	Curie point, °C
Fe	218.0	1,714	221.9	2.219	770
Co	161	1,422	162.5	1.715	1,131
Ni	54.39	484.1	57.50	0.604	358
Gd	0	0	253.5	7.12	16

$^a\sigma_s$ = saturation magnetic moment/gram.
$^b M_s$ = saturation magnetic moment/cm^3, in cgs units.
$^c n_B$ = magnetic moment per atom in Bohr magnetons.

10

Filter Devices and Circuits

10.1 Introduction

A *filter* is a multiport-network designed specifically to respond differently to signals of different frequency [1]. This definition excludes *networks*, which incidentally behave as filters, sometimes to the detriment of their main purpose. Passive filters are constructed exclusively with passive elements (i.e., resistors, inductors, and capacitors). Filters are generally categorized by the following general parameters:

- Type
- Alignment (or class)
- Order

10.2 Filter Type

Filters are categorized by type, according to the magnitude of the frequency response, as one of the following [1]:

- *Low-pass* (LP)
- *High-pass* (HP)
- *Band-pass* (BP)
- *Band-stop* (BS)

The terms *band-reject* or *notch* are also used as descriptions of the BS filter. The term *all-pass* is sometimes applied to a filter whose purpose is to alter the phase angle without affecting the magnitude of the frequency response. Ideal and practical interpretations of the types of filters and the associated terminology are illustrated in Figure 10.1.

In general, the voltage gain of a filter in the *stop band* (or *attenuation band*) is less than $\sqrt{2}/2\,(\approx\ 0.707)$ times the maximum voltage gain in the pass band. In logarithmic

(a)   (b)

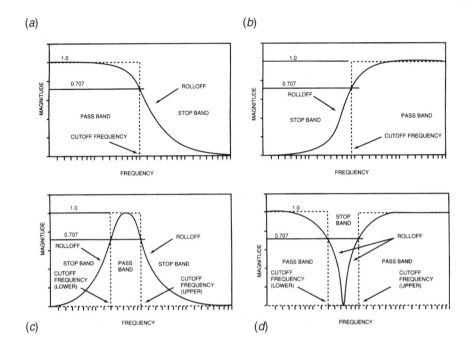

Figure 10.1 Filter characteristics by type: (a) low-pass, (b) high-pass, (c) bandpass, (d) bandstop. (*From* [1]. *Used with permission.*)

terms, the gain in the stop band is at least 3.01 dB less than the maximum gain in the pass band. The cutoff (*break* or *corner*) frequency separates the pass band from the stop band. In BP and BS filters, there are two cutoff frequencies, sometimes referred to as the *lower* and *upper* cutoff frequencies. Another expression for the cutoff frequency is *half-power frequency*, because the power delivered to a resistive load at cutoff frequency is one-half the maximum power delivered to the same load in the pass band. For BP and BS filters, the center frequency is the frequency of maximum or minimum response magnitude, respectively, and bandwidth is the difference between the upper and lower cutoff frequencies. *Rolloff* is the transition from pass band to stop band and is specified in gain unit per frequency unit (e.g., gain unit/Hz, dB/decade, dB/octave, etc.)

10.2.1 Filter Alignment

The *alignment* (or class) of a filter refers to the shape of the frequency response [1]. Fundamentally, filter alignment is determined by the coefficients of the filter network transfer function, so there are an indefinite number of filter alignments, some of which may not be realizable. The more common alignments are:

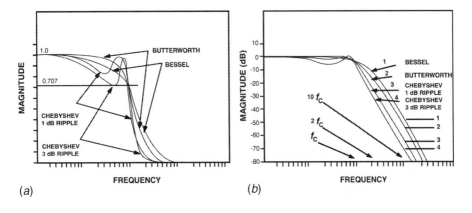

Figure 10.2 Filter characteristics by alignment, third-order, all-pole filters: (*a*) magnitude, (*b*) magnitude in decibels. (*From* [1]. *Used with permission.*)

- Butterworth
- Chebyshev
- Bessel
- Inverse Chebyshev
- Elliptic (or Cauer)

Each filter alignment has a frequency response with a characteristic shape, which provides some particular advantage. (See Figure 10.2.) Filters with Butterworth, Chebyshev, or Bessel alignment are called *all-pole filters* because their low-pass transfer functions have no zeros. Table 10.1 summarizes the characteristics of the standard filter alignments.

10.2.2 Filter Order

The *order* of a filter is equal to the number of poles in the filter network transfer function [1]. For a lossless LC filter with resistive (nonreactive) termination, the number of reactive elements (inductors or capacitors) required to realize a LP or HP filter is equal to the order of the filter. Twice the number of reactive elements are required to realize a BP or a BS filter of the same order. In general, the order of a filter determines the slope of the rolloff—the higher the order, the steeper the rolloff. At frequencies greater than approximately one octave above cutoff (i.e., $f > 2f_c$), the rolloff for all-pole filters is $20n$ dB/decade (or approximately $6n$ dB/octave), where n is the order of the filter (Figure 10.3). In the vicinity of f_c, both filter alignment and filter order determine rolloff.

Table 10.1 Summary of Standard Filter Alignments (*After* [1].)

Alignment	Pass Band Description	Stop Band Description	Comments
Butterworth	Monotonic	Monotonic	All-pole; maximally flat
Chebyshev	Rippled	Monotonic	All-pole
Bessel	Monotonic	Monotonic	All-pole; constant phase shift
Inverse Chebyshev	Monotonic	Rippled	
Elliptic (or Cauer)	Rippled	Rippled	

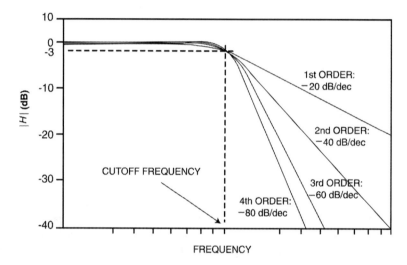

Figure 10.3 The effects of filter order on rolloff (Butterworth alignment). (*From* [1]. *Used with permission.*)

10.3 Filter Design Implementation

Conventional filter design techniques may be implemented using a number of different resonators. The principal available technologies are *inductor-capacitor* (LC) resonators, mechanical resonators, quartz crystal resonators, quartz monolithic filters, and ceramic resonators. The classical approach to radio frequency filtering involves cascading single- or dual-resonator filters separated by amplifier stages. Overall selectivity is provided by this combination of one- or two-pole filters. The disadvantages of this approach are alignment problems, and the possibility of intermodulation (IM) and overload in certain stages because of out-of-band signals. An advantage is the relatively low cost of using a large number of essentially identical two-resonator devices. This approach has been largely displaced in modern systems by the use of *multiresonator filters* inserted as appropriate in the signal chain to reduce nonlinear distortion, localize alignment and stability problems to a single assembly, and permit easy attainment of any of a variety of selectivity patterns. The simple single- or dual-resonator pairs are now used mainly for impedance matching between stages or to reduce noise between very broadband cascaded amplifiers.

10.3.1 LC Filters

LC resonators are limited to Q values on the order of a few hundred for reasonable sizes, and in most cases designers must be satisfied with lower Q values. The size of the structures depends strongly on the center frequency, which may range from the audio band to several hundred megahertz. Bandwidth below about 1 percent is not easily obtained. However, broader bandwidths can be obtained more easily than with other resonator types. Skirt selectivity depends on the number of resonators used; ultimate filter rejection can be made higher than 100 dB with careful design. The filter loss depends on the percentage bandwidth required and the resonator Q, and can be expected to be as high as 1 dB per resonator at the most narrow bandwidths. This type of filter does not generally suffer from nonlinearities unless the frequency is so low that very high permeability cores must be used. Frequency stability is limited by the individual components and cannot be expected to achieve much better than 0.1 percent of center frequency under extremes of temperature and aging.

10.3.2 Electrical Resonators

As frequencies increase into the VHF region, the construction of inductors for use in *LC* resonant circuits becomes more difficult. The *helical resonator* is an effective alternative for the VHF and lower UHF ranges. This type of resonator looks like a shielded coil (see Figure 10.4a). However, it acts as a resonant transmission line section. High Q can be achieved in reasonable sizes (Figure 10.4b). When such resonators are used singly, coupling in and out can be achieved by a tap on the coil, a loop in the cavity near the grounded end (high magnetic field), or a probe near the ungrounded end (high electric field). The length of the coil is somewhat shorter than the predicted open-circuit quarter-wave line because of the end capacity to the shield. A separate adjustable screw or vane can be inserted near the open end of the coil to provide tuning. Multiresonator filters are designed using a cascade of similar resonators, with coupling between them. The coupling may be of the types mentioned previously or may be obtained by locating an aperture in the common shield between two adjacent resonators. At still higher frequencies, coaxial transmission line resonators or resonant cavities are used for filtering (mostly above 1 GHz). The use of *striplines* to provide filtering is another useful technique for these frequency regions.

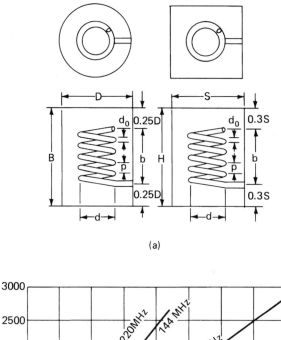

Figure 10.4 Helical resonators: (*a*) round and square shielded types, showing principal dimensions (diameter *D* or side *S* is determined by the desired unloaded *Q*), (*b*) unloaded *Q* versus shield diameter *D* for bands from 1.8 MHz to 1.3 GHz. (*From* [3]. *Used with permission.*)

10.3.3 Stripline Technology

Stripline typically utilizes a double-sided printed circuit board made of fiberglass. The board is usually 30- to 50-thousandths of an inch thick. The board is uniform over the entire surface and forms an electrical ground plane for the circuit. This ground plane serves as a return for the electrical fields built up on the component side of the board.

The shape and length of each trace of stripline on the component side dictates the impedance and reactance of the trace. The impedance is a function of the width of the trace, its height above the lower surface ground plane, and the dielectric constant of the circuit board material. The length of the trace is another important factor. At microwave frequencies, a quarter-wavelength can be as short as 0.5-in in air. Because all printed circuit boards have a dielectric constant that is greater than the dielectric constant of air, waves are slowed as they travel through the board-trace combination. This effect causes the wavelength on a circuit board to be dependent on the dielectric constant of the material of which the board is made. At a board dielectric constant of 5 to 10 (common with the materials typically used in printed circuit boards), a wavelength may be up to 33 percent shorter than in air. The wider the trace, the lower the RF impedance.

Traces that supply bias and require operating or control voltages are usually made thin so as to present a high RF impedance while maintaining a low dc resistance. Narrow bias and control traces are usually made to be a multiple of a quarter-wavelength at the operating frequency so that unwanted RF energy may be easily shunted to ground.

Figure 10.5 shows stripline technology serving several functions in a satellite-based communications system. The circuit includes the following elements:

A 3-section, low-pass filter

- Quarter-wave line used as half of a transmit-receive switch

- Bias lines to supply operating voltages to a transistor

- Impedance-matching strip segments that convert a high impedance (130 Ω) to 50 Ω

- Coupling lines that connect two circuit sections at one impedance

A wide variety of techniques can be used to synthesize filter and coupling networks in stripline. After an initial choice of one of the components is made, however, only a small number of solutions are practical. While it is apparent that all components must be kept within reasonable physical limits, the most critical parameter is usually the length of the stripline trace. This technique is popular with equipment designers because of the following benefits:

- **Low cost.** Stripline coupling networks are simply a part of the PC board layout. No assembly time is required during construction of the system.

- **Excellent repeatability.** Variations in dimensions, and therefore performance, are virtually eliminated.

Stripline also has the following drawbacks:

- **Potential for and/or susceptibility to radiation.** Depending on the design, shielding of stripline sections may be necessary to prevent excessive RF emis-

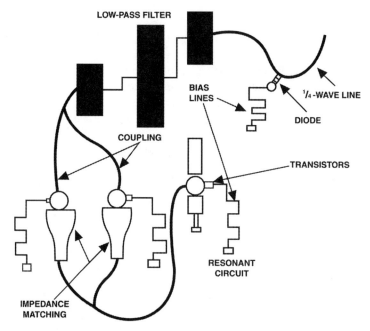

Figure 10.5 A typical application of stripline showing some of the components commonly used.

sions or to prevent emissions from nearby sources from coupling into the stripline filter.

- **Repair difficulties.** If a stripline section is damaged, it may be necessary to replace the entire PC board.

10.3.4 Electromechanical Filters

Most of the other resonators used in receiver design are electromechanical, where the resonance of acoustic waves is employed. During a period when quartz resonators were in limited supply, electromechanical filters were constructed from metals, using metal plates or cylinders as the resonant element and wires as the coupling elements. Filters can be machined from a single metal bar of the right diameter. This type of electromechanical filter is limited by the physical size of the resonators to center frequencies between about 60 and 600 kHz. Bandwidths can be obtained from a few tenths of a percent to a maximum of about 10 percent. A disadvantage of these filters is the loss encountered when coupling between the electrical and mechanical modes at input and output. This tends to result in losses of 6 dB or more. Also, spurious resonances can limit the ultimate out-of-band attenuation. Size and weight are somewhat lower, but are generally comparable to LC filters. Temperature and aging stability is about 10 times greater than for LC filters. Because of their limited frequency range, electromechanical filters have been largely superseded by quartz crystal filters, which have greater stability at comparable price.

10.3.5 Quartz Crystal Resonators

While other piezoelectric materials have been used for filter resonators, quartz crystals have proved most satisfactory. Filters are available from 5 kHz to 100 MHz, and bandwidths from less than 0.01 percent to about 1 percent. (The bandwidth, center frequency, and selectivity curve type are interrelated, so manufacturers should be consulted as to the availability of specific designs.) Standard filter shapes are available, and with modern computer design techniques, it is possible to obtain custom shapes. Ultimate filter rejection can be greater than 100 dB. Input and output impedances are determined by input and output matching networks in the filters, and typically range from a few tens to a few thousands of ohms. Insertion loss varies from about 1 to 10 dB, depending on filter bandwidth and complexity. While individual crystal resonators have spurious modes, these tend not to overlap in multiresonator filters, so that high ultimate rejection is possible. Nonlinearities can occur in crystal resonators at high input levels, and at sufficiently high input the resonator may even shatter. Normally these problems should not be encountered in common use, and in any event, the active devices preceding the filter are likely to fail prior to destruction of the filter. Frequency accuracy can be maintained to about 0.001 percent, although this is relatively costly; somewhat less accuracy is often acceptable. Temperature stability of 0.005 percent is achievable.

10.3.6 Monolithic Crystal Filters

In monolithic crystal filter technology, a number of resonators are constructed on a single quartz substrate, using the *trapped-energy* concept. The principal energy of each resonator is confined primarily to the region between plated electrodes, with a small amount of energy escaping to provide coupling. Usually these filters are constrained to about four resonators, but the filters can be cascaded using electrical coupling circuits if higher-order characteristics are required. The filters are available from 3 to more than 100 MHz, with characteristics generally similar to those of discrete crystal resonator filters, except that the bandwidth is limited to several tenths of a percent. The volume and weight are also much less than those of discrete resonator filters.

10.3.7 Ceramic Filters

Piezoelectric ceramics are also used for filter resonators, primarily to achieve lower cost than quartz. Such filters are comparable in size to monolithic quartz filters but are available over a limited center frequency range (100 to 700 kHz). The cutoff rate, stability, and accuracy are not as good as those of quartz, but are adequate for many applications. Selectivity designs available are more limited than for quartz filters. Bandwidths are 1 to 10 percent. Single- and double-resonator structures are manufactured, and multiple-resonator filters are available that use electrical coupling between sections.

10.4 References

1. Harrison, Cecil, "Passive Filters," in *The Electronics Handbook*, Jerry C. Whitaker (ed.), CRC Press, Boca Raton, FL, pp. 279–290, 1996.
2. Rohde, Ulrich L., Jerry C. Whitaker, and T. T. N. Bucher, *Communications Receivers*, 2nd ed., McGraw-Hill, New York, NY, 1996.
3. Fisk, J. R., "Helical Resonator Design Techniques," *QST*, July 1976.

Thermal Properties

11.1 Introduction

In the commonly used model for materials, heat is a form of energy associated with the position and motion of the molecules, atoms, and ions of the material [1]. The position is analogous with the state of the material and is *potential energy*, while the motion of the molecules, atoms, and ions is *kinetic energy*. Heat added to a material makes it hotter, and heat withdrawn from a material makes it cooler. Heat energy is measured in *calories* (cal), *British Thermal Units* (Btu), or *joules*. A calorie is the amount of energy required to raise the temperature of one gram of water one degree Celsius (14.5 to 15.5 °C). A Btu is the unit of energy necessary to raise the temperature of one pound of water by one degree Fahrenheit. A joule is an equivalent amount of energy equal to the work done when a force of one newton acts through a distance of one meter.

Temperature is a measure of the average kinetic energy of a substance. It can also be considered a relative measure of the difference of the heat content between bodies.

Heat capacity is defined as the amount of heat energy required to raise the temperature of one mole or atom of a material by one °C without changing the state of the material. Thus, it is the ratio of the change in heat energy of a unit mass of a substance to its change in temperature. The heat capacity, often referred to as *thermal capacity*, is a characteristic of a material and is measured in cal/gram per °C or Btu/lb per °F.

Specific heat is the ratio of the heat capacity of a material to the heat capacity of a reference material, usually water. Because the heat capacity of water is one Btu/lb and one cal/gram, the specific heat is numerically equal to the heat capacity.

Heat transfers through a material by conduction resulting when the energy of atomic and molecular vibrations is passed to atoms and molecules with lower energy. As heat is added to a substance, the kinetic energy of the lattice atoms and molecules increases. This, in turn, causes an expansion of the material that is proportional to the temperature change, over normal temperature ranges. If a material is restrained from expanding or contracting during heating and cooling, internal stress is established in the material.

Table 11.1 Thermal Conductivity of Common Materials

Material	Btu/(h·ft·°F)	W/(m·°C)
Silver	242	419
Copper	228	395
Gold	172	298
Beryllia	140	242
Phosphor bronze	30	52
Glass (borosilicate)	0.67	1.67
Mylar	0.11	0.19
Air	0.015	0.026

11.2 Heat Transfer Mechanisms

The process of heat transfer from one point or medium to another is a result of temperature differences between the two. Thermal energy can be transferred by any of three basic modes:

- Conduction

- Convection

- Radiation

A related mode is the convection process associated with the change of phase of a fluid, such as condensation or boiling.

11.2.1 Conduction

Heat transfer by conduction in solid materials occurs whenever a hotter region with more rapidly vibrating molecules transfers a portion of its energy to a cooler region with less rapidly vibrating molecules. Conductive heat transfer is the most common form of thermal exchange in electronic equipment. Thermal conductivity for solid materials used in electronic equipment spans a wide range of values, from excellent (high conductivity) to poor (low conductivity). Generally speaking, metals are the best conductors of heat, whereas insulators are the poorest. Table 11.1 lists the thermal conductivity of materials commonly used in the construction (and environment) of electronic systems. Table 11.2 compares the thermal conductivity of various substances as a percentage of the thermal conductivity of copper.

11.2.2 Convection

Heat transfer by natural convection occurs as a result of a change in the density of a fluid (including air), which causes fluid motion. Convective heat transfer between a

Table 11.2 Relative Thermal Conductivity of Various Materials As a Percentage of the Thermal Conductivity of Copper

Material	Relative Conductivity
Silver	105
Copper	100
Berlox high-purity BeO	62
Aluminum	55
Beryllium	39
Molybdenum	39
Steel	9.1
High-purity alumina	7.7
Steatite	0.9
Mica	0.18
Phenolics, epoxies	0.13
Fluorocarbons	0.05

heated surface and the surrounding fluid is always accompanied by a mixing of fluid adjacent to the surface. Electronic devices relying on convective cooling typically utilize forced air or water passing through a heat-transfer element [2]. This *forced convection* provides for a convenient and relatively simple cooling system. In such an arrangement, the temperature gradient is confined to a thin layer of fluid adjacent to the surface so that the heat flows through a relatively thin *boundary layer*. In the main stream outside this layer, isothermal conditions exist.

11.2.3 Radiation

Cooling by radiation is a function of the transfer of energy by electromagnetic wave propagation. The wavelengths between 0.1 and 100 m are referred to as *thermal radiation wavelengths*. The ability of a body to radiate thermal energy at any particular wavelength is a function of the body temperature and the characteristics of the surface of the radiating material. Figure 11.1 charts the ability to radiate energy for an ideal radiator, a *blackbody*, which, by definition, radiates the maximum amount of energy at any wavelength. Materials that act as perfect radiators are rare. Most materials radiate energy at a fraction of the maximum possible value. The ratio of the energy radiated by a given material to that emitted by a blackbody at the same temperature is termed *emissivity*. Table 11.3 lists the emissivity of various common materials.

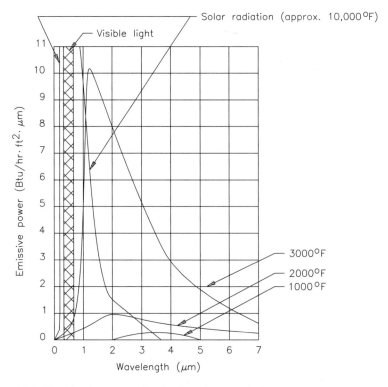

Figure 11.1 Blackbody energy distribution characteristics.

Table 11.3 Emissivity Characteristics of Common Materials at 80°F

Surface Type	Finish	Emissivity
Metal	Copper (polished)	0.018
Metal	Nickel	0.21
Metal	Silver	0.10
Metal	Gold	0.04–0.23
Glass	Smooth	0.9–0.95
Ceramic	Cermet[1]	0.58
[1] Ceramic containing sintered metal		

11.2.4 The Physics of Boiling Water

The *Nukiyama curve* shown in Figure 11.2 charts the heat-transfer capability (measured in watts per square centimeter) of a heated surface, submerged in water at vari-

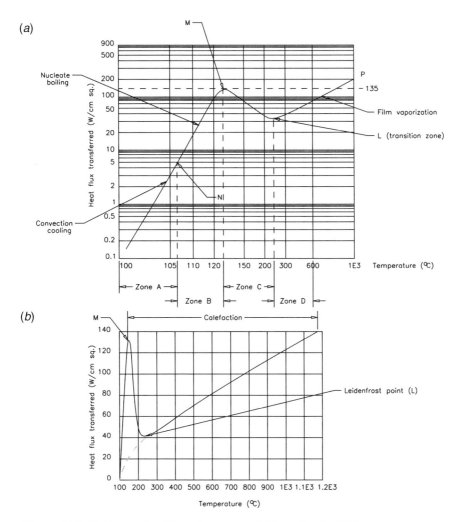

Figure 11.2 Nukiyama heat-transfer curves: (*a*) logarithmic, (*b*) linear.

ous temperatures [2]. The first portion of the curve—*zone A*—indicates that from 100 to about 108°C, heat transfer is a linear function of the temperature differential between the hot surface and the water, reaching a maximum of about 5 W/cm² at 108°C. This linear area is known as the *convection cooling zone*. Boiling takes place in the heated water at some point away from the surface.

From 108 to 125°C—*zone B*—heat transfer increases as the fourth power of ΔT until, at 125°C, it reaches 135 W/cm². This zone is characterized by *nucleate boiling*; that is, individual bubbles of vapor are formed at the hot surface, break away, and travel upward through the water to the atmosphere.

Above 125°C, an unstable portion of the Nukiyama curve is observed, where increasing the temperature of the heated surface actually reduces the unit thermal conductivity. At this area—*zone C*—the vapor partially insulates the heated surface from the water until a temperature of approximately 225°C is reached on the hot surface. At this point—called the *Leidenfrost point*—the surface becomes completely covered with a sheath of vapor, and all heat transfer is accomplished through this vapor cover. Thermal conductivity of only 30 W/cm^2 is realized at this region.

From the Leidenfrost point on through *zone D*—the *film vaporization zone*—heat transfer increases with temperature until at about 1000°C the value of 135 W/cm^2 again is reached.

The linear plot of the Nukiyama curve indicates that zones *A* and *B* are relatively narrow areas and that a heated surface with unlimited heat capacity will tend to pass from zone *A* to zone *D* in a short time. This irreversible superheating is known as *calefaction*. For a cylindrical vacuum tube anode, for example, the passing into total calefaction would not be tolerated, because any unit heat-transfer density above 135 W/cm^2 would result in temperatures above 1000°C, well above the safe limits of the device.

11.3 Application of Cooling Principles

Excessive operating temperature is perhaps the single greatest cause of catastrophic failure in electronic systems. Temperature control is important because the properties of many of the materials used to build individual devices change with increasing temperature. In some applications, these changes are insignificant. In others, however, such changes can result in detrimental effects, leading to—in the worst case—catastrophic failure. Table 11.4 details the variation of electrical and thermal properties with temperature for various substances. Figure 11.3 shows the temperature dependence of thermal parameters of selected electronic packaging materials. Figure 11.4 shows the operating ranges of a variety of heat removal/cooling techniques.

11.3.1 Forced-Air Cooling Systems

Air cooling is the simplest and most common method of removing waste heat from an electronic device or system [2]. The normal flow of cooling air is upward, making it consistent with the normal flow of convection currents. Attention must be given to airflow efficiency and turbulence in the design of a cooling system. Consider the case shown in Figure 11.5. Improper layout has resulted in inefficient movement of air because of circulating thermal currents. The cooling arrangement illustrated in Figure 11.6 provides for the uniform passage of cooling air over the device.

Long-term reliability of an electronic system requires regular attention to the operating environment. Periodic tests and preventive maintenance are important components of this effort. Optimum performance of the cooling system can be achieved only when all elements of the system are functioning properly.

Table 11.4 Variation of Electrical and Thermal Properties of Common Insulators As a Function of Temperature

Parameters		20°C	120°C	260°C	400°C	538°C
Thermal conductivity[1]	99.5% BeO	140	120	65	50	40
	99.5% Al_2O_3	20	17	12	7.5	6
	95.0% Al_2O_3	13.5				
	Glass	0.3				
Power dissipation[2]	BeO	2.4	2.1	1.1	0.9	0.7
Electrical resistivity[3]	BeO	10^{16}	10^{14}	5×10^{12}	10^{12}	10^{11}
	Al_2O_3	10^{14}	10^{14}	10^{12}	10^{12}	10^{11}
	Glass	10^{12}	10^{10}	10^{8}	10^{6}	
Dielectric constant[4]	BeO	6.57	6.64	6.75	6.90	7.05
	Al_2O_3	9.4	9.5	9.6	9.7	9.8
Loss tangent[4]	BeO	0.00044	0.00040	0.00040	0.00049	0.00080

[1] Heat transfer in $Btu/ft^2/hr/°F$
[2] Dissipation in $W/cm/°C$
[3] Resistivity in Ω-cm
[4] At 8.5 GHz

11.3.2 Air-Handling System

The temperature of the intake air supply for an electronic installation is a parameter that is usually under the control of the end user. The preferred cooling air temperature is typically no higher than 75°F, and no lower than the room dew point. The air temperature should not vary because of an oversized air-conditioning system or because of the operation of other pieces of equipment at the facility.

Another convenient method for checking the efficiency of the cooling system over a period of time involves documenting the back pressure that exists within the pressurized compartments of the equipment. This measurement is made with a *manometer*, a simple device that is available from most heating, ventilation, and air-conditioning (HVAC) suppliers. The connection of a simplified manometer to a transmitter output compartment is illustrated in Figure 11.7.

By charting the manometer readings, it is possible to accurately measure the performance of the cooling system over time. Changes resulting from the buildup of small dust particles (*microdust*) may be too gradual to be detected except through back-pressure charting. Deviations from the typical back-pressure value, either higher or lower, could signal a problem with the air-handling system. Decreased input or output com-

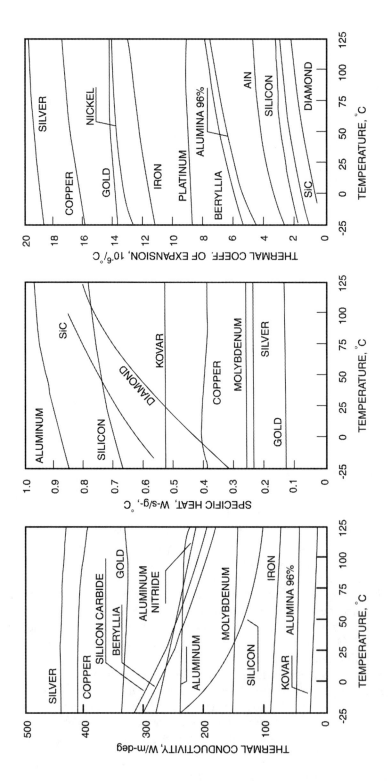

Figure 11.3 Temperature dependence of thermal conductivity k, specific heat c_p, and coefficient of thermal expansion (CTE) β of selected packaging materials. (*From* [3]. *Used with permission.*)

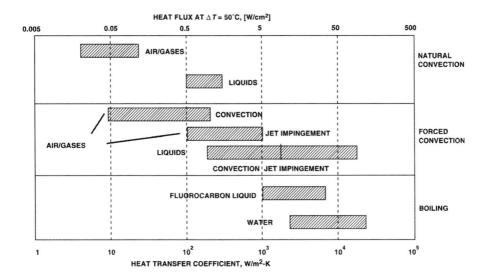

Figure 11.4 Heat transfer coefficient for various heat removal/cooling techniques. (*From* [3]. *Used with permission.*)

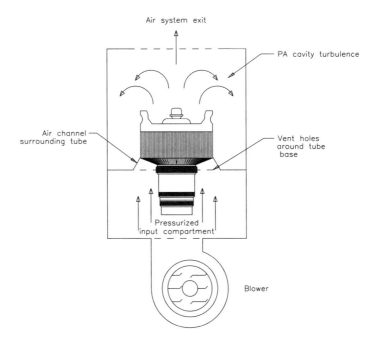

Figure 11.5 A poorly designed cooling system in which circulating air in the output compartment reduces the effectiveness of the heat-removal system.

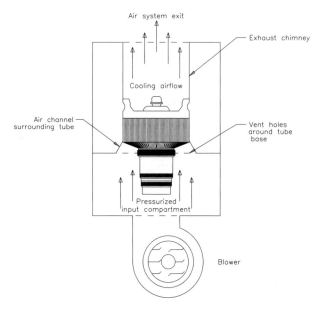

Figure 11.6 The use of a chimney to improve cooling of a power grid tube.

partment back pressure could indicate a problem with the blower motor or an accumu-

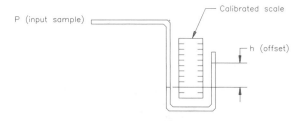

Figure 11.7 A manometer, used to measure air pressure.

lation of dust and dirt on the blades of the blower assembly. Increased back pressure, on the other hand, could indicate dirty or otherwise restricted cooling fins and/or exhaust ducting. Either condition is cause for concern. Cooling problems do not improve with time; they always get worse.

Failure of a pressurized compartment air-interlock switch to close reliably may be an early indication of impending trouble in the cooling system. This situation could be caused by normal mechanical wear or vibration of the switch assembly, or it may signal that the compartment air pressure has dropped. In such a case, documentation of manometer readings will show whether the trouble is caused by a failure of the air pressure switch or a decrease in the output of the air-handling system.

11.4 References

1 Besch, David F., "Thermal Properties," in *The Electronics Handbook*, Jerry C. Whitaker (ed.), CRC Press, Boca Raton, FL, pp. 127–134, 1996.
2. Laboratory Staff, *The Care and Feeding of Power Grid Tubes*, Varian Associates, San Carlos, CA, 1984.
3. Staszak, Zbigniew J., "Heat Management," in *The Electronics Handbook*, Jerry C. Whitaker (ed.), CRC Press, Boca Raton, FL, pp. 1133–1157, 1996.

Semiconductor Devices

12.1 Introduction

A *diode* is a passive electronic device that has a positive anode terminal and a negative cathode terminal and a nonlinear voltage-current characteristic. A *rectifier* is assembled from one or more diodes for the purpose of obtaining a direct current from an alternating current; this term also refers to large diodes used for this purpose. Many types of diodes exist.

Over the years, a great number of constructions and materials have been used as diodes and rectifiers. Rectification in electrolytes with dissimilar electrodes resulted in the *electrolytic rectifier*. The voltage-current characteristic of conduction from a heated cathode in vacuum or low-pressure noble gases or mercury vapor is the basis of vacuum tube diodes and rectifiers. Semiconductor materials such as germanium, silicon, selenium, copper-oxide, or gallium arsenide can be processed to form a pn junction that has a nonlinear diode characteristic. Although all these systems of rectification have seen use, the most widely used rectifier in electronic equipment is the silicon diode. The remainder of this section deals only with these and other silicon two-terminal devices.

12.1.1 The pn Junction

When biased in a reverse direction at a voltage well below breakdown, the diode reverse current is composed of two currents. One current is caused by leakage due to contamination and is proportional to voltage. The intrinsic diode reverse current is independent of voltage but doubles for every 10°C in temperature (approximately). The forward current of a silicon diode is approximately equal to the leakage current multiplied by e ($= 2.718$) raised to the power given by the ratio of forward voltage divided by 26 mV with the junction at room temperature. In practical rectifier calculations, the reverse current is considered to be important in only those cases where a capacitor must hold a charge for a time, and the forward voltage drop is assumed to be constant at 0.7 V, unless a wide range of currents must be considered.

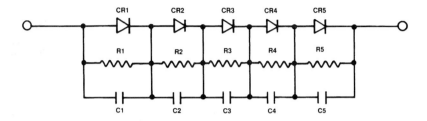

Figure 12.1 A high-voltage rectifier stack.

All diode junctions have a *junction capacitance* that is approximately inversely proportional to the square of the applied reverse voltage. This capacitance rises further with applied forward voltage. When a rectifier carries current in a forward direction, the junction capacitance builds up a charge. When the voltage reverses across the junction, this charge must flow out of the junction, which now has a lower capacitance, giving rise to a current spike in the opposite direction of the forward current. After the *reverse-recovery time*, this spike ends, but interference may be radiated into low-level circuits. For this reason, rectifier diodes are sometimes bypassed with capacitors of about 0.1 mF located close to the diodes. Rectifiers used in high-voltage assemblies use bypass capacitors and high value resistors to reduce noise and equalize the voltage distribution across the individual diodes (Figure 12.1).

Tuning diodes have a controlled reverse capacitance that varies with applied direct tuning voltage. This capacitance may vary over a 2-to-1 to as high as a 10-to-1 range and is used to change the resonant frequency of tuned RF circuits. These diodes find application in receiver circuits.

12.1.2 Zener Diodes and Reverse Breakdown

When the reverse voltage on a diode is increased to a certain *critical voltage*, the reverse leakage current will increase rapidly or *avalanche*. This breakdown or *zener voltage* sets the upper voltage limit a rectifier can experience in normal operation because the peak reverse currents may become as high as the forward currents. Rectifier and other diodes have a rated *peak reverse voltage*, and some rectifier circuits may depend on this reverse breakdown to limit high-voltage spikes that may enter the equipment from the power line. It should also be noted that diode dissipation is very high during these periods.

The reverse breakdown voltage can be controlled in manufacture to a few percent and used to advantage in a class of devices known as *zener diodes*, used extensively in voltage-regulator circuits. It should be noted that the voltage-current curve of a pn junction may go through a region where a *negative resistance* occurs and voltage decreases a small amount while current increases. This condition can give rise to noise and oscillation, which can be minimized by connecting a ceramic capacitor of about 0.02 μF and an electrolytic capacitor of perhaps 100 µF in parallel with the zener diode. Voltage-regulator diodes are available in more than 100 types, covering voltages from 2.4 to 200 V with rated dissipation between 1/4 and 10 W (typical). The forward characteristics of a zener diode usually are not specified but are similar to those of a conventional diode.

Precision voltage or *bandgap reference diodes* make use of the difference in voltage between two diodes carrying a precise ratio of forward currents. Packaged as a two-terminal device including an operational amplifier, these devices produce stable reference voltages of 1.2, 2.5, 5, and 10 V, depending on type.

12.1.3 Current Regulators

The *current regulator diode* is a special class of device used in many small signal applications where constant current is needed. These diodes are *junction field-effect transistors* (FETs) with the gate connected to the source and effectively operated at zero-volt bias. Only two leads are brought out. Current-regulator diodes require a minimum voltage of a few volts for good regulation. Ratings from 0.22 to 4.7 mA are commonly available.

12.1.4 Varistor

Varistors are symmetrical nonlinear voltage-dependent resistors, behaving not unlike two zener diodes connected back to back. The current in a varistor is proportional to applied voltage raised to a power N. These devices are normally made of zinc oxide, which can be produced to have an N factor of 12 to 40. In circuits at normal operating voltages, varistors are nearly open circuits shunted by a capacitor of a few hundred to a few thousand picofarads. Upon application of a high voltage pulse, such as a lightning discharge, they conduct a large current, thereby absorbing the pulse energy in the bulk of the material with only a relatively small increase in voltage, thus protecting the circuit. (See Figure 12.2.) Varistors are available for operating voltages from 10 to 1000 V rms and can handle pulse energies from 0.1 to more than 100 J and maximum peak currents from 20 to 2000 A. Typical applications include protection of power supplies and power-switching circuits, and the protection of telephone and data-communication lines.

12.2 Bipolar Transistors

A bipolar transistor has two pn junctions that behave in a manner similar to that of the diode pn junctions described previously. These junctions are the base-emitter junction and the base-collector junction. In typical use, the first junction would normally have a forward bias, causing conduction, and the second junction would have a reverse bias. If the material of the base were very thick, the flow of electrons into the p-material base junction (of an NPN transistor) would go entirely into the base junction and no current would flow in the reverse-biased collector-base junction.

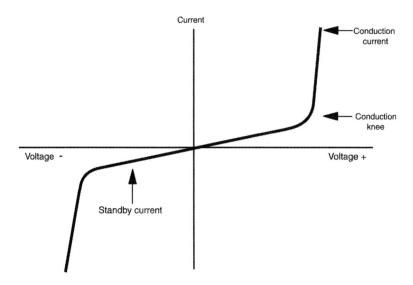

Figure 12.2 The current-vs.-voltage transfer curve for a varistor. Note the *conduction knee*.

If, however, the base junction were quite thin, electrons would diffuse in the semi-conductor crystal lattice into the base-collector junction, having been injected into the base material of the base-emitter junction. This diffusion occurs because an excess electron moving into one location will bump out an electron in the adjacent semicon-ductor molecule, which will bump its neighbor. Thus, a collector current will flow that is nearly as large as the injected emitter current.

The ratio of collector to emitter current is *alpha* (α)or the common-base current gain of the transistor, normally a value a little less than 1.000. The portion of the emitter current not flowing into the collector will flow as a base current in the same direction as the collector current. The ratio of collector current to base current, or *beta* (β), is the conventional current gain of the transistor and may be as low as 5 in power transistors operating at maximum current levels to as high as 5000 in super-beta transistors oper-ated in the region of maximum current gain.

12.2.1 NPN and PNP Transistors

Bipolar transistors are identified by the sequence of semiconductor material going from emitter to collector. NPN transistors operate normally with a positive voltage on the collector with respect to the emitter, with PNP transistors requiring a negative voltage at the collector and the flow of current being internally mostly a flow of *holes* or absent excess electrons in the crystal lattice at locations of flow. (See Figure 12.3.)

Because the diffusion velocity of holes is slower than that of pn electrons, PNP tran-sistors have more junction capacitance and slower speed than NPN transistors of the

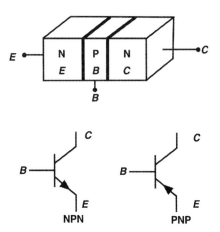

Figure 12.3 Bipolar junction transistor—basic construction and symbols. (*From* [1]. *Used with permission.*)

same size. Holes and electrons in pn junctions are *minority carriers* of electric current as opposed to electrons, which are *majority carriers* and which can move freely in resistors or in the conductive channel of field-effect transistors. Consequently, bipolar transistors are known as *minority carrier devices*.

The most common transistor material is silicon, which permits transistor junction temperatures as high as 200°C. The normal base-emitter voltage is about 0.7 V, and collector-emitter voltage ratings of up to hundreds of volts are available. At room temperature these transistors can dissipate from tens of milliwatts to hundreds of watts with proper heat removal.

Transistors made of gallium arsenide and similar materials are also available for use in microwave and high speed circuits, taking advantage of the high diffusion speeds and low capacitances characteristic of such materials.

12.2.2 Transistor Impedance and Gain

Transistor impedances and gain are normally referred to the *common-emitter* connection, which also results in the highest gain. It is useful to treat transistor parameters first as if the transistor were an ideal device and then to examine degradations resulting from nonideal behavior.

If we assume that the transistor has a fixed current gain, then the collector current is equal to the base current multiplied by the current gain of the transistor, and the emitter current is the sum of both of these currents. Because the collector-base junction is reverse-biased, the output impedance of the ideal transistor is very high.

Actual bipolar transistors suffer degradations from this ideal model. Each transistor terminal may be thought of having a resistor connected in series, although these resistors are actually distributed rather than lumped components. These resistors cause the transistor to have lower gain than predicted and to have a saturation voltage in both input and output circuits. In addition, actual transistors have resistances connected between terminals that cause further reductions in available gain, particularly at low currents and with high load resistances.

In addition to resistances, actual transistors also exhibit stray capacitance between terminals, causing further deviation from the ideal case. These capacitances are—in part—the result of the physical construction of the devices and also the finite diffusion velocities in silicon. The following effects result:

- Transistor current gain decreases with increasing frequency, with the transistor reaching unity current gain at a specific *transition frequency*.

- A feedback current exists from collector to base through the base-collector capacitance.

- Storage of energy in the output capacitance similar to energy storage in a rectifier diode. This stored energy limits the turn-off speed of transistors, a critical factor in certain applications.

12.2.3 Transistor Configurations

Table 12.1 summarizes the most common transistor operating modes. For stages using a single device, the common-emitter arrangement is by far the most common. Power output stages of push-pull amplifiers make use of the common-collector or *emitter-follower* connection. Here, the collector is directly connected to the supply voltage, and the load is connected to the emitter terminal with signal and bias voltage applied to the base terminal. The voltage gain of such a circuit is a little less than 1.0, and the load impedance at the emitter is reflected to the base circuit as if it were increased by the current gain of the transistor.

At high frequencies, the base of a transistor is often grounded for high-frequency signals, which are fed to the emitter of the transistor. With this arrangement, the input impedance of the transistor is low, which is easily matched to radio frequency transmission lines, assisted in part by the minimal capacitive feedback within the transistor.

So far in this discussion, transistor analysis has dealt primarily with the small signal behavior. For operation under large signal conditions, other limitations must be observed. When handling low-frequency signals, a transistor can be viewed as a variable-controlled resistor between the supply voltage and the load impedance. The *quiescent operating point* in the absence of ac signals is usually chosen so that the maximum signal excursions in both positive and negative directions can be handled without limiting resulting from near-zero voltage across the transistor at maximum output current or near-zero current through the transistor at maximum output voltage. This is most critical in class B push-pull amplifiers where first one transistor stage conducts current to the load during part of one cycle and then the other stage conducts during the other part. Similar considerations also apply for distortion reduction considerations.

Limiting conditions also constitute the maximum capabilities of transistors under worst-case conditions of supply voltage, load impedance, drive signal, and temperature

Table 12.1 Basic Amplifier Configurations of Bipolar Transistors (*From Rohde, U., et al., Communications Receivers, 2nd ed., McGraw-Hill, New York, NY, 1996. Reproduced with permission of the McGraw-Hill Companies.*)

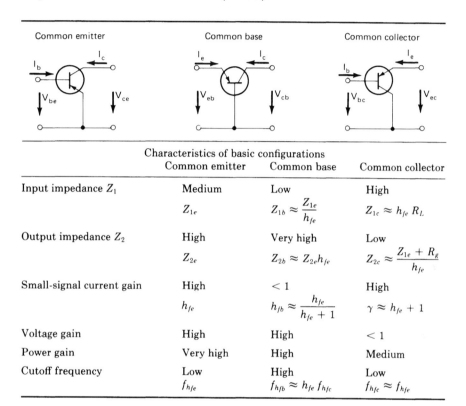

| | Characteristics of basic configurations | | |
	Common emitter	Common base	Common collector
Input impedance Z_1	Medium	Low	High
	Z_{1e}	$Z_{1b} \approx \dfrac{Z_{1e}}{h_{fe}}$	$Z_{1c} \approx h_{fe}\,R_L$
Output impedance Z_2	High	Very high	Low
	Z_{2e}	$Z_{2b} \approx Z_{2e}h_{fe}$	$Z_{2c} \approx \dfrac{Z_{1e}+R_g}{h_{fe}}$
Small-signal current gain	High	< 1	High
	h_{fe}	$h_{fb} \approx \dfrac{h_{fe}}{h_{fe}+1}$	$\gamma \approx h_{fe}+1$
Voltage gain	High	High	< 1
Power gain	Very high	High	Medium
Cutoff frequency	Low	High	Low
	f_{hfe}	$f_{hfb} \approx h_{fe}\,f_{hfc}$	$f_{hfc} \approx f_{hfe}$

consistent with safe operation. In no case should the maximum voltage across a transistor ever be exceeded.

12.2.4 Switching and Inductive-Load Ratings

When using transistors for driving relays, deflection yokes of cathode ray tubes, or any other inductive or resonant load, current in the inductor will tend to flow in the same direction, even if interrupted by the transistor. The resultant voltage spike caused by the collapse of the magnetic field can destroy the switching device unless it is designed to handle the energy of these voltage excursions. The manufacturers of power semiconductors have special transistor types and application information relating to inductive switching circuits. In many cases, the use of protection diodes are sufficient.

Transistors are often used to switch currents into a resistive load. The various junction capacitances are voltage-dependent in the same manner as the capacitance of tuning diodes that have maximum capacitance at forward voltages, becoming less at zero voltage and lowest at reverse voltages. These capacitances and the various resistances combine into the switching delay times for turn-on and turn-off functions. If the transistor is prevented from being saturated when turned on, shorter delay times will occur for nonsaturated switching than for saturated switching. These delay times are of importance in the design of switching amplifiers or D/A converters.

12.2.5 Noise

Every resistor creates noise with equal and constant energy for each hertz of bandwidth, regardless of frequency. A useful number to remember is that a 1000-Ω resistor at room temperature has an open-circuit output noise voltage of 4 nanovolts per root-hertz. This converts to 40 nV in a 100-Hz bandwidth or 400 μV in a 10-kHz bandwidth.

Bipolar transistors also create noise in their input and output circuits, and every resistor in the circuit also contributes its own noise energy. The noise of a transistor is effectively created in its input junction, and all transistor noise ratings are referred to it.

In an ideal bipolar transistor, the voltage noise at the base is created by an equivalent resistor that has a value of twice the transistor input conductance at its emitter terminal, and the current noise is created by a resistor that has the value of twice the transistor input conductance at its input terminal. This means that the current noise energy is less at the base terminal of a common-emitter stage by the current gain of the transistor when compared to the current noise at the input of a grounded-base stage.

The highest signal-to-noise ratio in an amplifier can be achieved when the resistance of the signal source is equal to the ratio of amplifier input noise voltage and input noise current, and the reactive impedances have been tuned to zero. Audio frequency amplifiers usually cannot be tuned, and minimum noise may be achieved by matching transformers or by bias current adjustment of the input transistor. With low source impedances, the optimum may not be reached economically, and the equipment must then be designed to have an acceptable input noise voltage.

Practical transistors are not ideal from the standpoint of noise performance. All transistors show a voltage and current noise energy that increases inversely with frequency. At a *corner frequency* this noise will become independent of frequency. Very low noise transistors may have a corner frequency as low as a few hertz, and ordinary high-frequency devices may have a corner frequency well above the audio frequency range. Transistor noise may also be degraded by operating a transistor at more than a few percent of its maximum current rating. Poor transistor design or manufacturing techniques can result in transistors that exhibit "popcorn" noise, so named after the audible characteristics of a random low-level switching effect.

The noise level produced by thermal noise sources is not necessarily large, however, because signal power may also be low, it is usually necessary to amplify the source signal. Because noise is combined with the source signal, both are then amplified, with more noise added at each successive stage of amplification. Noise can, thus, become a noticeable phenomenon (Figure 12.4).

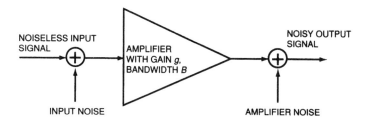

Figure 12.4 A block diagram modeling how noise is introduced to a signal during amplification. (*From* [1]. *Used with permission.*)

12.3 Field-Effect Transistors

Field-effect transistors (FETs) have a conducting channel terminated by source and drain electrodes and a gate terminal that effectively widens or narrows the channel by the electric field between the gate and each portion of the channel. No gate current is required for steady-state control.

Current flow in the channel is by majority carriers only, analogous to current flow in a resistor. The onset of conduction is not limited by diffusion speeds but by the electric field accelerating the charged electrons.

The input impedance of an FET is a capacitance. Because of this, electrostatic charges during handling may reach high voltages that are capable of breaking down gate insulation.

FETs for common applications use silicon as the semiconducting material. Field-effect transistors are made both in p-channel and n-channel configurations. An n-channel FET has a positive drain voltage with respect to the source voltage, and a positive increase in gate voltage causes an increase in channel current. Reverse polarities exist for p-channel devices. (See Figure 12.5).

An n-channel FET has a drain voltage that is normally positive, and a positive increase in gate-to-source voltage increases drain current and *transconductance*. In single-gate field-effect transistors, drain and source terminals may often be interchanged without affecting circuit performance; however, power handling and other factors may be different. Such an interchange is not possible when two FETs are interconnected internally to form a dual-gate cascode-connected FET, or matched pairs, or when channel conductance is controlled by gates on two sides of the channel as in insulated-gate FETs.

12.3.1 FET Impedance and Gain

The input impedance of a field-effect transistor is usually quite high, and is primarily capacitive. The input capacitance consists of the gate-source capacitance in parallel

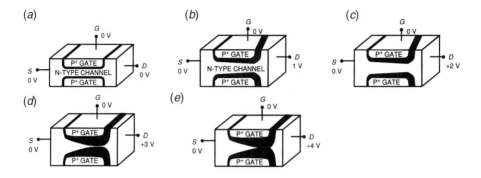

Figure 12.5 Junction FET (JFET) operational characteristics: (*a*) uniform channel from drain to source, (*b*) depletion region wider at the drain end, (*c*) depletion region significantly wider at the drain, (*d*) channel near pinchoff, (*e*) channel at pinchoff. (*From* [1]. *Used with permission.*)

with the gate-drain capacitance multiplied by the stage gain + 1, assuming the FET has its source at ac-ground potential.

The output impedance of a common-source FET is also primarily capacitive as long as the drain voltage is above a critical value, which, for a junction-gate FET, is equal to the sum of the *pinch-off voltage* and gate-bias voltage. When the pinch-off voltage is applied between the gate and source terminals, the drain current is nearly shut off (the channel is pinched off). Actual FETs have a high drain resistance in parallel with this capacitance. At low drain voltages near zero volts, the drain impedance of an ideal FET is a resistor reciprocal in value to the transconductance of the FET in series with the residual end resistances between the source and drain terminals and the conducting FET channel. This permits an FET to be used as a variable resistor in circuits controlling analog signals.

At drain voltages between zero and the critical voltage, the drain current will increase with both increasing drain voltage and increasing gate voltage. This factor will cause increased saturation voltages in power amplifier circuits when compared to circuits with bipolar transistors.

Table 12.2 summarizes the basic FET amplifier configurations.

12.4 Integrated Circuits

An *integrated circuit* (IC) is a combination of circuit elements that are interconnected and formed on and in a continuous substrate material. Usually, an integrated circuit is monolithic and formed by steps that produce semiconductor elements along with resistors and capacitors. A hybrid integrated circuit contains silicon chips along with circuit elements partially formed on the substrate.

Table 12.2 Basic Field-Effect Transistor Amplifier Configurations and Operating Characteristics (*From Rohde, U., et al., Communications Receivers, 2nd ed., McGraw-Hill, New York, NY, 1996. Reproduced with permission of the McGraw-Hill Companies.*)

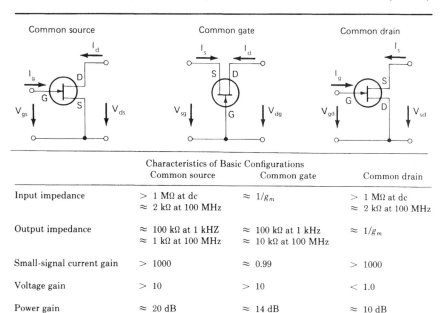

| Characteristics of Basic Configurations | | |
Common source	Common gate	Common drain
Input impedance $> 1 \text{ M}\Omega$ at dc $\approx 2 \text{ k}\Omega$ at 100 MHz	$\approx 1/g_m$	$> 1 \text{ M}\Omega$ at dc $\approx 2 \text{ k}\Omega$ at 100 MHz
Output impedance $\approx 100 \text{ k}\Omega$ at 1 kHZ $\approx 1 \text{ k}\Omega$ at 100 MHz	$\approx 100 \text{ k}\Omega$ at 1 kHz $\approx 10 \text{ k}\Omega$ at 100 MHz	$\approx 1/g_m$
Small-signal current gain > 1000	≈ 0.99	> 1000
Voltage gain > 10	> 10	< 1.0
Power gain $\approx 20 \text{ dB}$	$\approx 14 \text{ dB}$	$\approx 10 \text{ dB}$
Cutoff frequency $g_m/2\pi C_{gs}$	$g_m/2\pi C_{ds}$	$g_m/2\pi C_{gd}$

The circuit elements formed in integrated circuits are more closely matched to each other than separately selected components, and these elements are in intimate thermal contact with each other. The circuit configurations used in integrated circuits take advantage of this matching and thermal coupling.

ICs are classified according to their levels of complexity:

- Small-scale integration (SSI)

- Medium-scale integration (MSI)

- Large-scale integration (LSI)

- Very large-scale integration (VLSI)

Devices are further classified according to the technology employed for their fabrication: bipolar, N metal oxide semiconductor (NMOS), complementary metal oxide semiconductor (CMOS), and so on.

12.4.1 Digital Integrated Circuits

The basis of digital circuits is the logic gate that produces a high (or 1) or low (or 0) logic-level output with the proper combination of logic-level inputs. A number of these gates are combined to form a digital circuit that is part of the hardware of computers or controllers of equipment or other circuits. A digital circuit may be extremely complex, containing up to more than 1,000,000 gates.

Bipolar and field-effect transistors are the active elements of digital integrated circuits, divided into families such as *transistor-transistor logic* (TTL), *high-speed complementary metal-oxide-gate semiconductor* (HCMOS), and many others. Special families include memories, microprocessors, and interface circuits between transmission lines and logic circuits. Thousands of digital integrated circuit types in tens of families have been produced.

12.4.2 Linear Integrated Circuits

Linear integrated circuits are designed to process linear signals in their entirety or in part, as opposed to *digital circuits* that process logic signals only. Major classes of linear integrated circuits include operational amplifiers, voltage regulators, digital-to-analog and analog-to-digital circuits, circuits for consumer electronic equipment and communications equipment, power control circuits, and others not as easily classified.

12.5 References

1. Whitaker, Jerry C. (ed.), *The Electronics Handbook*, CRC Press, Boca Raton, FL, 1996.
2. Rhode, U., J. Whitaker, and T. Bucher, *Communications Receivers*, 2nd ed., McGraw-Hill, New York, NY, 1996.

12.6 Bibliography

Benson, K. Blair, and Jerry C. Whitaker, *Television and Audio Handbook for Technicians and Engineers*, McGraw-Hill, New York, NY, 1990.
Benson, K. Blair, *Audio Engineering Handbook*, McGraw-Hill, New York, NY, 1988.
Whitaker, Jerry C., and K. Blair Benson (eds.), *Standard Handbook of Video and Television Engineering*, McGraw-Hill, New York, NY, 2000.
Whitaker, Jerry C., *Television Engineers' Field Manual*, McGraw-Hill, New York, NY, 2000.

12.7 Tabular Data

Table 12.3 Semiconducting Properties of Selected Materials (*From* [1]. *Used with permission.*)

Substance	Minimum energy gap, eV R.T.	0 K	$\frac{dE_g}{dT}$ ×10⁴, eV/°C	$\frac{dE_g}{dP}$ ×10⁶, eV · cm²/kg	Density of states electron effective mass m_{d_n} (m_o)	Electron mobility and temperature dependence μ_n, cm²/V·s	$-x$	Density of states hole effective mass m_{d_p}, (m_o)	Hole mobility and temperature dependence μ_p, cm²/V · s	$-x$
Si	1.107	1.153	−2.3	−2.0	1.1	1,900	2.6	0.56	500	2.3
Ge	0.67	0.744	−3.7	±7.3	0.55	3,800	1.66	0.3	1,820	2.33
αSn	0.08	0.094	−0.5		0.02	2,500	1.65	0.3	2,400	2.0
Te	0.33				0.68	1,100		0.19	560	
III–V Compounds										
AlAs	2.2	2.3				1,200			420	
AlSb	1.6	1.7	−3.5	−1.6	0.09	200	1.5	0.4	500	1.8
GaP	2.24	2.40	−5.4	−1.7	0.35	300	1.5	0.5	150	1.5
GaAs	1.35	1.53	−5.0	+9.4	0.068	9,000	1.0	0.5	500	2.1
GaSb	0.67	0.78	−3.5	+12	0.050	5,000	2.0	0.23	1,400	0.9
InP	1.27	1.41	−4.6	+4.6	0.067	5,000	2.0		200	2.4
InAs	0.36	0.43	−2.8	+8	0.022	33,000	1.2	0.41	460	2.3
InSb	0.165	0.23	−2.8	+15	0.014	78,000	1.6	0.4	750	2.1
II–VI Compounds										
ZnO	3.2		−9.5	+0.6	0.38	180	1.5			
ZnS	3.54		−5.3	+5.7		180			5(400°C)	
ZnSe	2.58	2.80	−7.2	+6		540			28	
ZnTe	2.26			+6		340			100	
CdO	2.5 ± 0.1		−6		0.1	120				
CdS	2.42		−5	+3.3	0.165	400		0.8		
CdSe	1.74	1.85	−4.6		0.13	650	1.0	0.6		
CdTe	1.44	1.56	−4.1	+8	0.14	1,200		0.35	50	
HgSe	0.30				0.030	20,000	2.0			
HgTe	0.15		−1		0.017	25,000		0.5	350	
Halite Structure Compounds										
PbS	0.37	0.28	+4		0.16	800		0.1	1,000	2.2
PbSe	0.26	0.16	+4		0.3	1,500		0.34	1,500	2.2
PbTe	0.25	0.19	+4	−7	0.21	1,600		0.14	750	2.2
Others										
ZnSb	0.50	0.56			0.15	10				1.5
CdSb	0.45	0.57	−5.4		0.15	300			2,000	1.5
Bi₂S₃	1.3					200			1,100	
Bi₂Se₃	0.27					600			675	
Bi₂Te₃	0.13		−0.95		0.58	1,200	1.68	1.07	510	1.95
Mg₂Si		0.77	−6.4		0.46	400	2.5		70	
Mg₂Ge		0.74	−9			280	2		110	
Mg₂Sn	0.21	0.33	−3.5		0.37	320			260	
Mg₃Sb₂		0.32				20			82	
Zn₃As₂	0.93					10	1.1		10	
Cd₃As₂	0.55				0.046	100,000	0.88			
GaSe	2.05		3.8						20	
GaTe	1.66	1.80	−3.6			14	−5			
InSe	1.8					9000				
TlSe	0.57		−3.9		0.3	30		0.6	20	1.5
CdSnAs₂	0.23				0.05	25,000	1.7			
Ga₂Te₃	1.1	1.55	−4.8							
α-In₂Te₃	1.1	1.2			0.7				50	1.1
β-In₂Te₃	1.0								5	
Hg₅In₂Te₈	0.5								11,000	
SnO₂									78	

13

Analog Circuits

13.1 Introduction

Amplifiers are the functional building blocks of electronic systems, and each of these building blocks typically contains several amplifier stages coupled together. An amplifier may contain its own power supply or require one or more external sources of power. The active component of each amplifier stage is usually a transistor or an FET. Other amplifying components, such as vacuum tubes, can also be used in amplifier circuits if the operating power and/or frequency of the application demands it.

13.2 Single-Stage Transistor/FET Amplifier

The single-stage amplifier can best be described using a single transistor or FET connected as a *common-emitter* or *common-source* amplifier, using an npn transistor (Figure 13.1a) or an n-channel FET (Figure 13.1b) and treating pnp transistors or p-channel FET circuits by simply reversing the current flow and the polarity of the voltages.

At zero frequency (dc) and at low frequencies, the transistor or FET amplifier stage requires an input voltage E_1 equal to the sum of the input voltages of the device (the transistor V_{be} or FET V_{gs}) and the voltage across the resistance R_e or R_s between the common node (ground) and the emitter or source terminal. The input current I_1 to the amplifier stage is equal to the sum of the current through the external resistor connected between ground and the base or gate and the base current I_b or gate current I_g drawn by the device. In most FET circuits, the gate current may be so small that it can be neglected, while in transistor circuits the base current I_b is equal to the collector current I_c divided by the current gain beta of the transistor. The input resistance R_1 to the amplifier stage is equal to the ratio of input voltage E_1 to input current I_1.

The input voltage and the input resistance of an amplifier stage increases as the value of the emitter or source resistor becomes larger.

The output voltage E_2 of the amplifier stage, operating without any external load, is equal to the difference of supply voltage V+ and the product of collector or drain load

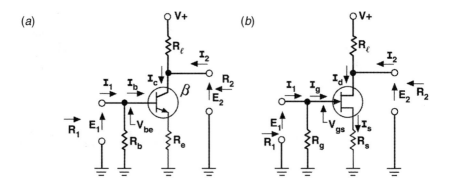

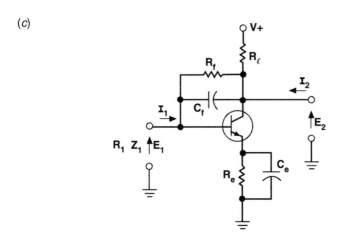

Figure 13.1 Single-stage amplifier circuits: (*a*) common-emitter NPN, (*b*) common-source n-channel FET, (*c*) single-stage with current and voltage feedback.

resistor R_1 and collector current I_c or drain current I_d. An external load will cause the device to draw an additional current I_2, which increases the device output current.

As long as the collector-to-emitter voltage is larger than the saturation voltage of the transistor, collector current will be nearly independent of supply voltage. Similarly, the drain current of an FET will be nearly independent of drain-to-source voltage as long as this voltage is greater than an equivalent saturation voltage. This saturation voltage is approximately equal to the difference between gate-to-source voltage and *pinch-off* voltage, the latter being the bias voltage that causes nearly zero drain current. In some FET data sheets, the pinch-off voltage is referred to as the *threshold voltage*. At lower supply voltages, the collector or drain current will become less until it reaches zero, when the drain-to-source voltage is zero or the collector-to-emitter voltage has a very small reverse value.

The output resistance R_2 of a transistor or FET amplifier stage is—in effect—the parallel combination of the collector or drain load resistance and the series connection of two resistors, consisting of R_e or R_s, and the ratio of collector-to-emitter voltage and collector current or the equivalent drain-to-source voltage and drain current. In actual devices, an additional resistor, the relatively large output resistance of the device, is connected in parallel with the output resistance of the amplifier stage.

The collector current of a single-stage transistor amplifier is equal to the base current multiplied by the current gain of the transistor. Because the current gain of a transistor may be specified as tightly as a two-to-one range at one value of collector current, or it may have just a minimum value, knowledge of the input current is usually not quite sufficient to specify the output current of a transistor.

13.2.1 Impedance and Gain

The input impedance is the ratio of input voltage to input current, and the output impedance is the ratio of output voltage to output current. As the input current increases, the output current into the external output load resistor will increase by the current amplification factor of the stage. The output voltage will decrease because the increased current flows from the collector or drain voltage supply source into the collector or drain of the device. Therefore, the voltage amplification is a negative number having the magnitude of the ratio of output voltage change to input voltage change.

The magnitude of voltage amplification is often calculated as the product of transconductance G_m of the device and the load resistance value. This can be done as long as the emitter or source resistance is zero or the resistor is bypassed with a capacitor that effectively acts as a short circuit for all signal changes of interest but allows the desired bias currents to flow through the resistor. In a bipolar transistor, the transconductance is approximately equal to the emitter current multiplied by 39, which is the charge of a single electron divided by the product of Boltzmann's constant and absolute temperature in degrees Kelvin. In a field-effect transistor, this value will be less and usually proportional to the input-bias voltage, with reference to the pinch-off voltage.

The power gain of the device is the ratio of output power to input power, often expressed in decibels. Voltage gain or current gain can be stated in decibels but must be so marked.

The resistor in series with the emitter or source causes negative feedback of most of the output current, which reduces the voltage gain of the single amplifier stage and raises its input impedance (Figure 13.1c). When this resistor R_e is bypassed with a capacitor C_e, the amplification factor will be high at high frequencies and will be reduced by approximately 3 dB at the frequency where the impedance of capacitor C_e is equal to the emitter or source input impedance of the device, which in turn is approximately equal to the inverse of the transconductance G_m of the device (Figure 13.2a). The gain of the stage will be approximately 3 dB higher than the dc gain at the frequency where the impedance of the capacitor is equal to the emitter or source resistor. These simplifica-

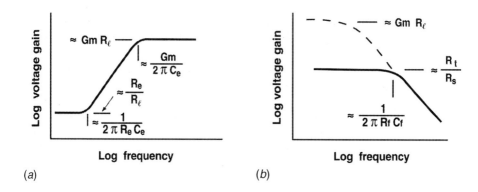

Figure 13.2 Feedback amplifier voltage gains: (a) current feedback, (b) voltage feedback.

tions hold in cases where the product of transconductance and resistance values are much larger than 1.

A portion of the output voltage may also be fed back to the input, which is the base or gate terminal. This resistor R_f will lower the input impedance of the single amplifier stage, reduce current amplification, reduce output impedance of the stage, and act as a supply voltage source for the base or gate. This method is used when the source of input signals, and internal resistance R_s, is coupled with a capacitor to the base or gate and a group of devices with a spread of current gains, transconductances, or pinch-off voltages must operate with similar amplification in the same circuit. If the feedback element is also a capacitor C_f, high-frequency current amplification of the stage will be reduced by approximately 3 dB when the impedance of the capacitor is equal to the feedback resistor R_f and voltage gain of the stage is high (Figure 13.2b). At still higher frequencies, amplification will decrease at the rate of 6 dB per octave of frequency. It should be noted that the base-collector or gate-drain capacitance of the device has the same effect of limiting high-frequency amplification of the stage; however, this capacitance becomes larger as the collector-base or drain-gate voltage decreases.

Feedback of the output voltage through an impedance lowers the input impedance of an amplifier stage. Voltage amplification of the stage will be affected only as this lowered input impedance loads the source of input voltage. If the source of input voltage has a finite source impedance and the amplifier stage has very high voltage amplification and reversed phase, the effective amplification for this stage will approach the ratio of feedback impedance to source impedance and also have reversed phase.

13.2.2 Common-Base or Common-Gate Connection

For the common-base or common-gate case, voltage amplification is the same as in the common-emitter or common-source connection; however, the input impedance is

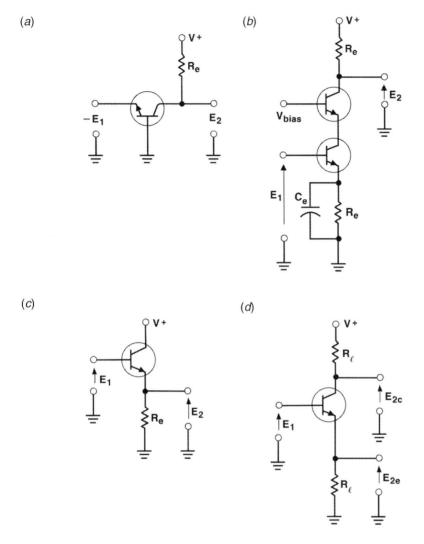

Figure 13.3 Transistor amplifier circuits: (a) common-base NPN, (b) cascode NPN, (c) common-collector NPN emitter follower, (d) split-load phase inverter.

approximately the inverse of the transconductance of the device. (See Figure 13.3a.) As a benefit, high-frequency amplification will be less affected because of the relatively lower emitter-collector or source-drain capacitance and the relatively low input impedance. This is the reason why the *cascade connection* (Figure 13.3b) of a common-emitter amplifier stage driving a common-base amplifier stage exhibits nearly the dc amplification of a common-emitter stage with the wide bandwidth of a common-base stage. Another advantage of the common-base or common-gate amplifier

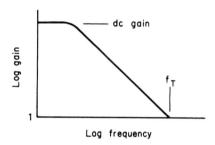

Figure 13.4 Amplitude-frequency response of a common-emitter or common-source amplifier.

stage is stable amplification at very high frequencies and ease of matching to RF transmission-line impedances, usually 50 to 75 Ω.

13.2.3 Common-Collector or Common-Drain Connection

The voltage gain of a transistor or FET is slightly below 1.0 for the common-collector or common-drain configuration. However, the input impedance of a transistor so connected will be equal to the value of the load impedance multiplied by the current gain of the device plus the inverse of the transconductance of the device (Figure 13.3c). Similarly, the output impedance of the stage will be the impedance of the source of signals divided by the current gain of the transistor plus the inverse of the transconductance of the device.

When identical resistors are connected between the collector or drain and the supply voltage and the emitter or source and ground, an increase in base or gate voltage will result in an increase of emitter or source voltage that is nearly equal to the decrease in collector or drain voltage. This type of connection is known as the *split-load phase inverter*, useful for driving push-pull amplifiers, although the output impedances at the two output terminals are unequal (Figure 13.3d).

The current gain of a transistor decreases at high frequencies as the emitter-base capacitance shunts a portion of the transconductance, thereby reducing current gain until it reaches a value of 1 at the transition frequency of the transistor (Figure 13.4). From this figure it can be seen that the output impedance of an emitter-follower or common-collector stage will increase with frequency, having the effect of an inductive source impedance when the input source to the stage is resistive. If the source impedance is inductive, as it might be with cascaded-emitter followers, the output impedance of such a combination can be a negative value at certain high frequencies and be a possible cause of amplifier oscillation. Similar considerations also apply to common-drain FET stages.

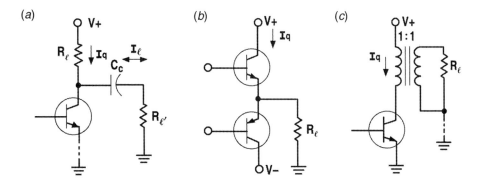

Figure 13.5 Output load-coupling circuits: (*a*) ac-coupled, (*b*) series-parallel ac, push-pull half-bridge, (*c*) single-ended transformer-coupled.

13.2.4 Bias and Large Signals

When large signals have to be handled by a single-stage amplifier, distortion of the signals introduced by the amplifier itself must be considered. Although feedback can reduce distortion, it is necessary to ensure that each stage of amplification operates in a region where normal signals will not cause the amplifier stage to operate with nearly zero voltage drop across the device or to operate the device with nearly zero current during any portion of the cycle of the signal. Although described primarily with respect to a single-device-amplifier stage, the same holds true for any amplifier stage with multiple devices, except that here at least one device must be able to control current flow in the load without being *saturated* (nearly zero voltage drop) or *cut off* (nearly zero current).

If the single-device-amplifier load consists of the collector or drain load resistor only, the best operating point should be chosen so that in the absence of a signal, one-half of the supply voltage appears as a *quiescent voltage* across the load resistor R_l. If an additional resistive load R_l is connected to the output through a coupling capacitor C_c (Figure 13.5a), the maximum peak load current I_l in one direction is equal to the difference between quiescent current I_l of the stage and the current that would flow if the collector resistor and the external load resistor were connected in series across the supply voltage. In the other direction, the maximum load current is limited by the quiescent voltage across the device divided by the load resistance. The quiescent current flows in the absence of an alternating signal and is caused by bias voltage or current only. Because most audio frequency signals (and others, depending upon the application) have positive and negative peak excursions of equal probability, it is usually advisable to have the two peak currents be equal. This can be accomplished by increasing the quiescent current as the external load resistance decreases.

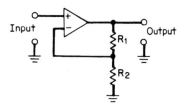

Figure 13.6 Operational amplifier with unbalanced input and output signals and a fixed level of feedback to set the voltage gain V_g, which is equal to $(1 + R)/R$.

When several devices contribute current into an external load resistor (Figure 13.5b), one useful strategy is to set bias currents so that the sum of all transconductances remains as constant as practical, which means a design for minimum distortion. This operating point for one device is near one-quarter the peak device current for push-pull FET stages and at a lesser value for bipolar push-pull amplifiers.

When the load resistance is coupled to the single-device-amplifier stage with a transformer (Figure 13.5c), the optimum bias current should be nearly equal to the peak current that would flow through the load impedance at the transformer with a voltage drop equal to the supply voltage.

13.3 Operational Amplifiers

An operational amplifier is a circuit (device) with a pair of differential input terminals that have very high gain to the output for differential signals of opposite phase at each input and relatively low gain for common-mode signals that have the same phase at each input (see Figure 13.6). An external feedback network between the output and the minus (–) input and ground or signal sets the circuit gain, with the plus (+) input at signal or ground level. Most operational amplifiers require a positive and a negative power supply voltage. One to eight operational amplifiers may be contained on one substrate mounted in a plastic, ceramic, or hermetically sealed metal-can package. Operational amplifiers may require external capacitors for circuit stability or may be internally compensated. Input stages may be field-effect transistors for high input impedance or bipolar transistors for low-offset voltage and low-voltage noise. Available types of operational amplifiers number in the hundreds. Precision operational amplifiers generally have more tightly controlled specifications than general-purpose types. Table 13.1 details the most common application and their functional parameters.

Table 13.1 Common Op-Amp Circuits (*From* [1]. *Used with permission.*)

No.	Type of Circuit	Schematic	Circuit Gain or Variable of Interest	Input Resistance or Input Currents or Voltages	Special Requirements or Remarks
1	Noninverting amplifier		$\dfrac{v_O}{v_i} = 1 + \dfrac{R_2}{R_1}$	$R_{in} = \infty$ (ideally)	
2	Buffer		$\dfrac{v_O}{v_i} = 1$	$R_{in} = \infty$ (ideally)	Special case of circuit 1
3	Difference amplifier		$v_O = \dfrac{R_2}{R_1}(v_2 - v_1)$	$i_1 = \dfrac{v_1 - v_2 \dfrac{R_b}{(R_a + R_b)}}{R_1}$ $i_2 = \dfrac{v_2}{R_a + R_b}$	$\dfrac{R_1}{R_2} = \dfrac{R_a}{R_b}$
4	Adder		$v_O = -\left\{ v_1 \dfrac{R_f}{R_1} + v_2 \dfrac{R_f}{R_2} + \cdots + v_n \dfrac{R_f}{R_n} \right\}$	$i_1 = \dfrac{v_1}{R_1}$ $i_2 = \dfrac{v_2}{R_2}$ $\vdots$ $i_n = \dfrac{v_n}{R_n}$	
5	Variable gain circuit		$\dfrac{v_O}{v_i} = (2Kx - K)$ $0 \le x \le 1,\ K > 1$	$i = \dfrac{v_i}{R_3} + \dfrac{Kv_i(1 - x)}{R}$	Potentiometer R_3 adjusts the gain over the range $-K$ to $+K$.

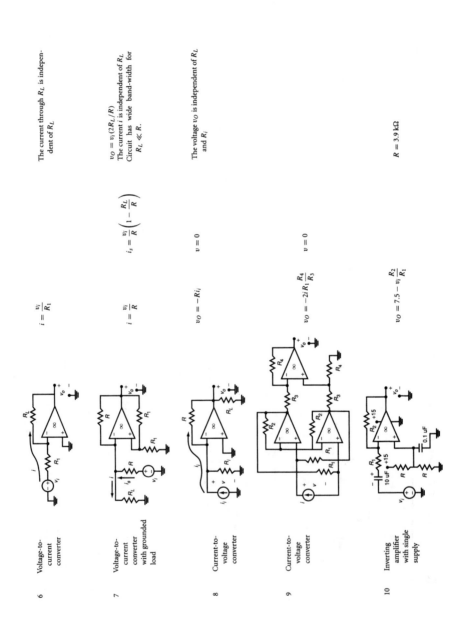

6 Voltage-to-current converter

$$i = \frac{v_i}{R_1}$$

The current through R_L is independent of R_L.

7 Voltage-to-current converter with grounded load

$$i = \frac{v_i}{R}$$

$$i_s = \frac{v_i}{R}\left(1 - \frac{R_L}{R}\right)$$

$$v_O = v_i(2R_L/R)$$

The current i is independent of R_L. Circuit has wide band-width for $R_L \ll R$.

8 Current-to-voltage converter

$$v_O = -Ri_i$$

$$v = 0$$

The voltage v_O is independent of R_L and R_i.

9 Current-to-voltage converter

$$v_O = -2iR_1\frac{R_4}{R_3}$$

$$v = 0$$

10 Inverting amplifier with single supply

$$v_O = 7.5 - v_i\frac{R_2}{R_1}$$

$$R = 3.9\text{ k}\Omega$$

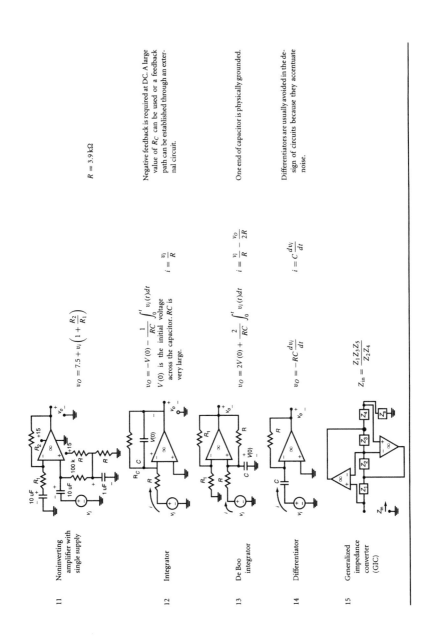

11 Noninverting amplifier with single supply

$$v_O = 7.5 + v_i\left(1 + \frac{R_2}{R_1}\right)$$

$R = 3.9\,k\Omega$

12 Integrator

$$v_O = -V(0) - \frac{1}{RC}\int_0^t v_i(t)\,dt$$

$V(0)$ is the initial voltage across the capacitor. RC is very large.

$$i = \frac{v_i}{R}$$

Negative feedback is required at DC. A large value of R_C can be used or a feedback path can be established through an external circuit.

13 De Boo integrator

$$v_O = 2V(0) + \frac{2}{RC}\int_0^t v_i(t)\,dt$$

$$i = \frac{v_i}{R} - \frac{v_O}{2R}$$

One end of capacitor is physically grounded.

14 Differentiator

$$v_O = -RC\frac{dv_i}{dt}$$

$$i = C\frac{dv_i}{dt}$$

Differentiators are usually avoided in the design of circuits because they accentuate noise.

15 Generalized impedance converter (GIC)

$$Z_{in} = \frac{Z_1 Z_3 Z_5}{Z_2 Z_4}$$

The input-bias current of an operational amplifier is the average current drawn by each of the two inputs, + and –, from the input and feedback circuits. Any difference in dc resistance between the circuits seen by the two inputs multiplied by the input-bias current will be amplified by the circuit gain and become an *output-offset voltage*. The *input-offset current* is the difference in bias current drawn by the two inputs, which, when multiplied by the sum of the total dc resistance in the input and feedback circuits and the circuit gain, becomes an additional output-offset voltage. The *input-offset voltage* is the internal difference in bias voltage within the operational amplifier, which, when multiplied by the circuit gain, becomes an additional output-offset voltage. If the normal input voltage is zero, the open-circuit output voltage is the sum of the three offset voltages.

13.4 References

1. Whitaker, Jerry C. (ed.), *The Electronics Handbook*, CRC Press, Boca Raton, FL, 1996.

13.5 Bibliography

Benson, K. Blair (ed.), *Audio Engineering Handbook*, McGraw-Hill, New York, NY, 1988.

Fink, Donald (ed.), *Electronics Engineers' Handbook*, McGraw-Hill, New York, NY, 1982.

Whitaker, Jerry C., and K. Blair Benson (eds.), *Standard Handbook of Video and Television Engineering*, 3rd ed., McGraw-Hill, New York, NY, 2000.

Whitaker, Jerry C. (ed.), *Video and Television Engineer's Field Manual*, McGraw-Hill, New York, NY, 2000.

14

Logic Concepts and Devices

14.1 Introduction

Digital signals differ from analog in that only two steady-state levels are used for the storage, processing, and/or transmission of information. The definition of a digital transmission format requires specification of the following parameters:

- The type of information corresponding to each of the binary levels

- The frequency or rate at which the information is transmitted as a bilevel signal

The digital coding of signals for most applications uses a scheme of binary numbers in which only two digits, 0 and 1, are used. This is called a *base*, or *radix*, of 2. It is of interest that systems of other bases are used for some more complex mathematical applications, the principal ones being *octal* (8) and *hexadecimal* (16). Table 14.1 compares the decimal, binary, and octal counting systems. Note that numbers in the decimal system are equal to the number of items counted, if used for a tabulation.

14.1.1 Analog-to-Digital (A/D) Conversion

Because the inputs and outputs of devices that interact with humans usually deal in analog values, the inputs must be represented as numbered sequences corresponding to the analog levels of the signal. This is accomplished by sampling the signal levels and assigning a binary code number to each of the samples. The rate of sampling must be substantially higher than the highest signal frequency in order to cover the bandwidth of the signal and to avoid spurious patterns (*aliasing*) generated by the interaction between the sampling signal and the higher signal frequencies. A simplified block diagram of an A/D converter (ADC) is shown in Figure 14.1. The *Nyquist law* for digital coding dictates that the sample rate must be at least twice the cutoff frequency of the signal of interest to avoid these effects.

The sampling rate, even in analog sampling systems, is crucial. Figure 14.2*a* shows the spectral consequence of a sampling rate that is too low for the input bandwidth; Figure 14.2*b* shows the result of a rate equal to the theoretical minimum value, which is im-

Table 14.1 Comparison of Counting in the Decimal, Binary, and Octal Systems

Decimal	Binary	Octal
0	0	0
1	1	1
2	10	2
3	11	3
4	100	4
5	101	5
6	110	6
7	111	7
8	1000	10
9	1001	11
10	1010	12
11	1011	13
12	1100	14
13	1101	15

practical; and Figure 14.2c shows typical practice. The input spectrum must be limited by a low-pass filter to greatly attenuate frequencies near one-half the sampling rate and above. The higher the sampling rate, the easier and simpler the design of the input filter becomes. An excessively high sampling rate, however, is wasteful of transmission bandwidth and storage capacity, while a low but adequate rate complicates the design and increases the cost of input and output analog filters.

14.1.2 Digital-to-Analog (D/A) Conversion

The digital-to-analog converter (DAC) is, in principle, quite simple. The digital stream of binary pulses is decoded into discrete, sequentially timed signals corresponding to the original sampling in the A/D. The output is an analog signal of varying levels. The time duration of each level is equal to the width of the sample taken in the A/D conversion process. The analog signal is separated from the sampling components by a low-pass filter. Figure 14.3 shows a simplified block diagram of a D/A. The deglitching sample-and-hold circuits in the center block set up the analog levels from the digital decoding and remove the unwanted high-frequency sampling components.

Each digital number is converted to a corresponding voltage and stored until the next number is converted. Figure 14.4 shows the resulting spectrum. The energy surrounding the sampling frequency must be removed, and an output low-pass filter is used to accomplish that task. One cost-effective technique used in compact disk players and other applications is called *oversampling*. A new sampling rate is selected that is a whole multiple of the input sampling rate. The new rate is typically two or four times

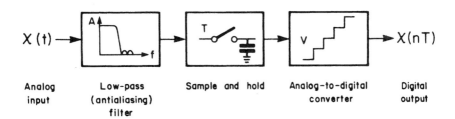

Figure 14.1 Analog-to-digital converter block diagram.

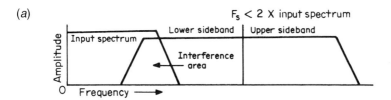

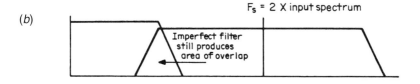

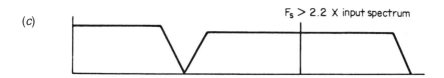

Figure 14.2 Relationship between sampling rate and bandwidth: (a) a sampling rate too low for the input spectrum, (b) the theoretical minimum sampling rate (F_s), which requires a theoretically perfect filter, (c) a practical sampling rate using a practical input filter.

the old rate. Every second or fourth sample is filled with the input value, while the others are set to zero. The result is passed through a digital filter that distributes the energy in the real samples among the empty ones and itself. The resulting spectrum (for a 4× oversampling system) is shown in Figure 14.5. The energy around the 4× sample fre-

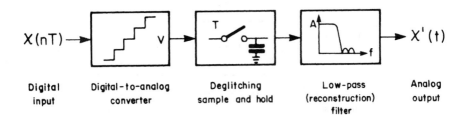

Figure 14.3 Digital-to-analog converter block diagram.

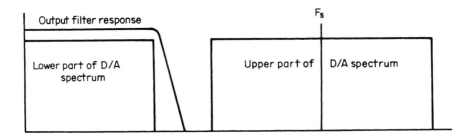

Figure 14.4 Output filter response requirements for a common D/A converter.

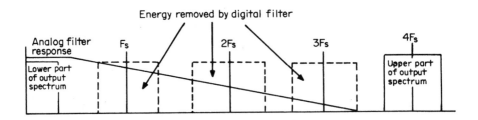

Figure 14.5 The filtering benefits of oversampling.

quency must be removed, which can be done simply because it is so distant from the upper band edge. The response of the output filter is chiefly determined by the digital processing and is therefore very stable with age, in contrast to a strictly analog filter, whose component values are susceptible to drift with age and other variables.

14.2 Combinational Logic

When the inputs to a logic circuit have only one meaning for each, the circuit is said to be *combinational*. These devices tend to have names reflecting the function they will perform, such as AND, OR, exclusive OR, latch, flip-flop, counter, and gate. Logic circuits are usually documented through the use of schematic diagrams. For simple devices, the shape of the symbol tells the function it performs, while the presence of small bubbles at the points of connection tell whether that point is high or low when the function is being performed. More complicated functions are shown as rectangular boxes. Figure 14.6 shows a collection of common logic symbols.

The clocking input to memory devices and counters is indicated by a small triangle at (usually) the inside left edge of the box. If the device is a transparent latch, the output follows the input while the clock input is active, and the output is "frozen" when the clock becomes inactive. A flip-flop, on the other hand, is an *edge-triggered* device. The output is allowed to change only upon a transition of the clock input from low to high (no bubble) or high to low (bubble present).

Three types of flip-flops are shown in Figure 14.6:

- *T* (toggle) flip-flop, which will reverse its output state when clocked while the T input is active.

- *D* flip-flop, which will allow the output to assume the state of the D input when clocked.

- *J-K* flip-flop. If both J and K inputs are inactive, the output does not change when clocked. If both are active, the output will toggle as in T. If J and K are different, the output will assume the state of the J input when clocked, similar to the D case.

Flip-flops, latches, and counters are often supplied with additional inputs used to force the output to a known state. An *active set* input will force the output into the active state, while a *reset input* will force the output into the inactive state. Counters also have inputs to force the output states; there are two types:

- Asynchronous, in which the function (preset or clear) is performed immediately

- Synchronous, in which the action occurs on the next clock transition

Usually, if both preset and clear are applied at once, the clear function outranks the preset function. Figure 14.7 shows some common logic stages and their truth tables. These gates and a few simple rules of *Boolean algebra*, the basics of which are shown in Table 14.2, facilitate the design of very complex circuits.

14.2.1 Boolean Algebra

Boolean algebra provides a means to analyze and design binary systems. It is based on the seven postulates given in Table 14.3. All other Boolean relationships are derived from these postulates. The OR and AND operations are normally designated by the arithmetic operator symbols + and • and are referred to as *sum* and *product* opera-

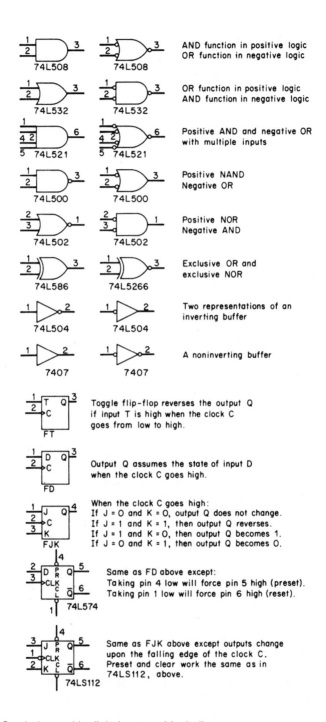

Figure 14.6 Symbols used in digital system block diagrams.

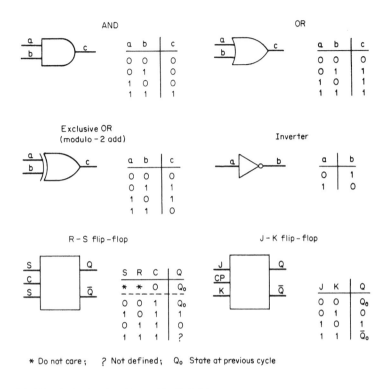

Figure 14.7 Basic logic circuits and truth tables.

tors in basic digital logic literature. A set of theorems derived from the postulates, given in Table 14.4, facilitate the development of more complex logic systems.

14.2.2 Logic Device Families

Resistor-transistor-logic (RTL) is mostly of historic interest only. It used a 3.6-V positive power supply, and was essentially incompatible with the logic families that came later. The packages were round with a circular array of wires (not pins) for circuit board mounting. Inputs were applied to the base of a transistor, and the transistor was turned on directly by the input signal if it was high. An open input could usually be considered as an "off" or "0."

Diode-Transistor Logic (DTL)

RTL was followed by the popular DTL, mounted in a DIP (dual in-line package). It had 14 or 16 stiff pins arranged in two parallel rows 0.3 in apart with the pins on 0.1-in centers. For simple devices, such as a two-input NAND gate, four gates were packaged into one DIP. The stiff pins made possible the use of sockets. An internal re-

Table 14.2 Fundamental Rules of Boolean Algebra

Rules and relations	$0 + 0 = 0$
	$0 \cdot 0 = 0$
	$1 + 1 = 1$
	$1 \cdot 1 = 1$
	$0 \cdot 1 = 0$
	$0 + 1 = 1$
	$\bar{0} = 1$
	$\bar{1} = 0$
	$a \cdot a = a$
	$a + a = a$
	$\bar{a} \cdot a = 0$
	$\bar{a} + a = 1$
	$0 + a = a$
	$0 \cdot a = 0$
	$1 + a = 1$
	$1 \cdot a = a$
	$a + a \cdot b = a$
	$a \cdot (a + b) = a$
	$a + a \cdot b = a + b$
Commutative law	$a + b = b + a$
	$a \cdot b = b \cdot a$
Associative law	$(a + b) + c = a + (b + c)$
	$(a \cdot b) \cdot c = a \cdot (b \cdot c)$
Distributive law	$a \cdot (b + c) = a \cdot b + a \cdot c$
DeMorgan's rules	$\overline{(a \cdot b)} = \bar{a} + \bar{b}$
	$\overline{(a + b)} = \bar{a} \cdot \bar{b}$

† $\cdot$ denotes AND, + denotes OR.

sistor attached to the positive 5.0-V supply turned on the input transistor. Input signals were applied through diodes such that if an input signal were low, it pulled down the resistor's current, and the transistor turned off. It is important to remember that a disconnected DTL or TTL input is a logic high. The DTL output circuit was pulled low by a transistor and pulled up to +5 V by an internal resistor. As a result, fall times were faster than rise times.

Transistor-Transistor Logic (TTL)

TTL, like DTL, supplies its own turn-on current but uses a transistor instead of a resistor. The inputs do not use diodes but instead use multiple emitters on an input transistor. The output is pulled down by one transistor and pulled up by another. There are a considerable number of family variations on this basic design. For example, the 7400 device (a two-input NAND gate) has the following common variations:

Table 14.3 Boolean Postulates (*From* [1]. *Used with permission.*)

Postulate	Name	Meaning	Forms (a) OR	(b) AND
1	Definition	$\exists$ a set $\{K\} = \{a, b, \ldots\}$ of two or more elements and two binary operators $\ni\{K\} = \{a, b, a+b, a \cdot b, \ldots\}$	$+$ $\vee$ $\cup$	$\cdot$ $\wedge$ $\cap$
2	Substitution Law	$\text{expression}_1 = \text{expression}_2$ If one replaced by the other does not alter the value		
3	Identity Element	$\exists$ identity elements for each operator	$a + 0 = a$	$a \cdot 1 = a$
4	Commutativity	For every a and b in K	$a + b = b + a$	$a \bullet b = b \bullet a$
5	Associativity	For every $a, b,$ and c in K	$a + (b + c) = (a + b) + c$	$a \cdot (b \cdot c) = (a \cdot b) \cdot c$
6	Distributivity	For every $a, b,$ and c in K	$a + (b \cdot c) = (a + b) \cdot (a + c)$	$a \cdot (b + c) = (a \cdot b) + (a \cdot c)$
7	Complement	For every a in K $\exists$ a complement in K $\ni$	$a + \bar{a} = 1$	$a \cdot \bar{a} = 0$

Table 14.4 Boolean Theorems (*From* [1]. *Used with permission.*)

Theorem		Forms (a)	(b)
8	Idempotency	$a + a = a$	$a \cdot a = a$
9	Complement Theorem	$a + 1 = 1$	$a \cdot 0 = 0$
10	Absorption	$a + ab = a$	$a(a + b) = a$
11	Extra Element Elimination	$a + \bar{a}b = a + b$	$a(\bar{a} + b) = ab$
12	De Morgan's Theorem	$\overline{a + b} = \bar{a} \cdot \bar{b}$	$\overline{ab} = \bar{a} + \bar{b}$
13	Concensus	$ab + \bar{a}c + bc = ab + \bar{a}c$	$(a + b)(\bar{a} + c)(b + c) = (a + b)(\bar{a} + c)$
14	Complement Theorem 2	$ab + a\bar{b} = a$	$(a + b)(a + \bar{b}) = a$
15	Concensus 2	$ab + a\bar{b}c = ab + ac$	$(a + b)(a + \bar{b} + c) = (a + b)(a + c)$
16	Concensus 3	$ab + \bar{a}c = (a + c)(\bar{a} + b)$	$(a + b)(\bar{a} + c) = ac + \bar{a}b$

- 7400—the prototype
- 74L00—a low-power version, but with relatively slow switching speed
- 74S00—(Schottky) fast but power-hungry
- 74LS00—low power and relatively slow speed
- 74AS00—advanced Schottky
- 74ALS00—similar to LS, but with improved performance
- 74F00—F for fast

All variants can be used in the presence of the others, but doing so complicates the design rules that determine how many inputs can be driven by one output. The dividing line between an input high and an input low in this example is about 1.8 V. A high output is guaranteed to be 2.4 V or greater, while an output low will be 0.8 V or less.

NMOS and PMOS

Metal-oxide semiconductor (MOS) logic devices use field-effect transistors as the switching elements. The initial letter tells whether the device uses – or p-type dopant on the silicon. At low frequencies, MOS devices are very frugal in power consumption. Early MOSs were fairly slow, but smaller conductor sizes reduced on-chip capacitance and therefore charging time.

Complementary MOS (CMOS)

A very popular logic family, CMOS devices use both p- and n-type transistors. At direct current, input currents are almost zero. Output current rises with frequency because the output circuit must charge and discharge the capacitance of the inputs it is driving. Early CMOS devices were fairly slow when powered with a 5-V supply, but performance improved when powered at 10 or 15 V. Modern microscopic geometry produces CMOS parts that challenge TTL speeds while using less power.

The input decision level of a CMOS device is nominally midway between the positive supply and ground. The logic state of an open input is indeterminate. It can and will wander around depending on which of the two input transistors is leaking the most. Unused inputs must be returned either to ground or the supply rail. CMOS outputs, unlike TTL, are very close to ground when low and very close to the supply rail when high. CMOS can drive TTL inputs; however, in a 5-V environment, the CMOS decision level of 2.5 V is too close to the TTL guaranteed output high for reliable operation. The solution is an external pull-up resistor between the output pin of the TTL part and the supply rail.

Early CMOS devices had their own numbering system (beginning at 4000) that was totally different from the one used for TTL parts. Improvements in speed and other performance metrics spawned subfamilies that tended toward a return to the use of the 7400 convention; for example, 74HC00 is a high-speed CMOS part.

Emitter-Coupled Logic (ECL)

ECL has almost nothing in common with the families previously discussed. Inputs and outputs are push-pull. The supply voltage is negative with respect to ground at –5.2 V. Certain advantages accrue from this configuration:

- Because of the push-pull input-output, inverters are not needed. To invert, simply reverse the two connections.

- The differential-amplifier construction of ECL input and output stages causes the total current through the device to be almost constant.

- The output voltage swing is small and, from a crosstalk standpoint, is opposed by the complementary output.

- Driving a balanced transmission line does not require a line-driver because an ECL output (with some resistors) is a line-driver.

Because the transistors in ECL are never saturated, they operate at maximum speed. Early ECL was power-hungry, but newer ECL gate-array products are available that will toggle well into the gigahertz range without running hot.

14.2.3 Scaling of Digital Circuit Packages

The term small-scale integration (SSI) includes those packages containing, for example, a collection of four gates, a 4-bit counter, a 4-bit adder, and any other item of less than about 100 gate equivalents. Large-scale integration (LSI) describes more complex circuitry, such as an asynchronous bit-serial transmitter-receiver, or a DMA (direct memory access) controller, involving a few thousand gate equivalents. Very large scale integration (VLSI) represents tens of thousands of gate equivalents or more, such as a microprocessor or a graphics controller. LSI and VLSI devices are typically packaged in a larger version of the DIP package, usually with the two rows spaced 0.6 in or more, and having 24 to 68 pins or more.

Many devices are available in dual in-line packages designed to be soldered to the surface of a circuit board rather than using holes in the circuit board. The pin spacing is 0.05 in or less. The *leadless chip carrier* is another surface-mount device with contact spacing of 0.05 in or less, and an equal number of contacts along each edge of the square. Well over 100 contacts can be accommodated in such packages. Sockets are available for these packages, but once the package is installed, a special tool is required to extract it. Yet another large-scale package is called the *pin-grid array*, with pins protruding from the bottom surface of a flat, square package in a row-and-column "bed-of-nails" array. The pin spacing is 0.1 in or less. For this device, more than 200 pins may be incorporated. Extraction tools are available for these packages as well.

14.2.4 Representation of Numbers and Numerals

A single bit, terminal, or flip-flop in a binary system can have only two states. When a single bit is used to describe numerals, by convention those two numerals are 0 and 1.

A group of bits, however, can describe a larger range of numbers. Conventional groupings are identified in the following sections.

Nibble

A *nibble* is a group of 4 bits. It is customary to show the binary representation with the *least significant bit* (LSB) on the right. The LSB has a decimal value of 1 or 0. The next most significant bit has a value of 2 or 0, and the next, 4 or 0, and the *most significant bit* (MSB), 8 or 0. The nibble can describe any value from binary 0000 (= 0 decimal) and 1111 (= 8 + 4 + 2 + 1 = 15 decimal), inclusive. The 16 characters used to signify the 16 values of a nibble are the ordinary numerals 0 through 9, followed by the letters of the alphabet A through F. The 4-bit "digit" is a *hexadecimal representation*.

Octal, an earlier numbering scheme, used groupings of 3 bits to describe the numerals 0 through 7. Used extensively by the Digital Equipment Corporation, it has fallen out of use, but is still included in some figures for reference.

Byte

A *byte* is a collection of 8 bits, or 2 nibbles. It can represent numbers (a number is a collection of numerals) in two ways:

- Two hexadecimal digits, the least significant representing the number of 1s, and the most significant the number of 16s. The total range of values is 0 through 255 (FF).

- Two decimal digits, the least significant representing the number of 1s, and limited to the range of numerals 0 through 9, and the most significant representing the number of 10s, again limited to the range 0 through 9.

The use of 4 bits to represent decimal numbers is called *binary-coded decimal* (BCD). The use of a byte to store two numerals is called *packed* BCD. The least significant nibble is limited to the range of 0 through 9, as is the upper nibble, thus representing 00 through 90. The maximum value of the byte is 99.

Word

A *word*, usually a multiple of 8 bits, is the largest array of bits that can be handled by a system in one action of its logic. In most personal computers, a word is 16 or 32 bits. Larger workstations use words of 32 and 64 bits in length. In all cases, the written and electronically mapped representation of the numeric value of the word is either hexadecimal or packed BCD.

Negative Numbers

When a byte or word is used to describe a *signed number* (one that may be less than zero), it is customary for the most significant bit to represent the sign of the number, 0 meaning positive and 1 negative. This representation is known as *two's complement*.

Table 14.5 Number and Letter Representations

Decimal	Binary	Octal	Hexadecimal
0	0	0	0
1	1	1	1
2	10	2	2
3	11	3	3
4	100	4	4
5	101	5	5
6	110	6	6
7	111	7	7
8	1000	10	8
9	1001	11	9
10	1010	12	A
11	1011	13	B
12	1100	14	C
13	1101	15	D
14	1110	16	E
15	1111	17	F
81	01010001	121	51
250	11111010	372	FA
+127	01111111 (signed)	177	7F
−1	11111111 (signed)	377	FF

To negate (make negative) a number, simply show the number in binary, make all the zeros into 1s, and all the 1s into zeros, and then add 1.

Floating Point

In engineering work, the range of numerical values is tremendous and can easily overflow the range of values offered by 64-bit (and smaller) systems. Where the accuracy of a computation can be tolerably expressed as a percentage of the input values and the result, *floating-point calculation* is used. One or two bytes are used to express the *characteristic* (a power of 10 by which to multiply everything), and the rest are used to express the *mantissa* (that fractional power of 10 to be multiplied by). This is commonly referred to as *engineering notation*. (See Table 14.5.)

Compare

A *comparison* involves negating one of the two numbers being compared, then adding them and testing the result. If the test shows zero, the two numbers are equal. If not, the test reveals which of the two is greater than or less than the other, and the appropriate bits in the status register are set.

Jump

The orderly progression of the program counter may be interrupted and instructions fetched from a new location in memory, usually based upon a test or a comparison. For example, "If the result is zero, jump to location X and begin execution there; if the result is positive, jump to Y and begin execution there; else keep on counting." This ability is probably the most powerful asset of a computer because it permits logic-based branching of a program.

14.3 Errors in Digital Systems

When a digital signal is transmitted through a noisy path, errors can occur. Early methods to deal with this problem included generating one or more digital words, using *check sums*, *cyclic redundancy checks*, and similar error-coding schemes, and appending the result at the end of a block of transmitted data. Upon reception, the same arithmetic was used to generate the same results, which were compared to the data appended to the transmission. If they were identical, it was unlikely that an error had occurred. If they differed, an error was assumed to have occurred, and a retransmission was requested. Such methods, thus, performed only error detection. In the case of many digital transmission systems, however, retransmission is not possible and methods must be employed that not only detect but correct errors.

14.3.1 Error Detection and Correction

Given a string of 8-bit bytes, additional bytes can be generated using *Galois field arithmetic* and appended to the end of the string. The length of the string and the appended bytes must be 256 or less, since 8 bits can have no more than 256 different states. If 2 bytes are generated, upon reconstruction 2 *syndrome* (symptom) bytes are generated. If they are zero, there was likely no error. If they are nonzero, then after arithmetic processing, 1 byte "points" to the location of the damaged byte in the string, while the other contains the 8-bit *error pattern*. The error pattern is used in a bit-wise exclusive OR function upon the offending byte, thus reversing the damaged bits and correcting the byte. With 4 check bytes, 2 flawed bytes can be pinpointed and corrected; with 6, 3 can be treated; and so on. If the number of bytes in the string is significantly less than 256, for example, 64, the error-detection function becomes more robust because, if the error pointer points to a nonexistent byte, it may be assumed that the error-detection system itself made a mistake.

Errors in digital recorders, for example, fall into two classes: random errors brought on by thermal random noise in the reproduce circuitry, and dropouts (long strings of lost signal resulting from media imperfections). The error detection and correction system of digital recorders is designed to cope with both types of errors. Figure 14.8 shows

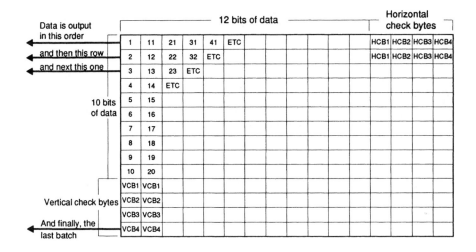

Figure 14.8 An example of row and column two-dimensional error-detection coding.

how data can be arranged in rows and columns, with separate check bytes generated for each row and each column in a two-dimensional array. The data is recorded (and reproduced) in row order. In the example given in the figure, it can be seen that a long interruption of signal will disrupt every tenth byte. The row corrector cannot cope with this, but it is likely that the column corrector can because it "sees" the burst error as being spread out over a large number of columns.

The column corrector, if taken alone, can correct $N/2$ errors, where N is the number of check bytes. Given knowledge of which rows are uncorrectable by the row corrector, then N errors can be corrected. Generally, the row (or "inner") corrector acts on errors caused by random noise, while the column (or "outer") corrector takes care of burst errors.

Generally, error detection and correction schemes have the following characteristics:

- Up to a threshold error rate, all errors are corrected.

- If the error rate is greater than the above first threshold, the system will flag the blocks of data it is unable to correct. This allows other circuits to attempt to conceal the error.

- Above an even higher error rate, the system will occasionally fail and either stop producing output data entirely, or simply pass along the data, correcting what it can and letting the rest pass through.

14.3.2 Error Concealment

When the error-correction system is overloaded and error-ridden samples are identified, it is typical practice in communications applications to calculate an estimation of the bad sample. In video applications, for example, samples that are visually nearby and that are not corrupted can be used to calculate an estimate of the damaged sample. The estimate is then substituted for the unusable sample. In the recording or transmission process, the video data samples are scrambled in a way that maximizes the chance that a damaged sample will be surrounded by good ones.

In the case of audio, the samples can be scrambled such that failure of the correction system is most likely to result in every alternate sample being in error. Replacement of a damaged audio sample can then consist of summing the previous (good) sample and the following (good) sample and dividing by 2. If the error rate becomes unreasonable, then the last good sample is simply repeated, or "held."

Video error concealment is roughly 10 times more effective than audio concealment, due in large part to differences in the way the eye and ear interpret and process input information.

14.4 References

1. Whitaker, Jerry C. (ed.), *The Electronics Handbook*, CRC Press, Boca Raton, FL, 1996.

14.5 Bibliography

Benson, K. Blair (ed.), *Audio Engineering Handbook*, McGraw-Hill, New York, NY, 1988.

Busby, E. Stanley, "Digital Fundamentals," in *Television and Audio Handbook for Technicians and Engineers*, K. Blair Benson and Jerry C. Whitaker (eds.), McGraw-Hill, New York, NY, 1990.

Fink, Donald (ed.), *Electronics Engineers' Handbook*, McGraw-Hill, New York, NY, 1982.

Texas Instruments, *2-mm CMOS Standard Cell Data Book*, Chapter 8, Texas Instruments, Dallas, TX, 1986.

Whitaker, Jerry C., and K. Blair Benson (eds.), *Standard Handbook of Video and Television Engineering*, third ed., McGraw-Hill, New York, NY, 2000.

Whitaker, Jerry C. (ed.), *Video and Television Engineer's Field Manual*, McGraw-Hill, New York, NY, 2000.

Amplitude Modulation

15.1 Introduction

In the simplest form of amplitude modulation, an analog carrier is controlled by an analog modulating signal. The desired result is an RF waveform whose amplitude is varied by the magnitude of the applied modulating signal and at a rate equal to the frequency of the applied signal. The resulting waveform consists of a carrier wave plus two additional signals:

- An upper-sideband signal, which is equal in frequency to the carrier *plus* the frequency of the modulating signal

- A lower-sideband signal, which is equal in frequency to the carrier *minus* the frequency of the modulating signal

This type of modulation system is referred to as *double-sideband amplitude modulation* (DSAM).

15.2 Fundamental Principles

The radio carrier wave signal onto which the analog amplitude variations are to be impressed is expressed as:

$$e(t) = A E_c \cos(\omega_c t) \tag{15.1}$$

Where:
$e(t)$ = instantaneous amplitude of carrier wave as a function of time (t)
A = a factor of amplitude modulation of the carrier wave
ω_c = angular frequency of carrier wave (radians per second)
E_c = peak amplitude of carrier wave

If A is a constant, the peak amplitude of the carrier wave is constant, and no modulation exists. Periodic modulation of the carrier wave results if the amplitude of A is caused to vary with respect to time, as in the case of a sinusoidal wave:

$$A = 1 + \left({}^{E_m}\!/\!_{E_c} \right) \cos(\omega_m t) \tag{15.2}$$

Where:
E_m/E_c = the ratio of modulation amplitude to carrier amplitude

The foregoing relationship leads to:

$$e(t) = E_c \left[1 + \left({}^{E_m}\!/\!_{E_c} \right) \cos(\omega_m t) \cos(\omega_c t) \right] \tag{15.3}$$

This is the basic equation for periodic (sinusoidal) amplitude modulation. When all multiplications and a simple trigonometric identity are performed, the result is:

$$e(t) = E_c \cos(\omega_c t) + \left({}^M\!/\!_2 \right) \cos(\omega_c t + \omega_m t) + \left({}^M\!/\!_2 \right) \cos(\omega_c t - \omega_m t) \tag{15.4}$$

Where:
M = the amplitude modulation factor (E_m/E_c)

Amplitude modulation is, essentially, a multiplication process in which the time functions that describe the modulating signal and the carrier are multiplied to produce a modulated wave containing *intelligence* (information or data of some kind). The frequency components of the modulating signal are translated in this process to occupy a different position in the spectrum.

The bandwidth of an AM transmission is determined by the modulating frequency. The bandwidth required for full-fidelity reproduction in a receiver is equal to twice the applied modulating frequency.

The magnitude of the upper sideband and lower sideband will not normally exceed 50 percent of the carrier amplitude during modulation. This results in an upper-sideband power of one-fourth the carrier power. The same power exists in the lower sideband. As a result, up to one-half of the actual carrier power appears additionally in the sum of the sidebands of the modulated signal. A representation of the AM carrier and its sidebands is shown in Figure 15.1. The actual occupied bandwidth, assuming pure sinusoidal modulating signals and no distortion during the modulation process, is equal to twice the frequency of the modulating signal.

The extent of the amplitude variations in a modulated wave is expressed in terms of the *degree of modulation* or *percentage of modulation*. For sinusoidal variation, the degree of modulation m is determined from the following:

$$m = \frac{E_{avg} - E_{min}}{E_{avg}} \tag{15.5}$$

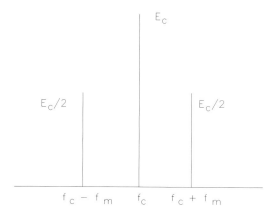

Figure 15.1 Frequency-domain representation of an amplitude-modulated signal at 100 percent modulation. E_c = carrier power, F_c = frequency of the carrier, and F_m = frequency of the modulating signal.

Where:
E_{avg} = average envelope amplitude
E_{min} = minimum envelope amplitude

Full (100 percent) modulation occurs when the peak value of the modulated envelope reaches twice the value of the unmodulated carrier, and the minimum value of the envelope is zero. The envelope of a modulated AM signal in the time domain is shown in Figure 15.2.

When the envelope variation is not sinusoidal, it is necessary to define the degree of modulation separately for the peaks and troughs of the envelope:

$$m_{pp} = \frac{E_{max} - E_{avg}}{E_{avg}} \times 100 \qquad (15.6)$$

$$m_{np} = \frac{E_{avg} - E_{min}}{E_{avg}} \times 100 \qquad (15.7)$$

Where:
m_{pp} = positive peak modulation (percent)
E_{max} = peak value of modulation envelope
m_{np} = negative peak modulation (percent)
E_{avg} = average envelope amplitude
E_{min} = minimum envelope amplitude

When modulation exceeds 100 percent on the negative swing of the carrier, spurious signals are emitted. It is possible to modulate an AM carrier asymmetrically; that is, to

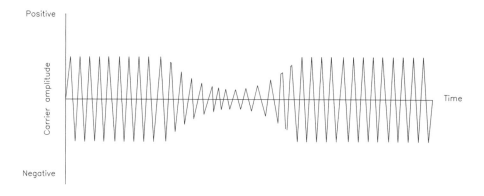

Figure 15.2 Time-domain representation of an amplitude-modulated signal. Modulation at 100 percent is defined as the point at which the peak of the waveform reaches twice the carrier level, and the minimum point of the waveform is zero.

restrict modulation in the negative direction to 100 percent, but to allow modulation in the positive direction to exceed 100 percent without a significant loss of fidelity. In fact, many modulating signals normally exhibit asymmetry, most notably human speech waveforms.

The carrier wave represents the average amplitude of the envelope and, because it is the same regardless of the presence or absence of modulation, the carrier transmits no information. The information is carried by the sideband frequencies. The amplitude of the modulated envelope may be expressed as follows [5]:

$$E = E_0 + E_1 \sin\left(2\pi f_1 t + \Phi_1\right) + E_2 \sin\left(2\pi f_2 t + \Phi_2\right) \tag{15.8}$$

Where:
E = envelope amplitude
E_0 = carrier wave crest value (volts)
E_1 = 2 × first sideband crest amplitude (volts)
f_1 = frequency difference between the carrier and the first upper/lower sidebands
E_2 = 2 × second sideband crest amplitude (volts)
f_2 = frequency difference between the carrier and the second upper/lower sidebands
Φ_1 = phase of the first sideband component
Φ_2 = phase of the second sideband component

15.2.1 High-Level AM Modulation

High-level anode modulation is the oldest and simplest method of generating a high-power AM signal. In this system, the modulating signal is amplified and com-

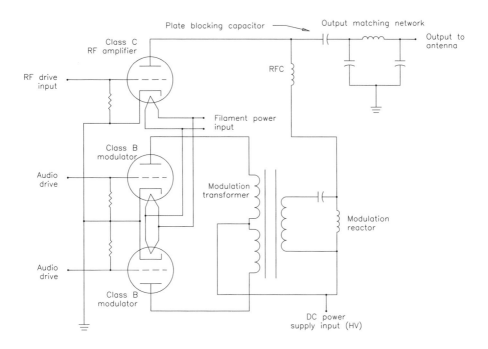

Figure 15.3 Simplified diagram of a high-level amplitude-modulated amplifier.

bined with the dc supply source to the anode of the final RF amplifier stage. The RF amplifier is normally operated class C. The final stage of the modulator usually consists of a pair of tubes operating class B in a push-pull configuration. A basic high-level modulator is shown in Figure 15.3.

The RF signal normally is generated in a low-level transistorized oscillator. It is then amplified by one or more solid-state or vacuum tube stages to provide final RF drive at the appropriate frequency to the grid of the final class C amplifier. The modulation input is applied to an intermediate power amplifier (usually solid-state) and used to drive two class B (or class AB) push-pull output devices. The final amplifiers provide the necessary modulating power to drive the final RF stage. For 100 percent modulation, this modulating power is equal to 50 percent of the actual carrier power.

The modulation transformer shown in Figure 15.3 does not usually carry the dc supply current for the final RF amplifier. The modulation reactor and capacitor shown provide a means to combine the signal voltage from the modulator with the dc supply to the final RF amplifier. This arrangement eliminates the necessity of having direct current flow through the secondary of the modulation transformer, which would result in magnetic losses and saturation effects. In some transmitter designs, the modulation reactor is eliminated from the system, thanks to improvements in transformer technology.

The RF amplifier normally operates class C with grid current drawn during positive peaks of the cycle. Typical stage efficiency is 75 to 83 percent. An RF tank following the amplifier resonates the output signal at the operating frequency and, with the assistance of a low-pass filter, eliminates harmonics of the amplifier caused by class C operation.

This type of system was popular in AM applications for many years, primarily because of its simplicity. The primary drawback is low overall system efficiency. The class B modulator tubes cannot operate with greater than 50 percent efficiency. Still, with inexpensive electricity, this was not considered to be a significant problem. As energy costs increased, however, more efficient methods of generating high-power AM signals were developed. Increased efficiency normally came at the expense of added technical complexity.

15.2.2 Vestigial-Sideband Amplitude Modulation (VSBAM)

Because the intelligence (modulating signal) of conventional AM transmission is identical in the upper *and* lower sidebands, it is possible to eliminate one sideband and still convey the required information. This scheme is implemented in *vestigial-sideband AM* (VSBAM). Complete elimination of one sideband (for example, the lower sideband) requires an ideal high-pass filter with infinitely sharp cutoff. Such a filter is difficult to implement in any practical design. VSBAM is a compromise technique wherein one sideband (typically the lower sideband) is attenuated significantly. The result is a savings in occupied bandwidth and transmitter power.

VSBAM is used for television broadcast transmission and other applications. A typical bandwidth trace for a VSBAM TV transmitter is shown in Figure 15.4.

15.2.3 Single-Sideband Amplitude Modulation (SSBAM)

The carrier in an AM signal does not convey any intelligence. All of the modulating information is in the sidebands. It is possible, therefore, to suppress the carrier upon transmission, radiating only one or both sidebands of the AM signal. The result is much greater efficiency at the transmitter (that is, a reduction in the required transmitter power). Suppression of the carrier may be accomplished with DSAM and SSBAM signals. *Single-sideband suppressed carrier* AM (SSB-SC) is the most spectrum- and energy-efficient mode of AM transmission. Figure 15.5 shows representative waveforms for suppressed carrier transmissions.

A waveform with carrier suppression differs from a modulated wave containing a carrier primarily in that the envelope varies at twice the modulating frequency. In addition, it will be noted that the SSB-SC wave has an apparent phase that reverses every time the modulating signal passes through zero. The wave representing a single sideband consists of a number of frequency components, one for each component in the original signal. Each of these components has an amplitude proportional to the amplitude of the corresponding modulating component and a frequency differing from that of the carrier by the modulating frequency. The result is that, in general, the envelope amplitude of the single sideband signal increases with the degree of modulation, and

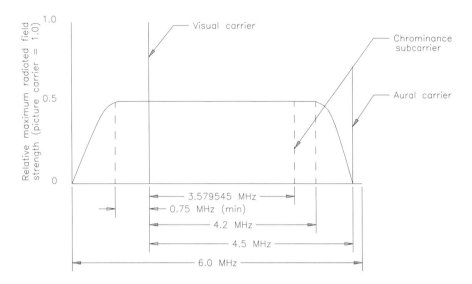

Figure 15.4 Idealized amplitude characteristics of the FCC standard waveform for monochrome and color TV transmission. (*Adapted from*: FCC Rules, Sec. 73.699.)

the envelope varies in amplitude in accordance with the difference frequencies formed by the various frequency components of the single sideband interacting with each other.

An SSB-SC system is capable of transmitting a given intelligence within a frequency band only half as wide as that required by a DSAM waveform. Furthermore, the SSB system saves more than two-thirds of the transmission power because of the elimination of one sideband and the carrier.

The drawback to suppressed carrier systems is the requirement for a more complicated receiver. The carrier must be regenerated at the receiver to permit demodulation of the signal. Also, in the case of SSBAM transmitters, it is usually necessary to generate the SSB signal in a low-power stage and then amplify the signal with a linear power amplifier to drive the antenna. Linear amplifiers generally exhibit low efficiency.

15.2.4 Quadrature Amplitude Modulation (QAM)

Single sideband transmission makes very efficient use of the spectrum; for example, two SSB signals can be transmitted within the bandwidth normally required for a single DSB signal. However, DSB signals can achieve the same efficiency by means of *quadrature amplitude modulation* (QAM), which permits two DSB signals to be transmitted and received simultaneously using the same carrier frequency [1]. A basic QAM DSB modulator is shown schematically in Figure 15.6.

Two DSB signals coexist separately within the same bandwidth by virtue of the 90° phase shift between them. The signals are, thus, said to be in *quadrature*. Demodulation

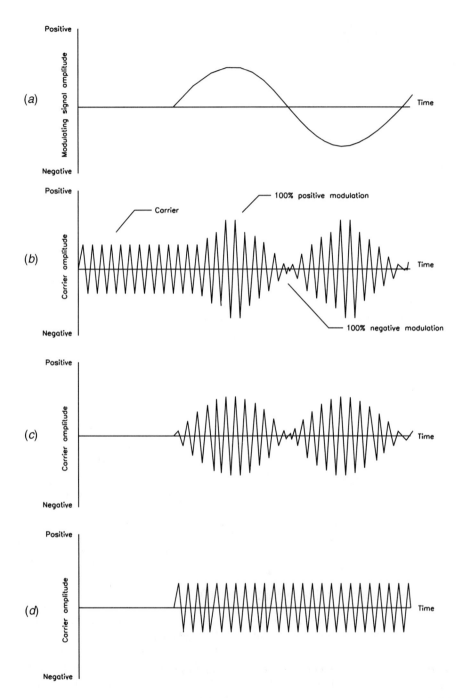

Figure 15.5 Types of suppressed carrier amplitude modulation: (*a*) the modulating signal, (*b*) double-sideband AM signal, (*c*) double-sideband suppressed carrier AM, (*d*) single-sideband suppressed carrier AM.

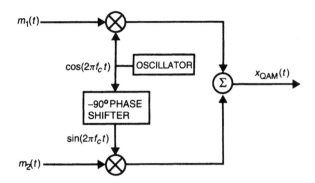

Figure 15.6 Simplified QAM modulator. (*From* [1]. *Used with permission.*)

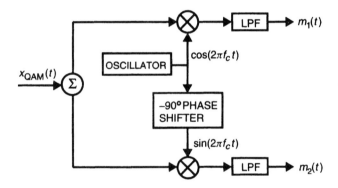

Figure 15.7 Simplified QAM demodulator. (*From* [1]. *Used with permission.*)

uses two local oscillator signals that are also in quadrature, i.e., a sine and a cosine signal, as illustrated in Figure 15.7.

The chief disadvantage of QAM is the need for a coherent local oscillator at the receiver exactly in phase with the transmitter oscillator signal. Slight errors in phase or frequency can cause both loss of signal and interference between the two signals (cochannel interference or crosstalk).

The relative merits of the various AM systems are summarized in Table 15.1.

15.3 References

1. Kubichek, Robert, "Amplitude Modulation," in *The Electronics Handbook*, Jerry C. Whitaker (ed.), CRC Press, Boca Raton, FL, pp. 1175–1187, 1996.

Table 15.1 Comparison of Amplitude Modulation Techniques (*After* [1].)

Modulation Scheme	Advantages	Disadvantages	Comments
DSB SC	Good power efficiency; good low-frequency response.	More difficult to generate than DSB+C; detection requires coherent local oscillator, pilot, or phase-locked loop; poor spectrum efficiency.	
DSB+C	Easier to generate than DSB SC, especially at high-power levels; inexpensive receivers using envelope detection.	Poor power efficiency; poor spectrum efficiency; poor low-frequency response; exhibits threshold effect in noise.	Used in commercial AM broadcasting.
SSB SC	Excellent spectrum efficiency.	Complex transmitter design; complex receiver design (same as DSB SC); poor low-frequency response.	Used in military communication systems, and to multiplex multiple phone calls onto long-haul microwave links.
SSB+C	Good spectrum efficiency; low receiver complexity.	Poor power efficiency; complex transmitters; poor low-frequency response; poor noise performance.	
VSB SC	Good spectrum efficiency; excellent low-frequency response; transmitter easier to build than for SSB.	Complex receivers (same as DSB SC).	
VSB+C	Good spectrum efficiency; good low-frequency response; inexpensive receivers using envelope detection.	Poor power efficiency; poor performance in noise.	Used in commercial TV broadcasting.
QAM	Good low-frequency response; good spectrum efficiency.	Complex receivers; sensitive to frequency and phase errors.	Two SSB signals may be preferable.

15.4 Bibliography

Benson, K. B., and Jerry. C. Whitaker (eds.), *Television Engineering Handbook*, McGraw-Hill, New York, NY, 1986.

Benson, K. B., and Jerry. C. Whitaker, *Television and Audio Handbook for Technicians and Engineers*, McGraw-Hill, New York, NY, 1989.

Crutchfield, E. B. (ed.), *NAB Engineering Handbook*, 8th Ed., National Association of Broadcasters, Washington, DC, 1991.

Fink, D., and D. Christiansen (eds.), *Electronics Engineers' Handbook*, 3rd ed., McGraw-Hill, New York, NY, 1989.

Fink, D., and D. Christiansen (eds.), *Electronics Engineers' Handbook*, 2nd ed., McGraw-Hill, New York, NY, 1982.

Jordan, Edward C., *Reference Data for Engineers: Radio, Electronics, Computer and Communications*, 7th ed., Howard W. Sams, Indianapolis, IN, 1985.

Laboratory Staff, *The Care and Feeding of Power Grid Tubes*, Varian Eimac, San Carlos, CA, 1984.

Whitaker, Jerry C., *Maintaining Electronic Systems*, CRC Press, Boca Raton, FL, 1992.

Whitaker, Jerry C., *Radio Frequency Transmission Systems: Design and Operation*, McGraw-Hill, New York, NY, 1991.

Frequency Modulation

16.1 Introduction

Frequency modulation is a technique whereby the phase angle or phase shift of a carrier is varied by an applied modulating signal. The *magnitude* of frequency change of the carrier is a direct function of the *magnitude* of the modulating signal. The *rate* at which the frequency of the carrier is changed is a direct function of the *frequency* of the modulating signal. In FM modulation, multiple pairs of sidebands are produced. The actual number of sidebands that make up the modulated wave is determined by the *modulation index* (MI) of the system.

16.1.1 Modulation Index

The modulation index is a function of the frequency deviation of the system and the applied modulating signal:

$$MI = \frac{F_d}{M_f} \tag{16.1}$$

Where:
MI = the modulation index
F_d = frequency deviation
M_f = modulating frequency

The higher the MI, the more sidebands produced. It follows that the higher the modulating frequency for a given deviation, the fewer number of sidebands produced, but the greater their spacing.

To determine the frequency spectrum of a transmitted FM waveform, it is necessary to compute a Fourier series or Fourier expansion to show the actual signal components involved. This work is difficult for a waveform of this type, because the integrals that must be performed in the Fourier expansion or Fourier series are not easily solved. The

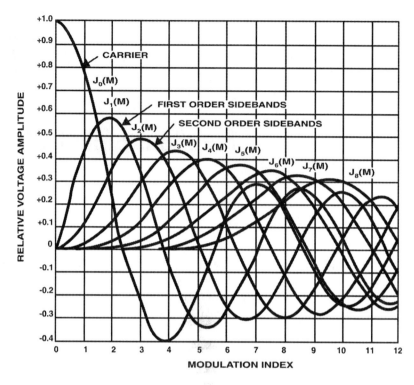

Figure 16.1 Plot of Bessel functions of the first kind as a function of modulation index.

result, however, is that the integral produces a particular class of solution that is identified as the *Bessel function*, illustrated in Figure 16.1.

The carrier amplitude and phase, plus the sidebands, can be expressed mathematically by making the modulation index the argument of a simplified Bessel function. The general expression is given from the following equations:

$$\text{RF output voltage} = E_1 = E_c + S_{1u} - S_{1l} + S_{2u} - S_{2l} + S_{3u} - S_{3l} + S_{nu} - S_{nl}$$

$$\text{Carrier amplitude} = E_c = A\left[J_0(M)\sin \omega c(t)\right]$$

$$\text{First-order upper sideband} = S_{1u} = J_1(M)\sin(\omega c + \omega m)t$$

$$\text{First-order lower sideband} = S_{1l} = J_1(M)\sin(\omega c - \omega m)t$$

$$\text{Second-order upper sideband} = S_{2u} = J_2(M)\sin(\omega c + 2\omega m)t$$

$$\text{Second-order lower sideband} = S_{2l} = J_2(M)\sin(\omega c - 2\omega m)t$$

Third-order upper sideband $= S_{3u} = J_3(M)\sin(\omega_c + 3\omega_m)t$

Third-order lower sideband $= S_{3l} = J_3(M)\sin(\omega_c - 3\omega_m)t$

Nth-order upper sideband $= S_{nu} = J_n(M)\sin(\omega_c + n\omega_m)t$

Nth-order lower sideband $= S_{nl} = J_n(M)\sin(\omega_c - n\omega_m)t$

Where:
A = the unmodulated carrier amplitude constant
J_0 = modulated carrier amplitude
$J_1, J_2, J_3...J_n$ = amplitudes of the nth-order sidebands
M = modulation index
$\omega_c = 2\pi F_c$, the carrier frequency
$\omega_m = 2\pi F_m$, the modulating frequency

Further supporting mathematics will show that an FM signal using the modulation indices that occur in a wideband system will have a multitude of sidebands. From the
purist point of view, *all* sidebands would have to be transmitted, received, and demodulated to reconstruct the modulating signal with complete accuracy. In practice, however, the channel bandwidths permitted FM systems usually are sufficient to reconstruct the modulating signal with little discernible loss in fidelity, or at least an acceptable loss in fidelity.

Figure 16.2 illustrates the frequency components present for a modulation index of
5. Figure 16.3 shows the components for an index of 15. Note that the number of significant sideband components becomes quite large with a high MI. This simple representation of a single-tone frequency-modulated spectrum is useful for understanding the
general nature of FM, and for making tests and measurements. When typical modulation signals are applied, however, many more sideband components are generated.
These components vary to the extent that sideband energy becomes distributed over the
entire occupied bandwidth, rather than appearing at discrete frequencies.

Although complex modulation of an FM carrier greatly increases the number of frequency components present in the frequency-modulated wave, it does not, in general,
widen the frequency band occupied by the energy of the wave. To a first approximation,
this band is still roughly twice the sum of the maximum frequency deviation at the peak
of the modulation cycle plus the highest modulating frequency involved.

FM is not a simple frequency translation, as with AM, but involves the generation of
entirely new frequency components. In general, the new spectrum is much wider than
the original modulating signal. This greater bandwidth may be used to improve the *signal-to-noise ratio* (S/N) of the transmission system. FM thereby makes it possible to
exchange bandwidth for S/N enhancement.

The power in an FM system is constant throughout the modulation process. The output power is increased in the amplitude modulation system by the modulation process,
but the FM system simply distributes the power throughout the various frequency components that are produced by modulation. During modulation, a wideband FM system

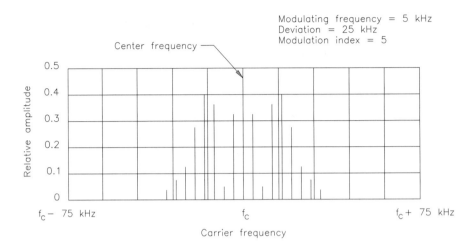

Figure 16.2 RF spectrum of a frequency-modulated signal with a modulation index of 5 and other operating parameters as shown.

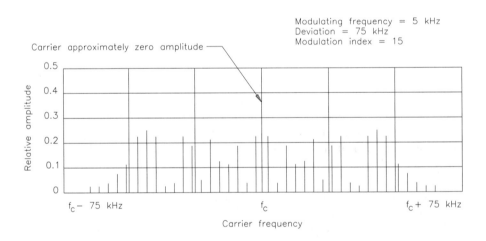

Figure 16.3 RF spectrum of a frequency-modulated signal with a modulation index of 15 and operating parameters as shown.

does not have a high amount of energy present in the carrier. Most of the energy will be found in the sum of the sidebands.

The constant-amplitude characteristic of FM greatly assists in capitalizing on the low noise advantage of FM reception. Upon being received and amplified, the FM signal normally is clipped to eliminate all amplitude variations above a certain threshold.

This removes noise picked up by the receiver as a result of man-made or atmospheric signals. It is not possible (generally speaking) for these random noise sources to change the frequency of the desired signal; they can affect only its amplitude. The use of *hard limiting* in the receiver will strip off such interference.

16.1.2 Phase Modulation

In a phase modulation (PM) system, intelligence is conveyed by varying the phase of the RF wave. Phase modulation is similar in many respects to frequency modulation, except in the interpretation of the modulation index. In the case of PM, the modulation index depends only on the amplitude of the modulation; MI is independent of the frequency of the modulating signal. It is apparent, therefore, that the phase-modulated wave contains the same sideband components as the FM wave and, if the modulation indices in the two cases are the same, the relative amplitudes of these different components also will be the same.

The modulation parameters of a PM system relate as follows:

$$\Delta f = m_p \times f_m \tag{16.2}$$

Where:
Δf = frequency deviation of the carrier
m_p = phase shift of the carrier
f_m = modulating frequency

In a phase-modulated wave, the phase shift m_p is independent of the modulating frequency; the frequency deviation Δf is proportional to the modulating frequency. In contrast, with a frequency-modulated wave, the frequency deviation is independent of modulating frequency. Therefore, a frequency-modulated wave can be obtained from a phase modulator by making the modulating voltage applied to the phase modulator inversely proportional to frequency. This can be readily achieved in hardware.

16.2 Modifying FM Waves

When a frequency-modulated wave is passed through a harmonic generator, the effect is to increase the modulation index by a factor equal to the frequency multiplication involved. Similarly, if the frequency-modulated wave is passed through a frequency divider, the effect is to reduce the modulation index by the factor of frequency division. Thus, the frequency components contained in the wave—and, consequently, the bandwidth of the wave—will be increased or decreased, respectively, by frequency multiplication or division. No distortion in the nature of the modulation is introduced by the frequency change.

When an FM wave is translated in the frequency spectrum by heterodyne action, the modulation index—hence the relative positions of the sideband frequencies and the bandwidths occupied by them—remains unchanged.

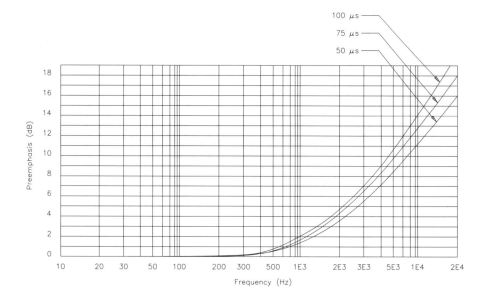

Figure 16.4 Preemphasis curves for time constants of 50, 75, and 100 μs.

16.2.1 Preemphasis and Deemphasis

The FM transmission/reception system offers significantly better noise-rejection characteristics than AM. However, FM noise rejection is more favorable at low modulating frequencies than at high frequencies because of the reduction in the number of sidebands at higher frequencies. To offset this problem, the input signal to the FM transmitter may be *preemphasized* to increase the amplitude of higher-frequency signal components in normal program material. FM receivers utilize complementary *deemphasis* to produce a flat overall system frequency response.

FM broadcasting, for example, uses a 75 μs preemphasis curve, meaning that the time constant of the resistance-inductance (*RL*) or resistance-capacitance (*RC*) circuit used to provide the boost of high frequencies is 75 μs. Other values of preemphasis are used in different types of FM communications systems. Figure 16.4 shows three common preemphasis curves.

16.2.2 Modulation Circuits

Early FM transmitters used *reactance modulators* that operated at a low frequency. The output of the modulator then was multiplied to reach the desired output frequency. This approach was acceptable for some applications and unsuitable for others. Modern FM systems utilize what is referred to as *direct modulation*; that is, the

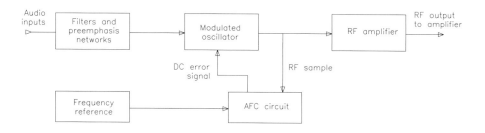

Figure 16.5 Block diagram of an FM exciter.

frequency modulation occurs in a modulated oscillator that operates on a center frequency equal to the desired transmitter output frequency.

Various techniques have been developed to generate the direct-FM signal. One of the most popular uses a variable-capacity diode as the reactive element in the oscillator. The modulating signal is applied to the diode, which causes the capacitance of the device to vary as a function of the magnitude of the modulating signal. Variations in the capacitance cause the frequency of the oscillator to change. The magnitude of the frequency shift is proportional to the amplitude of the modulating signal, and the rate of frequency shift is equal to the frequency of the modulating signal.

The direct-FM modulator is one element of an FM transmitter *exciter*, which generates the composite FM waveform. A block diagram of a complete FM exciter is shown in Figure 16.5. Input signals are buffered, filtered, and preemphasized before being summed to feed the modulated oscillator. Note that the oscillator is not normally coupled directly to a crystal, but to a free-running oscillator adjusted as closely as possible to the carrier frequency of the transmitter. The final operating frequency is maintained carefully by an automatic frequency control system employing a *phase locked loop* (PLL) tied to a reference crystal oscillator or frequency synthesizer.

A solid-state class C amplifier typically follows the modulated oscillator and raises the operating power of the FM signal to 20 to 30 W. One or more subsequent amplifiers in the transmitter raise the signal power to several hundred watts for application to the final power amplifier stage. Nearly all high-power FM transmitters use solid-state amplifiers up to the final RF stage, which is generally a vacuum tube for operating powers of 15 kW and above. All stages operate in the class C mode. In contrast to AM systems, each stage in an FM power amplifier can operate class C; no information is lost from the frequency-modulated signal because of amplitude changes. As mentioned previously, FM is a constant-power system.

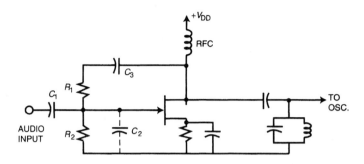

Figure 16.6 Simplified reactance modulator. (*From* [1]. *Used with permission.*)

Direct-FM Modulator

Many types of circuits have been used to produce a direct-FM signal [1]. In each case, a reactance device is used to shunt capacitive or inductive reactance across an oscillator. The value of capacitive or inductive reactance is made to vary as the amplitude of the modulating signal varies. Because the reactive load is placed across an oscillator's tuned circuit, the frequency of the oscillator will therefore shift by a predetermined amount, thereby creating an FM signal.

A typical example of a reactance modulator is illustrated in Figure 16.6. The circuit uses a field-effect transistor (FET), where the modulating signal is applied to the modulator through C_1. The actual components that affect the overall reactance consist of R_1 and C_2. Typically, the value of C_2 is small as this is the input capacitance to the FET, which may only be a few picofarads. However, this capacitance will generally be much larger by a significant amount as a result of the *Miller effect*. Capacitor C_3 has no significant effect on the reactance of the modulator; it is strictly a blocking capacitor that keeps dc from changing the gate bias of the FET.

To further understand the performance of the reactance modulator, the equivalent circuit of Figure 16.6 is represented in Figure 16.7. The FET is represented as a current source, gmV_g, with the internal drain resistance, r_d. The impedances Z_1 and Z_2 are a combination of resistance and capacitive reactance, which are designed to provide a 90° phase shift.

Using vector diagrams, we can analyze the phase relationship of the reactance modulator. Referring to Figure 16.6, the resistance of R_1 is typically high compared to the capacitive reactance of C_2. The R_1C_2 circuit is then resistive. Because this circuit is resistive, the current, I_{AB}, that flows through it is in phase with the voltage V_{AB}. Voltage V_{AB} is also across R_1C_2 (or Z_1Z_2 in Figure 16.7). This is true because current and voltage tend to be in phase in a resistive network. However, voltage V_{C2}, which is across C_2, is out of phase with I_{AB}. This is because the voltage that is across a capacitor lags behind its current by 90°. (See Figure 16.8.)

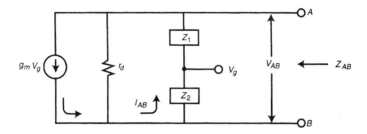

Figure 16.7 Equivalent circuit of the reactance modulator. (*From* [1]. *Used with permission.*)

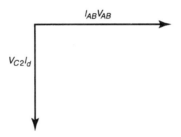

Figure 16.8 Vector diagram of a reactance modulator producing FM. Note: $V_g = V_{C2}$. (*From* [1]. *Used with permission.*)

VCO Direct-FM Modulator

One of the more common direct-FM modulation techniques uses an analog *voltage controlled oscillator* (VCO) in a phase locked loop arrangement [1]. In this configuration, shown in Figure 16.9, a VCO produces the desired carrier frequency that is—in turn—modulated by applying the input signal to the VCO input via a variactor diode. The variactor is used to vary the capacitance of an oscillator tank circuit. Therefore, the variactor behaves as a variable capacitor whose capacitance changes as the signal voltage across it changes. As the input capacitance of the VCO is varied by the variactor, the output frequency of the VCO is shifted, which produces a direct-FM modulated signal.

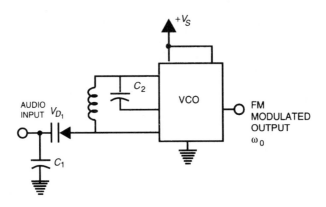

Figure 16.9 Voltage-controlled direct-FM modulator. (*From* [1]. *Used with permission.*)

16.3 References

1. Seymour, Ken, "Frequency Modulation," in *The Electronics Handbook*, Jerry C. Whitaker (ed.), CRC Press, Boca Raton, FL, pp. 1188–1200, 1996.

16.4 Bibliography

Benson, K. B., and Jerry C. Whitaker (eds.), *Television Engineering Handbook*, McGraw-Hill, New York, NY, 1986.

Benson, K. B., and Jerry C. Whitaker, *Television and Audio Handbook for Technicians and Engineers*, McGraw-Hill, New York, NY, 1989.

Crutchfield, E. B. (ed.), *NAB Engineering Handbook*, 8th ed., National Association of Broadcasters, Washington, DC, 1991.

Fink, D., and D. Christiansen (eds.), *Electronics Engineers' Handbook*, 3rd ed., McGraw-Hill, New York, NY, 1989.

Fink, D., and D. Christiansen (eds.), *Electronics Engineers' Handbook*, 2nd ed., McGraw-Hill, New York, NY, 1982.

Jordan, Edward C., *Reference Data for Engineers: Radio, Electronics, Computer and Communications*, 7th ed., Howard W. Sams, Indianapolis, IN, 1985.

Laboratory Staff, *The Care and Feeding of Power Grid Tubes*, Varian Eimac, San Carlos, CA, 1984.

Mendenhall, G. N., "Fine Tuning FM Final Stages," *Broadcast Engineering*, Intertec Publishing, Overland Park, KS, May 1987.

Whitaker, Jerry C., *Maintaining Electronic Systems*, CRC Press, Boca Raton, FL, 1992.

Whitaker, Jerry C., *Radio Frequency Transmission Systems: Design and Operation*, McGraw-Hill, New York, NY, 1991.

17

Pulse Modulation

17.1 Introduction

The growth of digital processing and communications has led to the development of modulation systems tailor-made for high-speed, spectrum-efficient transmission. In a *pulse modulation* system, the unmodulated carrier usually consists of a series of recurrent pulses. Information is conveyed by modulating some parameter of the pulses, such as amplitude, duration, time of occurrence, or shape. Pulse modulation is based on the *sampling principle*, which states that a message waveform with a spectrum of finite width can be recovered from a set of discrete samples if the sampling rate is higher than twice the highest sampled frequency (the Nyquist criteria). The samples of the input signal are used to modulate some characteristic of the carrier pulses.

17.2 Digital Modulation Systems

Because of the nature of digital signals (on or off), it follows that the amplitude of the signal in a pulse modulation system should be one of two heights (present or absent/positive or negative) for maximum efficiency. Noise immunity is a significant advantage of such a system. It is necessary for the receiving system to detect only the presence or absence (or polarity) of each transmitted pulse to allow complete reconstruction of the original intelligence. The pulse shape and noise level have minimal effect (to a point). Furthermore, if the waveform is to be transmitted over long distances, it is possible to regenerate the original signal exactly for retransmission to the next relay point. This feature is in striking contrast to analog modulation systems in which each modulation step introduces some amount of noise and signal corruption.

In any practical digital data system, some corruption of the intelligence is likely to occur over a sufficiently large span of time. Data encoding and manipulation schemes have been developed to detect and correct or conceal such errors. The addition of error-correction features comes at the expense of increased system overhead and (usually) slightly lower intelligence throughput.

17.2.1 Pulse Amplitude Modulation (PAM)

Pulse amplitude modulation (PAM) is one of the simplest forms of data modulation. PAM departs from conventional modulation systems in that the carrier exists as a series of pulses, rather than as a continuous waveform. The amplitude of the pulse train is modified in accordance with the applied modulating signal to convey intelligence, as illustrated in Figure 17.1. There are two primary forms of PAM sampling:

- *Natural sampling* (or *top sampling*), where the modulated pulses follow the amplitude variation of the sampled time function during the sampling interval.

- *Instantaneous sampling* (or *square-topped sampling*), where the amplitude of the pulses is determined by the instantaneous value of the sampled time function corresponding to a single instant of the sampling interval. This "single instant" may be the center or edge of the sampling interval.

There are two common methods of generating a PAM signal:

- Variation of the amplitude of a pulse sequence about a fixed nonzero value (or *pedestal*). This approach constitutes double-sideband amplitude modulation.

- Double-polarity modulated pulses with no pedestal. This approach constitutes double-sideband suppressed carrier modulation.

17.2.2 Pulse Time Modulation (PTM)

A number of modulating schemes have been developed to take advantage of the noise immunity afforded by a constant amplitude modulating system. *Pulse time modulation* (PTM) is one of those systems. In a PTM system, instantaneous samples of the intelligence are used to vary the time of occurrence of some parameter of the pulsed carrier. Subsets of the PTM process include:

- *Pulse duration modulation* (PDM), where the time of occurrence of either the leading or trailing edge of each pulse (or both pulses) is varied from its unmodulated position by samples of the input modulating waveform. PDM also may be described as *pulse length* or *pulse width* modulation (PWM).

- *Pulse position modulation* (PPM), where samples of the modulating input signal are used to vary the position in time of pulses, relative to the unmodulated waveform. Several types of pulse time modulation waveforms are shown in Figure 17.2.

- *Pulse frequency modulation* (PFM), where samples of the input signal are used to modulate the frequency of a series of carrier pulses. The PFM process is illustrated in Figure 17.3.

It should be emphasized that all of the pulse modulation systems discussed thus far may be used with both analog and digital input signals. Conversion is required for either signal into a form that can be accepted by the pulse modulator.

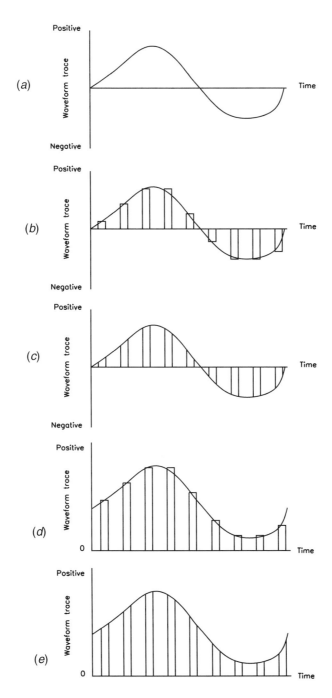

Figure 17.1 Pulse amplitude modulation waveforms: (*a*) modulating signal; (*b*) square-topped sampling, bipolar pulse train; (*c*) topped sampling, bipolar pulse train; (*d*) square-topped sampling, unipolar pulse train; (*e*) top sampling, unipolar pulse train.

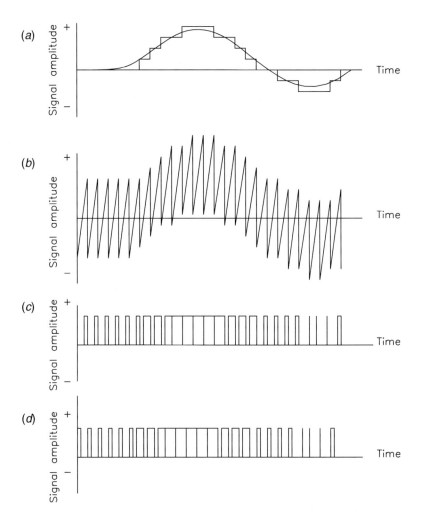

Figure 17.2 Pulse time modulation waveforms: (*a*) modulating signal and sample-and-hold (S/H) waveforms, (*b*) sawtooth waveform added to S/H, (*c*) leading-edge PTM, (*d*) trailing-edge PTM.

17.2.3 Pulse Code Modulation (PCM)

The pulse modulation systems discussed previously are *unencoded* systems. *Pulse code modulation* (PCM) is a scheme wherein the input signal is *quantized* into discrete steps and then sampled at regular intervals (as in conventional pulse modulation). In the *quantization* process, the input signal is sampled to produce a code representing the instantaneous value of the input within a predetermined range of values.

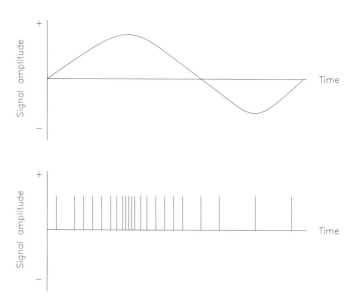

Figure 17.3 Pulse frequency modulation.

Figure 17.4 illustrates the concept. Only certain discrete levels are allowed in the quantization process. The code is then transmitted over the communications system as a pattern of pulses.

Quantization inherently introduces an initial error in the amplitude of the samples taken. This *quantization error* is reduced as the number of quantization steps is increased. In system design, tradeoffs must be made regarding low quantization error, hardware complexity, and occupied bandwidth. The greater the number of quantization steps, the wider the bandwidth required to transmit the intelligence or, in the case of some signal sources, the slower the intelligence must be transmitted.

In the classic design of a PCM encoder, the quantization steps are equal. The quantization error (or *quantization noise*) usually can be reduced, however, through the use of nonuniform spacing of levels. Smaller quantization steps are provided for weaker signals, and larger steps are provided near the peak of large signals. Quantization noise is reduced by providing an encoder that is matched to the *level distribution* (*probability density*) of the input signal.

Nonuniform quantization typically is realized in an encoder through processing of the input (analog) signal to compress it to match the desired nonuniformity. After compression, the signal is fed to a uniform quantization stage.

17.2.4 Delta Modulation (DM)

Delta modulation (DM) is a coding system that measures changes in the direction of the input waveform, rather than the instantaneous value of the wave itself. Figure 17.5

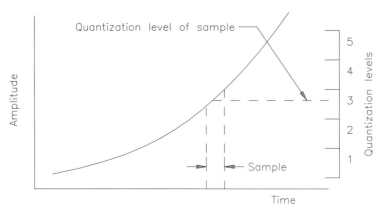

Figure 17.4 The quantization process.

illustrates the concept. The clock rate is assumed to be constant. Transmitted pulses from the pulse generator are positive if the signal is changing in a positive direction; they are negative if the signal is changing in a negative direction.

As with the PCM encoding system, quantization noise is a parameter of concern for DM. Quantization noise can be reduced by increasing the sampling frequency (the pulse generator frequency). The DM system has no fixed maximum (or minimum) signal amplitude. The limiting factor is the slope of the sampled signal, which must not change by more than one level or step during each pulse interval.

17.3 Digital Coding Systems

A number of methods exist to transmit digital signals over long distances in analog transmission channels. Some of the more common systems include:

- *Binary on-off keying* (BOOK), a method by which a high-frequency sinusoidal signal is switched on and off corresponding to 1 and 0 (on and off) periods in the input digital data stream. In practice, the transmitted sinusoidal waveform does not start or stop abruptly, but follows a predefined ramp up or down.

- *Binary frequency-shift keying* (BFSK), a modulation method in which a continuous wave is transmitted that is shifted between two frequencies, representing 1s and 0s in the input data stream. The BFSK signal may be generated by switching between two oscillators (set to different operating frequencies) or by applying a binary baseband signal to the input of a voltage-controlled oscillator (VCO). The transmitted signals often are referred to as a *mark* (binary digit 1) or a *space* (binary digit 0). Figure 17.6 illustrates the transmitted waveform of a BFSK system.

- *Binary phase-shift keying* (BPSK), a modulating method in which the phase of the transmitted wave is shifted 180° in synchronism with the input digital signal.

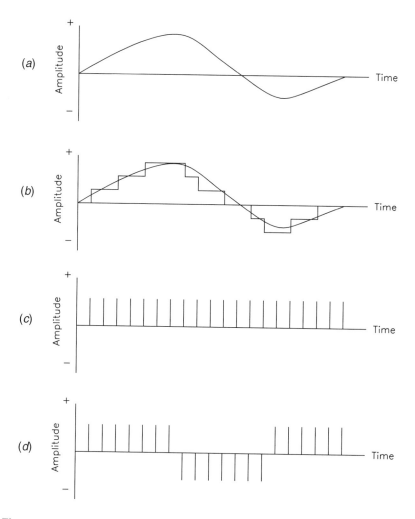

Figure 17.5 Delta modulation waveforms: (*a*) modulating signal, (*b*) quantized modulating signal, (*c*) pulse train, (*d*) resulting delta modulation waveform.

The phase of the RF carrier is shifted by $\pi/2$ radians or $-\pi/2$ radians, depending upon whether the data bit is a 0 or a 1. Figure 17.7 shows the BPSK transmitted waveform.

- *Quadriphase-shift keying* (QPSK), a modulation scheme similar to BPSK except that quaternary modulation is employed, rather than binary modulation. QPSK requires half the bandwidth of BPSK for the same transmitted data rate.

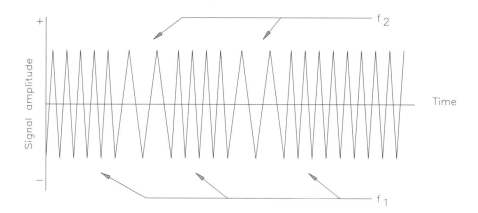

Figure 17.6 Binary FSK waveform.

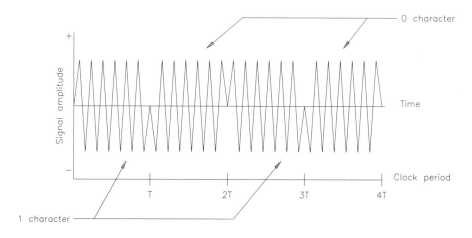

Figure 17.7 Binary PSK waveform.

17.3.1 Baseband Digital Pulse Modulation

After the input samples have been quantized, they are transmitted through a channel, received, and converted back to their approximate original form [1]. The format (modulation scheme) applied to the quantized samples is determined by a number of factors, not the least of which is the channel through which the signal passes. A number of different formats are possible and practical.

Several common digital modulation formats are shown in Figure 17.8. The first (a) is referred to as *non-return-to-zero* (NRZ) polar because the waveform does not return to zero during each signaling interval, but switches from $+V$ to $-V$, or vice versa, at the

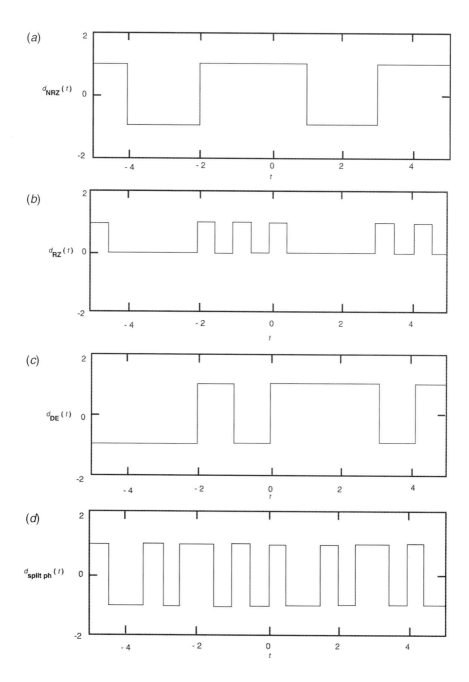

Figure 17.8 Various baseband modulation formats: (*a*) non-return-to zero, (*b*) unipolar return-to-zero, (*c*) differential encoded (NRZ-mark), (*d*) split phase. (*From* [1]. *Used with permission.*)

end of each signaling interval (NRZ unipolar uses the levels V and 0). On the other hand, the *unipolar return-to-zero* (RZ) format, shown in (*b*) returns to zero in each signaling interval. Because bandwidth is inversely proportional to pulse duration, it is apparent that RZ requires twice the bandwidth that NRZ does. Also, RZ has a nonzero dc component, whereas NRZ does not necessarily have a nonzero component (unless there are more 1s than 0s or vice versa). An advantage of RZ over NRZ is that a pulse transition is guaranteed in each signaling interval, whereas this is not the case for NRZ. Thus, in cases where there are long strings of 1s or 0s, it may be difficult to synchronize the receiver to the start and stop times of each pulse in NRZ-based systems. A very important modulation format from the standpoint of synchronization considerations is NRZ-*mark*, also known as *differential encoding*, where an initial reference bit is chosen and a subsequent 1 is encoded as a change from the reference and a 0 is encoded as no change. After the initial reference bit, the current bit serves as a reference for the next bit, and so on. An example of this modulation format is shown in (*c*).

Manchester is another baseband data modulation format that guarantees a transition in each signaling interval and does not have a dc component. Also known as *biphase* or *split phase*, this scheme is illustrated in (*d*). The format is produced by *OR*ing the data clock with an NRZ-formatted signal. The result is a + to − transition for a logic 1, and a − to + zero crossing for a logic 0.

A number of other data formats have been proposed and employed in the past, but further discussion is beyond the scope of this chapter.

17.4 References

1. Ziemer, Rodger E., "Pulse Modulation," in *The Electronics Handbook*, Jerry C. Whitaker (ed.), CRC Press, Boca Raton, FL, pp. 1201–1212, 1996.

17.5 Bibliography

Benson, K. B., and Jerry C. Whitaker (eds.), *Television Engineering Handbook*, McGraw-Hill, New York, NY, 1986.

Benson, K. B., and Jerry C. Whitaker, *Television and Audio Handbook for Technicians and Engineers*, McGraw-Hill, New York, NY, 1989.

Crutchfield, E. B. (ed.), *NAB Engineering Handbook*, 8th ed., National Association of Broadcasters, Washington, DC, 1991.

Fink, D., and D. Christiansen (eds.), *Electronics Engineers' Handbook*, 3rd ed., McGraw-Hill, New York, NY, 1989.

Fink, D., and D. Christiansen (eds.), *Electronics Engineers' Handbook*, 2nd ed., McGraw-Hill, New York, NY, 1982.

Jordan, Edward C., *Reference Data for Engineers: Radio, Electronics, Computer and Communications*, 7th ed., Howard W. Sams, Indianapolis, IN, 1985.

Whitaker, Jerry C., *Maintaining Electronic Systems*, CRC Press, Boca Raton, FL, 1992.

Whitaker, Jerry C., *Radio Frequency Transmission Systems: Design and Operation*, McGraw-Hill, New York, NY, 1991.

Network Communications

18.1 Introduction

The *open system interconnections* (OSI) model is the most broadly accepted explanation of LAN transmissions in an open system. The reference model was developed by the International Organization for Standardization (ISO) to define a framework for computer communication. The OSI model divides the process of data transmission into the following steps:

- Physical layer
- Data-link layer
- Network layer
- Transport layer
- Session layer
- Presentation layer
- Application layer

An overview of the OSI model is illustrated in Figure 18.1.

18.1.1 Physical Layer

Layer 1 of the OSI model is responsible for carrying an electrical current through the computer hardware to perform an exchange of information. The physical layer is defined by the following parameters:

- Bit transmission rate
- Type of transmission medium (twisted-pair, coaxial cable, or fiber-optic cable)
- Electrical specifications, including voltage- or current-based, and balanced or unbalanced

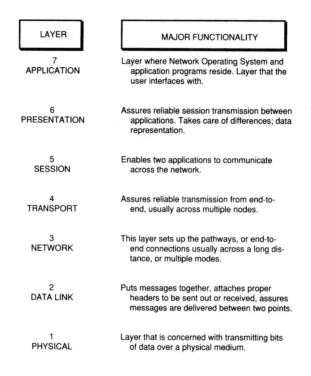

LAYER	MAJOR FUNCTIONALITY
7 APPLICATION	Layer where Network Operating System and application programs reside. Layer that the user interfaces with.
6 PRESENTATION	Assures reliable session transmission between applications. Takes care of differences; data representation.
5 SESSION	Enables two applications to communicate across the network.
4 TRANSPORT	Assures reliable transmission from end-to-end, usually across multiple nodes.
3 NETWORK	This layer sets up the pathways, or end-to-end connections usually across a long distance, or multiple modes.
2 DATA LINK	Puts messages together, attaches proper headers to be sent out or received, assures messages are delivered between two points.
1 PHYSICAL	Layer that is concerned with transmitting bits of data over a physical medium.

Figure 18.1 The OSI reference model.

- Type of connectors used (normally RJ-45 or DB-9)
- Many different implementations exist at the physical layer

Installation Considerations

Layer 1 can exhibit error messages as a result of overusage. For example, if a file server is being burdened with requests from workstations, the results may show up in error statistics that reflect the server's inability to handle all incoming requests. An overabundance of response timeouts may also be noted in this situation. A response timeout (in this context) is a message sent back to the workstation stating that the waiting period allotted for a response from the file server has passed without action from the server.

Error messages of this sort, which can be gathered by any number of commercially available software diagnostic utilities, can indicate an overburdened file server or a hardware flaw within the system. Intermittent response timeout errors can be caused by a corrupted network interface card (NIC) in the server. A steady flow of timeout errors throughout all nodes on the network may indicate the need for another server or bridge. Hardware problems are among the easiest to locate. In simple configurations, where something has suddenly gone wrong, the physical and data-link layers are usually the first suspects.

18.1.2 Data Link Layer

Layer 2 of the OSI model, the data-link layer, describes hardware that enables data transmission (NICs and cabling systems). This layer integrates data packets into messages for transmission and checks them for integrity. Sometimes layer 2 will also send an "arrived safely" or "did not arrive correctly" message back to the transport layer (layer 4), which monitors this communications layer. The data-link layer must define the frame (or package) of bits that is transmitted down the network cable. Incorporated within the frame are several important fields:

- Addresses of source and destination workstations

- Data to be transmitted between workstations

- Error control information, such as a *cyclic redundancy check* (CRC), which assures the integrity of the data

The data-link layer must also define the method by which the network cable is accessed, because only one workstation may transmit at a time on a baseband LAN. The two predominant schemes are:

- *Token passing*, used with the ARCnet and token-ring networks

- *Carrier sense multiple access with collision detection* (CSMA/CD), used with Ethernet and starLAN networks

At the data-link layer, the true identity of the LAN begins to emerge.

Installation Considerations

Because most functions of the data-link layer (in a PC-based system) take place in integrated circuits on NICs, software analysis is generally not required in the event of an installation problem. As mentioned previously, when something happens on the network, the data-link layer is among the first to be suspect. Because of the complexities of linking multiple topologies, cabling systems, and operating systems, the following failure modes may be experienced:

- RF disturbance. Transmitters, ac power controllers, and other computers can all generate energy that may interfere with data transmitted on the cable. RF interference (RFI) is usually the single biggest problem in a broadband network. This problem can manifest itself through excessive checksum errors and/or garbled data.

- Excessive cable run. Problems related to the data-link layer may result from long cable runs. Ethernet runs can stretch to 1,000 ft, depending on the cable. A typical token-ring system can stretch 600 ft, with the same qualification. The need for additional distance can be accommodated by placing a bridge, gateway, active hub, equalizer, or amplifier on the line.

The data-link layer usually includes some type of routing hardware with one or more of the following:

- Active hub

- Passive hub

- Multiple access units (for token-ring, starLAN, and ARCnet networks)

18.1.3 Network Layer

Layer 3 of the OSI model guarantees the delivery of transmissions as requested by the upper layers of the OSI. The network layer establishes the physical path between the two communicating endpoints through the *communications subnet*, the common name for the physical, data-link, and network layers taken collectively. As such, layer 3 functions (routing, switching, and network congestion control) are critical. From the viewpoint of a single LAN, the network layer is not required. Only one route—the cable—exists. Internetwork connections are a different story, however, because multiple routes are possible. The internet protocol (IP) and internet packet exchange (IPX) are two examples of layer 3 protocols.

Installation Considerations

The network layer confirms that signals get to their designated targets, and then translates logical addresses into physical addresses. The physical address determines where the incoming transmission is stored. Lost data errors can usually be traced back to the network layer, in most cases incriminating the network operating system. The network layer is also responsible for statistical tracking and communications with other environments, including gateways. Layer 3 decides which route is the best to take, given the needs of the transmission. If router tables are being corrupted or excessive time is required to route from one network to another, an operating system error on the network layer may be involved.

18.1.4 Transport Layer

Layer 4, the transport layer, acts as an interface between the bottom three and the upper three layers, ensuring that the proper connections are maintained. It does the same work as the network layer, only on a local level. The network operating system driver performs transport layer tasks.

Installation Considerations

Connection flaws between computers on a network can sometimes be attributed to the *shell driver*. The transport layer may be able to save transmissions that were en route in the case of a system crash, or reroute a transmission to its destination in case of a primary route failure. The transport layer also monitors transmissions, checking to make sure that packets arriving at the destination node are consistent with the build specifications given to the sending node in layer 2. The data-link layer in the sending node builds a series of packets according to specifications sent down from higher levels, then transmits the packets to a destination node. The transport layer monitors these packets to ensure that they arrive according to specifications indicated in the original build order. If they do not, the transport layer calls for a retransmission. Some operating systems refer to this technique as a *sequenced packet exchange* (SPX) transmission, meaning that the operating system guarantees delivery of the packet.

18.1.5 Session Layer

Layer 5 is responsible for turning communications on and off between communicating parties. Unlike other levels, the session layer can receive instructions from the application layer through the network basic input/output operation system (netBIOS), skipping the layer directly above it. The netBIOS protocol allows applications to "talk" across the network. The session layer establishes the session, or logical connection, between communicating host processors. Name-to-address translation is another important function; most communicating processors are known by a common name, rather than a numerical address.

Installation Considerations

Multi-vendor problems can often arise in the session layer. Failures relating to gateway access usually fall into layer 5 for the OSI model, and are often related to compatibility issues.

18.1.6 Presentation Layer

Layer 6 translates application layer commands into syntax that is understood throughout the network. It also translates incoming transmissions for layer 7. The presentation layer masks other devices and software functions. Reverse video, blinking cursors, and graphics also fall into the domain of the presentation layer. Layer 6 software controls printers and plotters, and may handle encryption and special file formatting. Data compression, encryption, and ASCII translations are examples of presentation layer functions.

Installation Considerations

Failures in the presentation layer are often the result of products that are not compatible with the operating system, an interface card, a resident protocol, or another application.

18.1.7 Application Layer

At the top of the seven-layer stack is the application layer. It is responsible for providing protocols that facilitate user applications. Print spooling, file sharing, and e-mail are components of the application layer, which translates local application requests into network application requests. Layer 7 provides the first layer of communications into other open systems on the network.

Installation Considerations

Failures at the application layer usually center on software quality and compatibility issues. The program for a complex network may include latent faults that will manifest only when a specific set of conditions are present. The compatibility of the network software with other programs is another source of potential complications.

18.2 Transmission System Options

A variety of options beyond the traditional local serial interface are available for linking intelligent devices. The evolution of wide area network (WAN) technology has permitted efficient two-way transmission of data between distant computer systems. High-speed facilities are cost-effective and widely available from the telephone company (telco) central office to the customer premises. Private communications companies also provide interconnection services.

LANs have proliferated and integrated with WANs through *bridges* and *gateways*. Interconnections via fiber-optic cable are common. Further extensions of the basic LAN include the following:

- *Campus area network* (CAN)—designed for communications within an industrial or educational campus.

- *Metropolitan area network* (MAN)—designed for communications among different facilities within a certain metropolitan area. MANs generally operate over common-carrier-owned switched networks installed in and over public rights of way.

- *Regional area network* (RAN)—interconnecting MANs within a unified geographical area, generally installed and owned by interexchange carriers (IECs).

- *Wide area network* (WAN)—communications systems operating over large geographic areas. Common carrier networks interconnect MANs and RANs within a contiguous land mass, generally within a country's political boundaries.

- *Global area network* (GAN)—networks interconnecting WANs, both across national borders and ocean floors, including between continents.

These network systems can carry a wide variety of multiplexed analog and/or digital signal transmissions on a single piece of coax or fiber.

18.2.1 System Design Alternatives

The signal form at the input and/or output interface of a large cable or fiber system may be either analog or digital, and the number of independent electrical signals transmitted may be one or many. Independent electrical signals may be combined into one signal for optical transmission by virtually unlimited combinations of electrical analog frequency division multiplexing (using analog AM and/or FM carriers) and digital bit stream multiplexing. Frequency division multiplexing involves the integration of two or more discrete signals into one complex electrical signal.

With the current availability of fiber-optic transmission lines, fiber interconnection of data networks is the preferred route for new systems. Three primary multiplexing schemes are used for fiber transmission:

- *Frequency division multiplexing* (FDM)

- *Time division multiplexing* (TDM)

- *Wave(length) division multiplexing* (WDM)

Frequency Division Multiplexing

The FDM technique of summing multiple AM or FM carriers is widely used in coaxial cable distribution. Unfortunately, nonlinearity of optical devices operated in the intensity-modulation mode can result in substantial—and often unacceptable—noise and intermodulation distortion in the delivered signal channels. Wide and selective spacing of carriers ameliorates this problem to some degree.

Time Division Multiplexing

TDM involves sampling the input signals at a high rate, converting the samples to high-speed digital codes, and interleaving the codes into pre-determined time slots. The principles of digital TDM are straightforward. Specific-length bit groups in a high-speed digital bit stream are repetitively allocated to carry the digital representations of individual analog signals and/or the outputs of separate digital devices.

Wave(length) Division Multiplexing

This multiplexing technique, illustrated in Figure 18.2, reduces the number of optical fibers required to meet a specific transmission requirement. Two or more complete and independent fiber-transmission systems operating at different optical wavelengths can be transported over a single fiber by combining them in a passive optical multiplexer. This device is an assembly in which pigtails from multiple optical transmitters are fused together and spliced into the transporting fiber. Demultiplexing the optical signals at the receiver end of the circuit is accomplished in an opposite-oriented passive optical multiplexer. The pigtails are coupled into photodetectors through wavelength-selective optical filters.

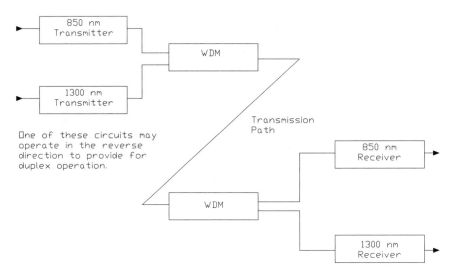

Figure 18.2 Basic operation of a wave division multiplexer. This type of passive assembly is created by fusing optical-fiber pigtails.

18.2.2 Selecting Cable for Digital Signals

Cable for the transmission of digital signals is selected on the basis of its electrical performance: the ability to transmit the required number of pulses at a specified bit rate over a specified distance, and its conformance to appropriate industry or government standards. A wide variety of data cables are available from manufacturers. Figure 18.3 illustrates some of the more common types. The type of cable chosen for an application is determined by the following:

- Type of network involved. Different network designs require different types of cable.

- Distance to be traveled. Long runs require low-loss cable.

- Physical environment. Local and national safety codes require specific types of cable for certain indoor applications. Outdoor applications require a cable suitable for burial or exposure to the elements.

- Termination required. The choice of cable type may be limited by the required connector termination on one or both ends.

18.2.3 Data Patch Panel

The growth of LANs has led to the development of a variety of interconnection racks and patch panels. Figure 18.4 shows two common types. Select data patch panels that offer many cycles of repeated insertion and removal. Use components specifically designed for network interconnection. Such components include the following:

(a)

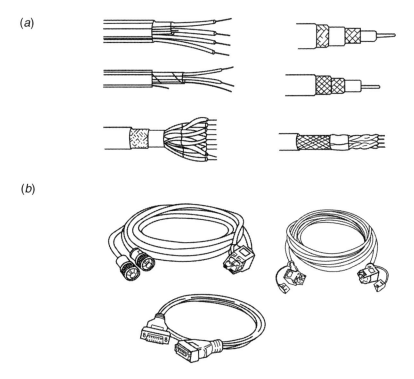

(b)

Figure 18.3 Common types of data cable: (a) shielded pair, multi-pair shielded, and coax, (b) data cable with various terminations.

- Twisted-pair network patch panels
- Coax-based network patch panels
- Fiber-based network patch panels
- Modular feed-through (normalled) patch panels
- Pre-assembled patch cables of various lengths
- Pre-assembled "Y" patch cables
- Patch cables offering different connectors on each end
- Media filter cables
- Balanced-to-unbalanced (balun) cable assemblies

Connector termination options for patch hardware include the following:

- Insulation displacement (punch-block) for twisted-pair cable
- Screw terminal (for twisted-pair)

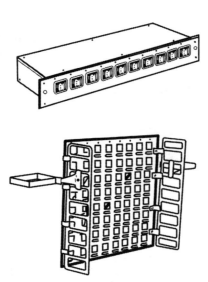

Figure 18.4 Data network patch-panel hardware.

- BNC connectors for coax
- Fiber-termination hardware

Although the cost of pre-assembled network patch panels and patch cables is higher than purchasing the individual components and then assembling them, most system engineers should specify factory-assembled hardware. Reliability is greater with pre-assembled elements, and installation is considerably faster.

18.3 Bibliography

Benson, K. B., and J. Whitaker, *Television and Audio Handbook for Engineers and Technicians*, McGraw-Hill, New York, NY, 1989.

Crutchfield, E. B. (ed.), *NAB Engineering Handbook*, 8th ed., National Association of Broadcasters, Washington, DC, 1992.

Dahlgren, Michael W., "Servicing Local Area Networks," *Broadcast Engineering*, Intertec Publishing, Overland Park, KS, November 1989.

International Organization for Standardization, "Information Processing Systems—Open Systems Interconnection—Basic Reference Model," ISO 7498, 1984.

Pearson, Eric, *How to Specify and Choose Fiber-Optic Cables,* Pearson Technologies, Acworth, GA, 1991.

Whitaker, Jerry C., *AC Power Systems*, 2nd ed., CRC Press, Boca Raton, FL, 1998.

Whitaker, Jerry C., *Maintaining Electronic Systems*, CRC Press, Boca Raton, FL, 1991.

Optical Devices and Systems

19.1 Introduction

Fiber-optic (FO) technology offers the end-user a number of benefits over metallic cable, including:

- **Signal-carrying ability.** The bandwidth information-carrying capacity of a communications link is directly related to the operating frequency. Light carrier frequencies are several orders of magnitude higher than the highest radio frequencies. Fiber-optic systems easily surpass the information-carrying capacity of microwave radio and coaxial cable alternatives; and fiber's future carrying capacity has only begun to be used. Fiber provides bandwidths in excess of several gigahertz per kilometer, which allows high-speed transfer of all types of information. Multiplexing techniques allow many signals to be sent over a single fiber.

- **Low loss.** A fiber circuit provides substantially lower attenuation than copper cables and twisted pairs. It also requires no equalization. Attenuation below 0.5 dB/km is available for certain wavelengths.

- **Electrical isolation.** The fiber and its coating are dielectric material, and the transmitter and receiver in each circuit are electrically isolated from each other. Isolation of separated installations from respective electrical grounds is assured if the strength material (messenger) in the cable is also a dielectric. Lightwave transmission is free of spark hazards and creates no EMI. All-dielectric fiber cable may also be installed in hazardous or toxic environments.

- **Size and weight.** An optical waveguide is less than the diameter of a human hair. A copper cable is many times larger, stiffer, and heavier than a fiber that carries the same quantity of signals. Installation, duct, and handling costs are much lower for a fiber installation than for a similar coaxial system. Fiber cable is the only alternative for circuit capacity expansion when ducts are full of copper.

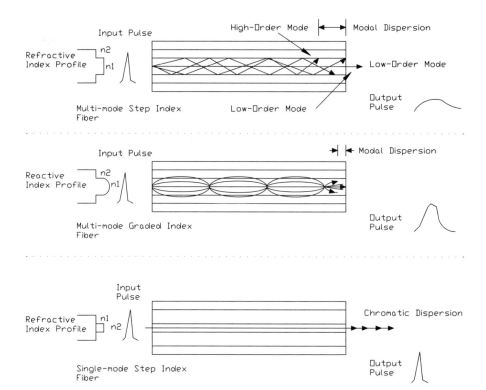

Figure 19.1 Modal dispersion in an FO cable. The core diameter and its refractive index characteristics determine the light propagation path(s) within the fiber core.

19.2 Types of Fibers

Of the many ways to classify fibers, the most informative is by *refractive index profile* and number of modes supported. The two main types of index profiles are *step* and *graded*. In a *step index* fiber, the core has a uniform index with a sharp change at the boundary of the cladding. In a *graded index* fiber, the core index is not uniform; it is highest at the center and decreases until it matches the cladding.

19.2.1 Step Index Multi-mode Fiber

A multi-mode step index fiber typically has a core diameter in the 50 to 1,000 micron range. The large core permits many modes of propagation. Because light will reflect differently for different modes, the path of each ray is a different length. The lowest-order mode travels down the center; higher-order modes strike the core-cladding interface at angles near the critical angle. As a result, a narrow pulse of light spreads out as it travels through this type of fiber. This spreading is called *modal dispersion* (Figure 19.1).

19.2.2 Step Index Single (Mono) -mode Fiber

Modal dispersion can be reduced by making the fiber core small, typically 5 to 10 microns (1/6 the diameter of a human hair). At this diameter, only one mode propagates efficiently. The small size of the core makes it difficult to splice. Single mode of propagation permits high-speed, long-distance transmission.

19.2.3 Graded Index Multi-mode Fiber

Like the step index single-mode fiber, a graded index fiber also limits modal dispersion. The core is essentially a series of concentric rings, each with a lower refractive index. Because light travels faster in a lower-index medium, light further from the axis travels faster. Because high-order modes have a faster average velocity than low-order modes, all modes tend to arrive at a given point at nearly the same time. Rays of light are not sharply reflected by the core-cladding interface; they are refracted successively by differing layers in the core.

19.2.4 Characteristics of Attenuation

Attenuation represents a loss of power. During transit, some of the light in a fiber-optic system is absorbed into the fiber or scattered by impurities. Attenuation for a fiber cable is usually specified in decibels per kilometer (dB/km). For commonly available fibers, attenuation ranges from approximately 0.5 dB/km for premium single-mode fibers to 1,000 dB/km for large-core plastic fibers. Because emitted light represents power, 3 dB represents a doubling or halving of any reference power level.

Attenuation and light wavelength are uniquely related in fiber-transmission systems. This is illustrated in Figure 19.2. Most fibers have a medium loss region in the 800–900 nm wavelength range (3–5 dB/km), a low loss region in the 1,150–1,350 nm range (0.6–1.5 dB/km), and a very low loss region (less than 0.5 dB/km) in the 1,550 nm range. As a result, optimum performance is achieved by careful balancing of fiber, light source wavelength, and distance requirements.

Light intensity attenuation has no direct effect on the bandwidth of the electrical signals being transported. There is a direct correlation, however, between the S/N of the fiber receiver electronic circuits and the usable recovered optical signal.

19.2.5 Types of Cable

The first step in packaging an optical fiber into a cable is the extrusion of a layer of plastic around the fiber. This layer of plastic is called a *buffer tube*; it should not be confused with the buffer coating. The buffer coating is placed on the fiber by the fiber manufacturer. The buffer tube is placed on the fiber by the cable manufacturer. This extrusion process can produce two different cable designs:

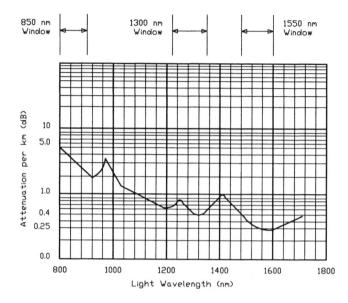

Figure 19.2 Fiber attenuation vs. light wavelength characteristics. Attenuation has been reduced steadily in the last two decades through improved fiber drawing techniques and a reduction in impurities. It has now approached the theoretical limits of silica-based glass at the 1,300 and 1,550 nm wavelengths.

- *Tight tube design*—The inner diameter of the plastic (buffer tube) is the same size as the outer diameter of the fiber, and is in contact with the fiber around its circumference.

- *Loose tube design*—The layer of plastic is significantly larger than the fiber, and, therefore, the plastic is not in contact with the fiber around the circumference of the fiber.

The two types of fiber cable are illustrated in Figure 19.3. Note that the loose tube design is available configured either as a *single-fiber-per-tube* (SFPT) or *multiple-fibers-per-tube* (MFPT) design. The six-fibers-per-tube MFPT design is often used for data communications.

After a fiber (or group of fibers) has been surrounded by a buffer tube, it is called an *element*. The cable manufacturer uses elements to build up the desired type of cable. In building the cable from elements, the manufacturer can create six distinct designs:

- Breakout design

- MFPT, central loose tube design

- MFPT, stranded loose tube design

- SFPT, stranded loose tube design

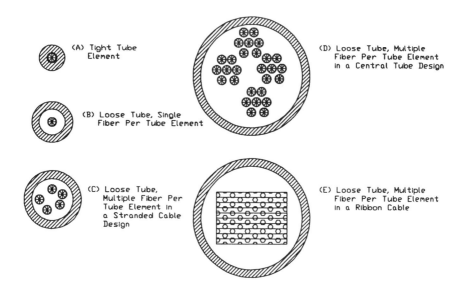

(A) Tight Tube Element

(B) Loose Tube, Single Fiber Per Tube Element

(C) Loose Tube, Multiple Fiber Per Tube Element In a Stranded Cable Design

(D) Loose Tube, Multiple Fiber Per Tube Element In a Central Tube Design

(E) Loose Tube, Multiple Fiber Per Tube Element In a Ribbon Cable

Figure 19.3 Loose-tube cables are available in either single-fiber-per-tube (SFPT) or multiple-fibers-per-tube (MFPT) designs. In both cases, the diameter of the plastic tube surrounding the core is larger than the outside diameter of the core. In a tight tube cable; the inner diameter tube is the same as the outer diameter of the fiber.

- *Star*, or *slotted core*, design
- *Tight tube*, or *stuffed*, design

Breakout Design

In the breakout design, shown in Figure 19.4, the element or buffered fiber is surrounded with a flexible-strength member, often Kevlar. The strength member is surrounded by an inner jacket to form a subcable, as shown. Multiple subcables are stranded around a central strength member or filler to form a cable core. This cable core is held together by a binder thread or Mylar wrapping tape. The core is surrounded by an extruded jacket to form the final cable.

Optional steps for this design include additional strength members, jackets, or armor. The additional jackets may be extruded directly on top of one another or separated by additional external strength members.

MFPT, Central Loose Tube Design

Fibers are placed together to form groups. Sometimes, the fibers are laid along a ribbon in groups of 12. These ribbons are then stacked up to 12 high and twisted. This

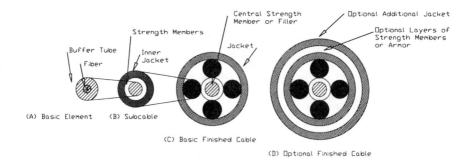

Figure 19.4 In the breakout type of cable, each element is surrounded by a flexible strength member, which is then surrounded by an inner jacket. This forms a subcable, which is incorporated into a larger cable. Optional additional jackets or armor can be applied.

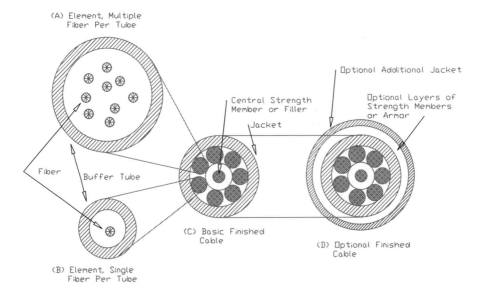

Figure 19.5 The MFPT stranded loose tube design relies on a center strength member to form the cable core. Multiple elements are then added to build up the desired cable capacity.

version of the central loose tube design is referred to as a *ribbon design*, and was developed by AT&T. The space between the fibers and the tube can be filled with a water-blocking compound.

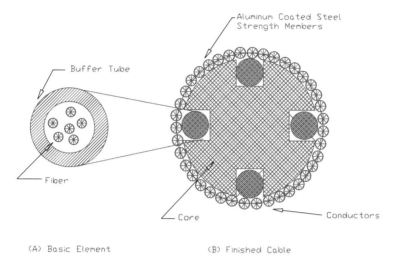

Aluminum Coated Steel
Strength Members

Buffer Tube

Fiber

Core

Conductors

(A) Basic Element

(B) Finished Cable

Figure 19.6 Utility companies sometimes use an optical power ground wire type of cable because it incorporates a metallic power ground wire within the design. The cable is based on a slotted-core configuration, but with the addition of helically wrapped wires around the outside for strength and conductivity.

MFPT, Stranded Loose Tube Design

Multiple buffer tubes are stranded around a central strength member or filler to form a core, as illustrated in Figure 19.5. This cable core is held together by a binder thread or Mylar wrapping tape. The core is surrounded by an extruded jacket to form a finished cable. Optional jacketing, strength members, or armor can be added.

SFPT, Stranded Loose Tube Design

This type of cable is manufactured similarly to MFPT cable. The primary difference is that the cable has one fiber per tube and smaller-diameter buffer tubes.

Star, or Slotted Core, Design

This design is seldom used in the U.S. In this scheme, the buffer tubes (usually MFPT) are laid in helical grooves, which are formed in the filler in the center of the cable. The core is then surrounded by an extruded jacket to form a finished cable. One variation of this design, shown in Figure 19.6, is used by power utility companies. This optical cable provides a conductive ground path from end to end. Instead of a jacket, the cable has helically wrapped wires, some of which are conductors and strength members.

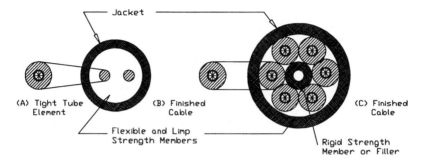

Figure 19.7 The tight tube stuffed design relies on a core filled with flexible strength members, usually Kevlar. Typically, two or more fibers are contained within the cable.

Tight Tube, or Stuffed, Design

This design is based on the tight tube element. The designs are common in that the core is filled, or stuffed, with flexible strength members, usually Kevlar. The design usually incorporates two or more fibers, as illustrated in Figure 19.7.

Application Considerations

Performance advantages exist for all designs, depending on what parameter is considered. For example, the tight tube design can force the ends of a broken fiber to remain in contact even after the fiber has broken. The result is that transmission may still be possible. When reliability is paramount, this feature may be important.

Loose tube designs have a different performance advantage. They offer a mechanical *dead zone*, which is not available in tight tube designs. The effect is that stress can be applied to the cable without that stress being transferred to the fiber. This dead zone exists for all mechanical forces, including tensile and crush loads, and bend strains. Tight tube designs do not have this mechanical dead zone. In the tight tube design, any force applied to the cable is also applied to the fibers. Loose tube designs also offer smaller size, lower cost, and smaller bend radii than tight tube designs.

When cable cost alone is considered, loose tube designs have the advantage over tight tube, breakout designs in long-length applications. However, when total installation cost is considered, the loose tube designs may or may not have a cost advantage. This is because loose tube designs have higher connector installation costs. The cost factor is composed of two parts:

- Labor cost
- Equipment cost

All designs, other than the breakout design, require handling of bare fibers or fibers with tight tubes. During this handling, fibers can be broken, especially where inexperienced personnel are involved.

19.2.6 Specifying Fiber-Optic Cable

In order to completely specify a fiber-optic cable, four primary performance categories must be quantified:

- Installation specifications

- Environmental specifications

- Fiber specifications

- Optical specifications

These criteria are outlined in Table 19.1. Note that not all specifications apply to all situations. The system engineer must review the specific application to determine which of the specifications are applicable. For example, cable installed in conduit or in protected locations will not need to meet a crush load specification.

Installation Specifications

The installation specifications are those that must be met to ensure successful cable installation. There are six:

- Maximum installation load in kilograms-force or pounds-force. This is the maximum tensile load that can be applied to a cable without causing fiber breakage or a permanent change in attenuation. This characteristic must always be specified. Load values for some typical installations are shown in Table 19.2. If the application requires a strength higher than those listed, specify a higher-strength cable. The increased cost of specifying a higher-strength cable is small, typically 5 to 10 percent of the cable cost.

- Minimum installation bend radius in inches or millimeters. This is the minimum radius to which the cable can be bent while loaded at the maximum installation load. This bending can be done without causing a permanent change in attenuation, fiber breakage, or breakage of any portion of the cable structure. The bend radius is usually specified as no less than 20 times the cable diameter. To determine this value, examine the locations where the cable will be installed, and identify the smallest bend the cable will encounter. Conversely, the system engineer can choose the cable and then specify that the conduits or ducts not violate this radius. The radius is actually limited more by the cabling materials than by the bend radius of the fiber.

- Diameter of the cable. Despite the space-effective nature of FO cable, it still must reside in the available space. This is especially true if the cable is to be installed in a partially filled conduit.

Table 19.1 Fiber Cable Specification Considerations

Installation Specifications:
 Maximum recommended installation load
 Minimum installation bend radius
 Cable diameter
 Diameter of subcables
 Maximum installation temperature range
 Maximum storage temperature range
Environmental Specifications:
 Temperature range of operation
 Minimum recommended unloaded bend radius
 Minimum long-term bend radius
 Maximum long-term use load
 Vertical rise
 National Electric Code or local electrical code requirements
 Flame resistance
 UV stability or UV resistance
 Resistance to rodent damage
 Resistance to water damage
 Crushing characteristics
 Resistance to conduction under high-voltage fields
 Toxicity
 High flexibility: static vs. dynamic applications
 Abrasion resistance
 Resistance to solvents, petrochemicals, and other substances
 Hermetically sealed fiber
 Radiation resistance
 Impact resistance
 Gas permeability
 Stability of filling compounds
Fiber Specifications:
Dimensional considerations:
 Core diameter
 Clad diameter
 Buffer coating diameter
 Mode field diameter
Optical Specifications:
Power considerations:
 Core diameter
 Numerical aperture
 Attenuation rate
 Cut-off wavelength
Capacity considerations:
 Bandwidth-distance product (dispersion)
 Zero-dispersion wavelength

Table 19.2 Maximum Installation Loads that Fiber Cable can be Exposed to in Various Applications

Application	Typical Maximum Recommended Installation Load Pounds force
1 fiber in raceway or tray	67 lb
1 fiber in duct or conduit	125 lb
2 fibers in duct or conduit	250–500 lb
Multi-fiber (6–12) cables	500 lb
Direct burial cables	600–800 lb
Lashed aerial cables	300 lb
Self-supported aerial cables	600 lb

- Diameter of subcables or elements. The diameter of the subcable or the cable elements can become a limiting factor. In the case of a breakout-style cable, the diameter of the subcable must be smaller than the maximum diameter of the connector boot so that the boot will fit on the subcable. Also, the diameter of the element must be less than the maximum diameter acceptable to the backshell of the connector. Most breakout cables have tight-tube elements, usually with a diameter of 1 mm or less.

- Recommended temperature range for installation (°C). All cables have a temperature range within which they can be installed without damage to either the cable materials or the fibers. Generally, the temperature range is affected more by the cable materials than the fibers. Not all cable manufacturers include this parameter in their data sheets. If the parameter is not specified, select a conservative temperature range of operation.

- Recommended temperature range for storage (°C). In severe climates, such as deserts and the Arctic, the system engineer must specify a recommended temperature range for storage in °C. This range will strongly influence the materials used in the cable.

Environmental Specifications

Environmental specifications are those that must be met to ensure successful long-term cable operation. Most of the items listed in Table 19.1 are self-explanatory. However, some environmental specifications deserve special attention.

The temperature range of operation is that range in which the attenuation remains less than the specified value. There are few applications where FO cable cannot be used because of temperature considerations. FO cables composed of plastic materials have maximum and minimum temperature points. If these are exceeded, the materials will not maintain their mechanical properties. After long exposures to high temperatures, plastics deteriorate and become soft. Some materials will begin to crack. After exposures to low temperatures, plastics become brittle and crack when flexed or moved. Under such conditions, the cable coverings will cease to protect the fiber.

Another reason for considering the temperature range of operation is the increase in attenuation that occurs when fibers are exposed to temperature extremes. This sensitivity occurs when the fibers are bent. When a cable is subject to extreme temperatures, the plastic materials will expand and contract. The rates at which the expansion and contraction take place are much greater (perhaps 100 times) than the rates of glass fibers. This movement results in the fiber being bent at a microscopic level. The fiber is either forced against the inside of the plastic tube as the plastic contracts, or the fiber is stretched against the inside of the tube as the plastic expands. In either case, the fiber is forced to conform to the microscopically uneven surface of the plastic. On a microscopic level, this is similar to placing the fiber against sandpaper. The bending results in light escaping from the core of the fiber. The result is referred to as a *microbend-induced* increase in attenuation.

19.2.7 Fiber Optic Connectors

The purpose of a fiber optic connector is to efficiently convey the optical signal from one link or element to the next. Most connectors share a design similar to the assembly shown in Figure 19.8. Typically, connectors are plugs (male) and are mated to precision couplers or sleeves (female). While the specific mechanical design of each connector type varies from one manufacturer to the next, the basic concept is the same: provide precise alignment of the optical fiber cores through a ferrule in the coupler. Some connectors are designed to keep the fiber ends separated, while other designs permit the fiber ends to touch in order to reduce reflections resulting from the glass-to-air-to-glass transition.

The fiber is prepared and attached to the connector, usually with an adhesive or epoxy cement, and polished flush with the connector tip. Factory-installed connectors typically use heat-cured epoxy and hand or machine polishing. Field installable connectors include epoxy-and-polish types, and crimp-on types. The crimp-on connector simplifies field assembly considerably.

The ferrule is a critical element of the connector. The ferrule functions to hold the fiber in place for optimum transmission of light energy. Ferrule materials include ceramics, stainless steel alloys, and glass.

Connector Properties

There are many types of fiber optic connectors. Each design has evolved to fill a specific application, or class of applications. Figure 19.9 shows three common fiber optic connectors.

The selection of a connector should take into consideration the following issues:

- Insertion loss

- Allowable loss budget for the fiber system

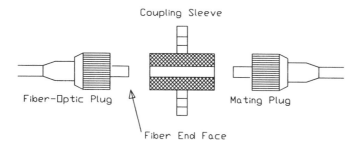

Figure 19.8 The mechanical arrangement of a simple fiber-optic connector.

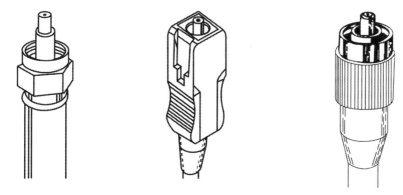

Figure 19.9 Common types of fiber-optic connectors.

- Consistent loss characteristics over a minimum number of connect/disconnect cycles
- Sufficiently high return loss for proper system operation
- Ruggedness
- Compatibility with fiber connectors of the same type
- High tensile strength
- Stable thermal characteristics

As with any system that transports energy, the fewer number of connectors and/or splices, the better. Pigtail leads are often required between a fiber termination panel and the transmission/reception system; however, keep such links to a minimum. Figure 19.10 shows how the fiber-optic light source may be terminated as a panel-mounted connector in order to minimize the number of pigtail links.

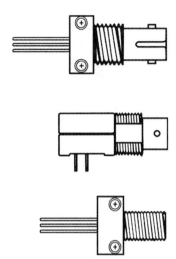

Figure 19.10 Circuit board mounted LED optical sources with connector terminations.

Performance Considerations

The optical loss in a fiber optic connector is the primary measure of device quality. Connector loss specifications are derived by measuring the optical power through a length of fiber. Next, the fiber is cut in the center of its length and the connectors are attached and mated with a coupler. The power is then measured again at the end of the fiber. The additional loss in the link represents the loss in the connector.

Return loss is another important measurement of connector quality. Return loss is the optical power that is reflected toward the source by a connector. Connector return loss in a single-mode link, for example, can diffuse back into the laser cavity, degrading its stability. In a multi-mode link, return loss can cause extraneous signals, reducing overall performance.

19.3 Bibliography

Ajemian, Ronald G., "Fiber Optic Connector Considerations for Professional Audio," *Journal of the Audio Engineering Society*, Audio Engineering Society, New York, NY, June 1992.

Benson, K. B., and J. Whitaker, *Television and Audio Handbook for Engineers and Technicians*, McGraw-Hill, New York, NY, 1989.

Pearson, Eric, *How to Specify and Choose Fiber-Optic Cables,* Pearson Technologies, Acworth, GA, 1991.

Whitaker, Jerry C., *AC Power Systems*, 2nd ed., CRC Press, Boca Raton, FL, 1998.

Whitaker, Jerry C., *Maintaining Electronic Systems*, CRC Press, Boca Raton, FL, 1991.

System Reliability

20.1 Introduction

The ultimate goal of any design engineer or maintenance department is zero downtime. This is an elusive goal, but one that can be approximated by examining the vulnerable areas of plant operation and taking steps to prevent a sequence of events that could result in system failure. In cases where failure prevention is not practical, a reliability assessment should encompass the stocking of spare parts, circuit boards, or even entire systems. A large facility may be able to cost-justify the purchase of backup gear that can be used as spares for the entire complex. Backup hardware is expensive, but so is downtime.

Failures can, and do, occur in electronic systems. The goal of product quality assurance at every step in the manufacturing and operating chain is to ensure that failures do not produce a systematic or repeatable pattern. The ideal is to eliminate failures altogether. Short of that, the secondary goal is to end up with a random distribution of failure modes. This indicates that the design of the system is fundamentally optimized and that failures are caused by random events that cannot be predicted. In an imperfect world, this is often the best that end users can hope for. Reliability and maintainability must be built into products or systems at every step in the design, construction, and maintenance process. They cannot be treated as an afterthought.

20.1.1 Terminology

To understand the principles of reliability engineering, the following basic terms must be defined:

- **Availability**. The probability that a system subject to repair will operate satisfactorily on demand.

- **Average life**. The mean value for a normal distribution of product or component lives. This term is generally applied to mechanical failures resulting from "wear-out."

- **Burn-in**. The initially high failure rate encountered when a component is placed on test. Burn-in failures usually are associated with manufacturing defects and the debugging phase of early service.

- **Defect**. Any deviation of a unit or product from specified requirements. A unit or product may contain more than one defect.

- **Degradation failure**. A failure that results from a gradual change, over time, in the performance characteristics of a system or part.

- **Downtime**. Time during which equipment is not capable of doing useful work because of malfunction. This does not include preventive maintenance time. Downtime is measured from the occurrence of a malfunction to its correction.

- **Failure**. A detected cessation of ability to perform a specified function or functions within previously established limits. A failure is beyond adjustment by the operator by means of controls normally accessible during routine operation of the system.

- **Failure mode and effects analysis (FMEA)**. An iterative documented process performed to identify basic faults at the component level and determine their effects at higher levels of assembly.

- **Failure rate**. The rate at which failure occurs during an interval of time as a function of the total interval length.

- **Fault tree analysis (FTA)**. An iterative documented process of a systematic nature performed to identify basic faults, determine their causes and effects, and establish their probabilities of occurrence.

- **Lot size**. A specific quantity of similar material or a collection of similar units from a common source; in inspection work, the quantity offered for inspection and acceptance at any one time. This may be a collection of raw material, parts, subassemblies inspected during production, or a consignment of finished products to be sent out for service.

- **Maintainability**. The probability that a failure will be repaired within a specified time after it occurs.

- **Mean time between failure (MTBF)**. The measured operating time of a single piece of equipment divided by the total number of failures during the measured period of time. This measurement normally is made during that period between early life and wear-out failures.

- **Mean time to repair (MTTR)**. The measured repair time divided by the total number of failures of the equipment.

- **Mode of failure**. The physical description of the manner in which a failure occurs and the operating condition of the equipment or part at the time of the failure.

- **Part failure rate**. The rate at which a part fails to perform its intended function.

- **Quality assurance (QA)**. All those activities, including surveillance, inspection, control, and documentation, aimed at ensuring that a product will meet its performance specifications.

- **Reliability**. The probability that an item will perform satisfactorily for a specified period of time under a stated set of use conditions.

- **Reliability growth**. Actions taken to move a hardware item toward its reliability potential, during development, subsequent manufacturing, or operation.

- **Reliability predictions**. Compiled failure rates for parts, components, subassemblies, assemblies, and systems. These generic failure rates are used as basic data to predict a value for reliability.

- **Sample**. One or more units selected at random from a quantity of product to represent that product for inspection purposes.

- **Sequential sampling**. Sampling inspection in which, after each unit is inspected, the decision is made to accept, reject, or inspect another unit. (Note: Sequential sampling as defined here is sometimes called *unit sequential sampling* or *multiple sampling*.)

- **System**. A combination of parts, assemblies, and sets joined together to perform a specific operational function or functions.

- **Test to failure**. Testing conducted on one or more items until a predetermined number of failures have been observed. Failures are induced by increasing electrical, mechanical, and/or environmental stress levels, usually in contrast to *life tests*, in which failures occur after extended exposure to predetermined stress levels. A life test can be considered a test to failure using age as the stress.

20.2 Quality Assurance

Electronic component and system manufacturers design and implement quality assurance procedures for one fundamental reason: Nothing is perfect. The goal of a QA program is to ensure, for both the manufacturer and the customer, that all but some small, mutually acceptable percentage of devices or systems shipped will be as close to perfection as economics and the state of the art allow. There are tradeoffs in this process. It would be unrealistic, for example, to perform extensive testing to identify potential failures if the cost of that testing exceeded the cost savings that would be realized by not having to replace the devices later in the field.

The focal points of any QA effort are *quality* and *reliability*. These terms are not synonymous. They are related, but they do not provide the same measure of a product:

- Quality is the measure of a product's performance relative to some established criteria.

- Reliability is the measure of a product's life expectancy.

Stated from a different perspective, quality answers the question of whether the product meets applicable specifications *now*; reliability answers the question of *how long* the product will continue to meet its specifications.

20.2.1 Inspection Process

Quality assurance for components normally is performed through sampling rather than through 100 percent inspection. The primary means used by QA departments for controlling product quality at the various processing steps include:

- *Gates.* A mandatory sampling of every lot passing through a critical production stage. Material cannot move on to the next operation until QA has inspected and accepted the lot.

- *Monitor points.* A periodic sampling of some attribute of the component. QA personnel sample devices at a predetermined frequency to verify that machines and operators are producing material that meets preestablished criteria.

- *Quality audit.* An audit carried out by a separate group within the QA department. This group is charged with ensuring that all production steps throughout the manufacturer's facility are in accordance with current specifications.

- *Statistical quality control.* A technique, based on computer modeling, that incorporates data accumulated at each gate and monitor point to construct statistical profiles for each product, operation, and piece of equipment within the plant. Analysis of this data over time allows QA engineers to assess trends in product performance and failure rates.

Quality assurance for a finished subassembly or system may range from a simple go/no-go test to a thorough operational checkout that may take days to complete.

20.2.2 Reliability Evaluation

Reliability prediction is the process of quantitatively assessing the reliability of a component or system during development, before large-scale fabrication and field operation. During product development, predictions serve as a guide by which design alternatives can be judged for reliability. To be effective, the prediction technique must relate engineering variables to reliability variables.

A prediction of reliability is obtained by determining the reliability of each critical item at the lowest system level and proceeding through intermediate levels until an estimate of overall reliability is obtained. This prediction method depends on the availability of accurate evaluation models that reflect the reliability of lower-level components. Various formal prediction procedures are used, based on theoretical and statistical concepts.

Parts-Count Method

The parts-count approach to reliability prediction provides an estimate of reliability based on a count by part type (ICs, transistors, vacuum tube devices, resistors, capacitors, and other components). This method is useful during the early design stage of a product, when the amount of available data is limited. The technique involves counting the number of components of each type, multiplying that number by a generic failure rate for each part type, and summing the products to obtain the failure rate of each functional circuit, subassembly, assembly, and/or block depicted in the system block diagram. The parts-count method is useful in the design phase because it provides rapid estimates of reliability, permitting assessment of the feasibility of a given concept.

Stress-Analysis Method

The stress-analysis technique is similar to the parts-count method, but utilizes a detailed parts model plus calculation of circuit stress values for each part before determining the failure rate. Each part is evaluated in its electric circuit and mechanical assembly application based on an electrical and thermal stress analysis. After part failure rates have been established, a combined failure rate for each functional block is determined.

20.2.3 Failure Analysis

Failure mode and effects analysis can be performed with data taken from actual failure modes observed in the field, or from hypothetical failure modes derived from one of the following:

- Design analysis
- Reliability prediction activities
- Experience with how specific parts fail

In the most complete form of FMEA, failure modes are identified at the component level. Failures are induced analytically into each component, and failure effects are evaluated and noted, including the severity and frequency (or probability) of occurrence. Using this approach, the probability of various system failures can be calculated, based on the probability of lower-level failure modes.

Fault tree analysis is a tool commonly used to analyze failure modes found during design, factory test, or field operations. The approach involves several steps, including the development of a detailed logic diagram that depicts basic faults and events that can lead to system failure and/or safety hazards. These data are used to formulate corrective suggestions that, when implemented, will eliminate or minimize faults considered critical. An example FTA chart is shown in Figure 20.1.

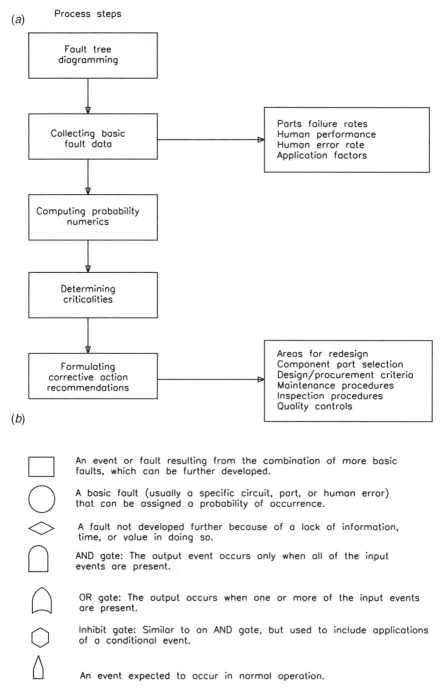

Figure 20.1 Example fault tree analysis diagram: (*a*) process steps, (*b*) fault tree symbols, (*c*) example diagram (next page).

(*c*)

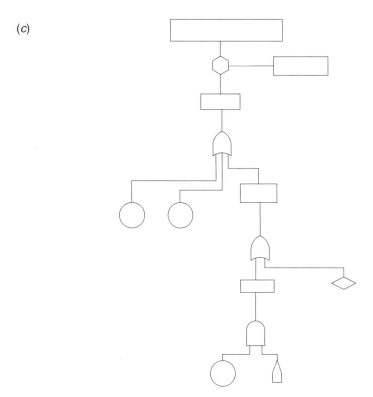

Figure 20.1c Example diagram.

20.2.4 Standardization

Standardization and reliability go hand in hand. Standardization of electronic components began with military applications in mind; the first recorded work was performed by the U.S. Navy with vacuum tubes. The navy recognized that some control
at the component level was essential to the successful incorporation of electronics
into naval systems.

Standardization and reliability are closely related, although there are many aspects
of standardization whose reliability implications are subtle. The primary advantages of
standardization include:

- *Total product interchangeability.* Standardization ensures that products of the
  same part number provide the same physical and electrical characteristics. There
  have been innumerable instances of a replacement device bearing the same part
  number as a failed device, but not functioning identically to it. In some cases, the
  differences in performance were so great that the system would not function at all
  with the new device.

- *Consistency and configuration control.* Component manufacturers constantly re-define their products to improve yields and performance. Consistency and con-figuration control assure the user that product changes will not affect the interchangeability of the part.

- *Efficiencies of volume production.* Standardization programs usually result in production efficiencies that reduce the costs of parts, relative to components with the same level of reliability screening and control.

- *Effective spares management.* The use of standardized components makes the stocking of spare parts a much easier task. This aspect of standardization is not a minor consideration. For example, the costs of placing, expediting, and receiving material against one Department of Defense purchase order may range from $300 to $1100 (or more). Accepting the lowest estimate, the conversion of 10 separate part numbers to one standardized component could effect immediate savings of $3000 just in purchasing and receiving costs.

- *Multiple product sources.* Standardization encourages second-sourcing. Multi-ple sources help hold down product costs and encourage manufacturers to strive for better product performance.

20.2.5 Reliability Analysis

The science of reliability and maintainability matured during the 1960s with the de-velopment of sophisticated computer systems and complex military and spacecraft electronics. Components and systems never fail without a reason. That reason may be difficult to find, but determination of failure modes and weak areas in system design or installation is fundamental to increasing the reliability of any component or sys-tem, whether it is a power vacuum tube, integrated circuit, aircraft autopilot, or radio transmitter.

All equipment failures are logical; some are predictable. A system failure usually is related to poor-quality components or to abuse of the system or a part within, either be-cause of underrating or environmental stress. Even the best-designed components can be badly manufactured. A process can go awry, or a step involving operator interven-tion may result in an occasional device that is substandard or likely to fail under normal stress. Hence, the process of screening and/or *burn-in* to weed out problem parts is a universally accepted quality control tool for achieving high reliability.

20.2.6 Statistical Reliability

Figure 20.2 illustrates what is commonly known as the *bathtub curve*. It divides the expected lifetime of a class of parts into three segments: *infant mortality, useful life,* and *wear-out.* A typical burn-in procedure consists of the following steps:

- The parts are electrically biased and loaded; that is, they are connected in a circuit representing a typical application.

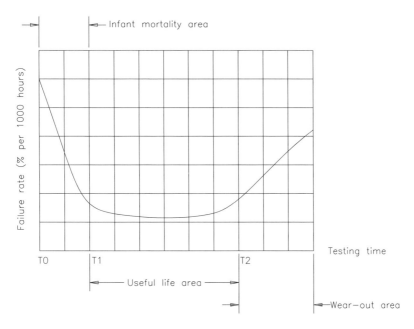

Figure 20.2 The statistical distribution of equipment or component failures vs. time for electronic systems and devices.

- The parts are cycled on and off (power applied, then removed) for a predetermined number of times. The number of cycles may range from 10 to several thousand during the burn-in period, depending on the component under test.

- The components under load are exposed to high operating temperatures for a selected time (typically 72 to 168 hours). This constitutes an accelerated life test for the part.

An alternative approach involves temperature shock testing, in which the component product is subjected to temperature extremes, with rapid changes between the *hot-soak* and *cold-soak* conditions. After the stress period, the components are tested for adherence to specifications. Parts meeting the established specifications are accepted for shipment to customers. Parts that fail to meet them are discarded.

Figure 20.3 illustrates the benefits of temperature cycling to product reliability. The charts compare the distribution of component failures identified through steady-state high-temperature burn-in vs. temperature cycling. Note that cycling screened out a significant number of failures. The distribution of failures under temperature cycling usually resembles the distribution of field failures. Temperature cycling simulates real-world conditions more closely than steady-state burn-in. The goal of burn-in testing is to ensure that the component lot is advanced beyond the infant mortality stage (*T1*

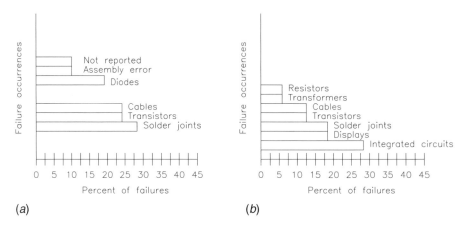

Figure 20.3 Distribution of component failures identified through burn-in testing: (*a*) steady-state high-temperature burn-in, (*b*) temperature cycling.

on the bathtub curve). This process is used not only for individual components, but for entire systems as well.

Such a systems approach to reliability is effective, but not foolproof. The burn-in period is a function of statistical analysis; there are no absolute guarantees. The natural enemies of electronic parts are heat, vibration, and excessive voltage. Figure 20.4 documents failures vs. hours in the field for a piece of avionics equipment. The conclusion is made that a burn-in period of 200 hours or more will eliminate 60 percent of the expected failures. However, the burn-in period for another system using different components may well be a different number of hours.

The goal of burn-in testing is to catch system problems and potential faults before the device or unit leaves the manufacturer. The longer the burn-in period, the greater the likelihood of catching additional failures. The problems with extended burn-in, however, are time and money. Longer burn-in translates to longer delivery delays and additional costs for the equipment manufacturer, which are likely to be passed on to the end user. The point at which a product is shipped is based largely on experience with similar components or systems and the financial requirement to move products to customers.

Roller-Coaster Hazard Rate

The bathtub curve has been used for decades to represent the failure rate of an electronic system. More recent data, however, has raised questions regarding the accuracy of the curve shape. A number of reliability scientists now believe that the probability of failure, known in the trade as the *hazard rate*, is more accurately represented as a roller-coaster track, as illustrated in Figure 20.5. Hazard rate calculations require analysis of the number of failures of the system under test, as well as the number of

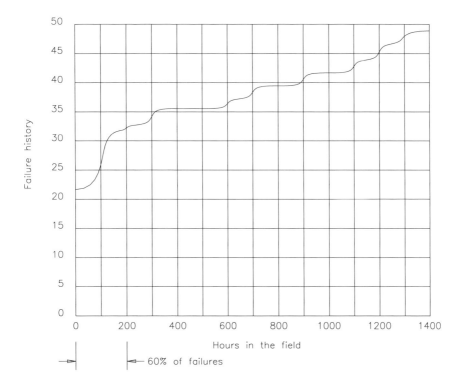

Figure 20.4 The failure history of a piece of avionics equipment vs. time. Note that 60 percent of the failures occurred within the first 200 hours of service. (*After* [1].)

survivors. Advocates of this approach point out that previous estimating processes and averaging tended to smooth the roller-coaster curve so that the humps were less pronounced, leading to an incorrect conclusion insofar as the hazard rate was concerned. The testing environment also has a significant effect on the shape of the hazard curve, as illustrated in Figure 20.6. Note that at the higher operating temperature (greater environmental stress), the roller-coaster hump has moved to an earlier age.

20.2.7 Environmental Stress Screening

The science of reliability analysis is rooted in the understanding that there is no such thing as a random failure; every failure has a cause. For reasonably designed and constructed electronic equipment, failures not caused by outside forces result from built-in flaws or *latent defects*. Because different flaws are sensitive to different stresses, a variety of environmental forces must be applied to a unit under test to identify any latent defects. This is the underlying concept behind *environmental stress screening* (ESS).

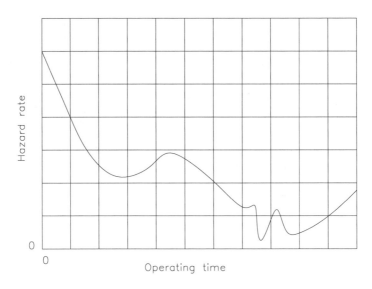

Figure 20.5 The roller-coaster hazard rate curve for electronic systems. (*After* [2].)

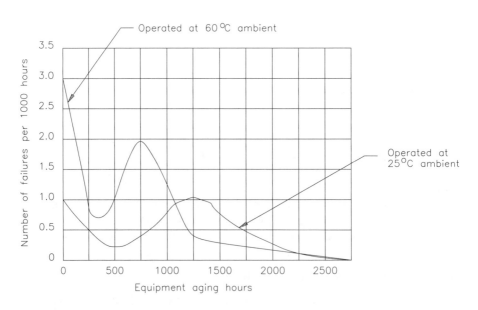

Figure 20.6 The effects of environmental conditions on the roller-coaster hazard rate curve. (*After* [2].)

ESS, which has come into widespread use for aeronautics and military products, takes the burn-in process a step further by combining two of the major environmental

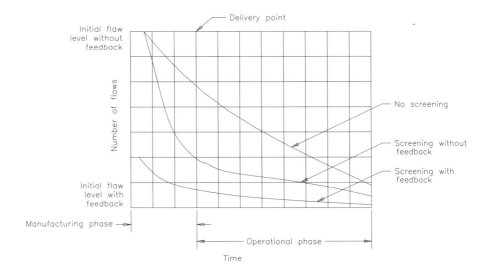

Figure 20.7 The effects of environmental stress screening on the reliability *bathtub* curve. (*After*[3].)

factors that cause parts or units to fail: heat and vibration. Qualification testing for products at a factory practicing ESS involves a carefully planned series of checks for each unit off the assembly line. Units are subjected to random vibration and temperature cycling during production (for subassemblies and discrete components) and upon completion (for systems). The goal is to catch product defects at the earliest possible stage of production. ESS also can lead to product improvements in design and manufacture if feedback from the qualification stage to the design and manufacturing stages is implemented. Figure 20.7 illustrates the improvement in reliability that typically can be achieved through ESS over simple burn-in screening, and through ESS with feedback to earlier production stages. Significant reductions in equipment failures in the field can be gained. Table 20.1 compares the merits of conventional reliability testing and ESS.

Designing an ESS procedure for a given product is no easy task. The environmental stresses imposed on the product must be great enough to cause fallout of marginal components during qualification testing. The stresses must not be so great, however, as to cause failures in good products. Any unit that is stressed beyond its design limits eventually will fail. The proper selection of stress parameters—generally, random vibration on a vibration generator and temperature cycling in an environmental chamber—can, in minutes, uncover product flaws that might take weeks or months to manifest themselves in the field. The result is greater product reliability for the user.

The ESS concept requires that every product undergo qualification testing before integration into a larger system for shipment to an end user. The flaws uncovered by

Table 20.1 Comparison of Conventional Reliability Testing and Environmental Stress Screening (*After* [2].)

Parameter	Conventional Testing	Environmental Stress Screening
Test condition	Simulates operational equipment profile	Accelerated stress condition
Test sample size	Small	100 percent of production
Total test time	Limited	High
Number of failures	Small	Large
Reliability growth	Potential for gathering useful data small	Potential for gathering useful data good

ESS vary from one unit to the next, but types of failures tend to respond to particular environmental stresses. Available data clearly demonstrate that the burn-in screens must match the flaws sought; otherwise, the flaws will probably not be found.

The concept of flaw-stimulus relationships can be presented in Venn diagram form. Figure 20.8 shows a Venn diagram for a hypothetical, but specific, product. The required screen would be different for a different product. For clarity, not all stimuli are shown. Note that there are many latent defects that will not be uncovered by any one stimulus. For example, a solder splash that is just barely clinging to a circuit element probably would not be broken loose by high-temperature burn-in or voltage cycling, but vibration or thermal cycling probably would break the particle loose. Remember also that the defect may be observable only during stimulation and not during a static bench test.

The levels of stress imposed on a product during ESS should be greater than the stress to which the product will be subjected during its operational lifetime, but still be below the maximum design parameters. This rule of thumb is pushed to the limits under an *enhanced screening* process. Enhanced screening places the component or system at well above the expected field environmental levels. This process has been found to be useful and cost-effective for many programs and products. Enhanced screening, however, requires the development of screens that are carefully identified during product design and development so that the product can survive the qualification tests. Enhanced screening techniques often are required for cost-effective products on a cradle-to-grave basis; that is, early design changes for screenability save tremendous costs over the lifetime of the product.

The types of products that can be checked economically through ESS break down into two categories: high-dollar items and mass-produced items. Units that are physically large in size, such as RF generators, are difficult to test in the finished state. Still, qualification tests using more primitive methods, such as cam-driven truck-bed shakers, are practical. Because most large systems generate a large amount of heat, subjecting the equipment to temperature extremes also may be accomplished. Sophisticated

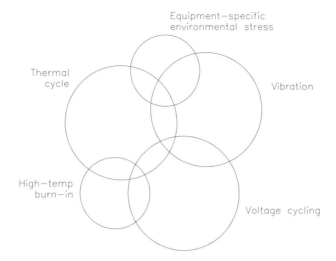

Figure 20.8 Venn diagram representation of the relationship between flaw precipitation and applied environmental stress. (*After* [4].)

ESS for large systems, however, must rely on qualification testing at the subassembly stage.

The basic hardware complement for an ESS test station includes a thermal chamber shaker and controller/monitor. A typical test sequence includes 10 minutes of exposure to random vibration, followed by 10 cycles between temperature minimum and maximum. To save time, the two tests may be performed simultaneously.

20.2.8 Latent Defects

The cumulative failure rate observed during the early life of an electronic system is dominated by the latent defect content of the product, not its inherent failure rate. Product design is the major determinant of inherent failure rate. A product design will show a higher-than-expected inherent rate if the system contains components that are marginally overstressed, have inadequate functional margin, or contain a subpopulation of components that exhibit a wear-out life shorter than the useful life of the product. Industry has grown to expect the high instantaneous failure rate observed when a new product is placed into service. The burn-in process, whether ESS or conventional, is aimed at shielding customers from the detrimental effects of infant mortality. The key to reducing early-product-life failures lies in reducing the number of latent defects.

A latent defect is some abnormal characteristic of the product or its parts that is likely to result in failure at some point, depending on:

- The degree of abnormality

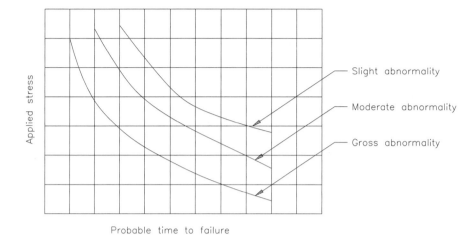

Figure 20.9 Estimation of the probable time to failure from an abnormal solder joint. (*After* [5].)

- The magnitude of applied stress
- The duration of applied stress

For example, consider a solder joint on the connecting pin of a device. One characteristic of the joint is the degree to which the pin hole is filled with solder, characterized as "percent fill." All other characteristics being acceptable, a joint that is 100 percent filled offers the maximum mechanical strength, minimum resistance, and greatest current carrying capacity. Conversely, a joint that is zero percent filled has no mechanical strength, and only if the lead is touching the barrel does it have any significant electrical properties. Between these two extreme cases are degrees of abnormality. For a fixed magnitude of applied stress:

- A grossly abnormal solder joint probably will fail in a short time.
- A moderately abnormal solder joint probably will fail, but after a longer period of time than a grossly abnormal joint.
- A mildly abnormal solder joint probably will fail, but after a much longer period of time than in either of the preceding conditions.

Figure 20.9 illustrates this concept. A similar time-stress relationship holds for a fixed degree of abnormality and variable applied stress.

A latent defect eventually will advance to a *patent defect* when exposed to environmental, or other, stimuli. A patent defect is a flaw that has advanced to the point at which an abnormality actually exists. To carry on the solder example, a cold solder joint represents a flaw (latent defect). After vibration and/or thermal cycling, the joint (it is assumed) will crack. The joint will now have become a detectable (patent) defect. Some latent defects can be stimulated into patent defects by thermal cycling, some by vibration, and some by voltage cycling. Not all flaws respond to all stimuli.

There is strong correlation between the total number of physical and functional defects found per unit of product during the manufacturing process, and the average latent defect content of shipping product. Product- and process-design changes aimed at reducing latent defects not only improve the reliability of shipping product, but also result in substantial manufacturing cost savings.

20.2.9 Operating Environment

The operating environment of an electronic system, either because of external environmental conditions or unintentional component underrating, may be significantly more stressful than the system manufacturer or the component supplier anticipated. Unintentional component underrating represents a design fault, but unexpected environmental conditions are possible for many applications, particularly in remote locations.

Conditions of extreme low or high temperatures, high humidity, and vibration during transportation may have a significant impact on long-term reliability of the system. For example, it is possible—and more likely, probable—that the vibration stress of the truck ride to a remote communications site will represent the worst-case vibration exposure of the radio equipment and all components within it during the lifetime of the product.

Manufacturers report that most of the significant vibration and shock problems for land-based products arise from the shipping and handling environment. Shipping tends to be an order of magnitude more severe than the operating environment with respect to vibration and shock. Early testing for these problems involved simulation of actual shipping and handling events, such as end-drops, truck trips, side impacts, and rolls over curbs and cobblestones. Although unsophisticated by today's standards, these tests are capable of improving product resistance to shipping-induced damage.

20.2.10 Failure Modes

Latent failures aside, the circuit elements most vulnerable to failure in any piece of electronic hardware are those exposed to the outside world. In most systems, the greatest threat typically involves one or more of the following components or subsystems:

- The ac-to-dc power supply
- Sensitive signal-input circuitry
- High-power output stages and devices
- Circuitry operating into an unpredictable load, or into a load that may be exposed to lightning and other transient effects (such as an antenna)

Derating of individual components is a key factor in improving the overall reliability of a given system. The goal of derating is the reduction of electrical, mechanical, thermal, and other environmental stresses on a component to decrease the degradation rate and prolong expected life. Through derating, the margin of safety between the operating stress level and the permissible stress level for a given part is increased. This adjustment provides added protection from system overstress, unforeseen during design.

20.2.11 Maintenance Considerations

The reliability and operating costs over the lifetime of a device or system can be affected significantly by the effectiveness of the preventive maintenance program designed and implemented by the engineering staff. In the case of a *critical-system* unit that must be operational continuously or during certain periods, maintenance can have a major impact—either positive or negative—on downtime.

The reliability of any electronic system may be compromised by an *enabling event phenomenon*. This is an event that does not cause a failure by itself, but sets up (or enables) a second event that can lead to failure of the system. Such a phenomenon is insidious because the enabling event may not be self-revealing. Examples include the following:

- A warning system that has failed or has been disabled for maintenance
- One or more controls that are set incorrectly, providing false readouts for operations personnel
- Redundant hardware that is out of service for maintenance
- Remote metering that is out of calibration

Common-Mode Failure

A *common-mode failure* is one that can lead to the failure of all paths in a redundant configuration. In the design of redundant systems, therefore, it is important to identify and eliminate sources of common-mode failures, or to increase their reliability to at least an order of magnitude above the reliability of the redundant system. Common-mode failure points include the following:

- Switching circuits that activate standby or redundant hardware
- Sensors that detect a hardware failure
- Indicators that alert personnel to a hardware failure
- Software that is common to all paths in a redundant system

The concept of software reliability in control and monitoring has limited meaning in that a good program will always run, and copies of a good program will always run. On the other hand, a program with one or more errors will always fail, and so will the copies, given the same input data. The reliability of software, unlike hardware, cannot be

Table 20.2 ISO 9000 Series Levels

Standard	Use
ISO-9000: quality management and assurance standards, guidelines for selection and use	Like a road map, this standard is to be used as a guideline to facilitate decisions with respect to selection and use of the other standards in the ISO 9000 Series.
ISO 9001: model for quality assurance in design, development, production, installation, and servicing	This standard is to be used when conformance to specified requirements is to be assured by the supplier during several stages: design, development, production, installation, and servicing.
ISO 9002: model for quality assurance in products and installation	This standard is to be used when conformance to specified requirements is to be assured by the supplier during production and installation.
ISO 9003: model for quality assurance in final inspection and testing	This standard is to be used when conformance to specified requirements is to be assured by the supplier solely at final inspection and test.
ISO 9004: quality management and quality system elements and guidelines	This standard is to be used as a model to develop and implement a quality management system. Basic elements of a quality management system are described. There is a heavy emphasis on meeting customer needs.

improved through redundancy if the software in the parallel path is identical to that in the primary path.

Spare Parts

The spare parts inventory is a key aspect of any successful equipment maintenance program. Having adequate replacement components on hand is important not only to correct equipment failures, but to identify those failures as well. Many parts—particularly in the high-voltage power supply and RF chain—are difficult to test under static conditions. The only reliable way to test the component may be to substitute one of known quality. If the system returns to normal operation, then the original component is defective. Substitution is also a valuable tool in troubleshooting intermittent failures caused by component breakdown under peak power conditions.

20.3 ISO 9000 Series

At its core, the ISO 9000 Series defines what a total quality system should do in order to guarantee product and service consistency. (See Table 20.2.) To that end, the ISO

9000 Series philosophically supports the age-old argument that form follows function; if a system's processes are defined and held within limits, consistent products and services will follow.

The ISO 9000 Series are documents that pertain to quality management standards. Individually titled and defined, they are listed in the table.

20.4 References

1. Capitano, J., and J. Feinstein, "Environmental Stress Screening Demonstrates Its Value in the Field," *Proceedings of the IEEE Reliability and Maintainability Symposium*, IEEE, New York, NY, 1986.
2. Wong, Kam L., "Demonstrating Reliability and Reliability Growth with Environmental Stress Screening Data," *Proceedings of the IEEE Reliability and Maintainability Symposium*, IEEE, New York, NY, 1990.
3. Tustin, Wayne, "Recipe for Reliability: Shake and Bake," *IEEE Spectrum*, IEEE, New York, NY, December 1986.
4. Hobbs, Gregg K., "Development of Stress Screens," *Proceedings of the IEEE Reliability and Maintainability Symposium*, IEEE, New York, NY, 1987.
5. Smith, William B., "Integrated Product and Process Design to Achieve High Reliability in Both Early and Useful Life of the Product," *Proceedings of the IEEE Reliability and Maintainability Symposium*, IEEE, New York, NY, 1987.

Glossary of Terms

A

absolute delay The amount of time a signal is delayed. The delay may be expressed in time or number of pulse events.

absolute zero The lowest temperature theoretically possible, −273.16°C. *Absolute zero* is equal to zero degrees Kelvin.

absorption The transference of some or all of the energy contained in an electromagnetic wave to the substance or medium in which it is propagating or upon which it is incident.

absorption auroral The loss of energy in a radio wave passing through an area affected by solar auroral activity.

ac coupling A method of coupling one circuit to another through a capacitor or transformer so as to transmit the varying (ac) characteristics of the signal while blocking the static (dc) characteristics.

ac/dc coupling Coupling between circuits that accommodates the passing of both ac and dc signals (may also be referred to as simply dc coupling).

accelerated life test A special form of reliability testing performed by an equipment manufacturer. The unit under test is subjected to stresses that exceed those typically experienced in normal operation. The goal of an *accelerated life test* is to improve the reliability of products shipped by forcing latent failures in components to become evident before the unit leaves the factory.

accelerating electrode The electrode that causes electrons emitted from an electron gun to accelerate in their journey to the screen of a cathode ray tube.

accelerating voltage The voltage applied to an electrode that accelerates a beam of electrons or other charged particles.

acceptable reliability level The maximum number of failures allowed per thousand operating hours of a given component or system.

acceptance test The process of testing newly purchased equipment to ensure that it is fully compliant with contractual specifications.

access The point at which entry is gained to a circuit or facility.

acquisition time In a communication system, the amount of time required to attain synchronism.

active Any device or circuit that introduces gain or uses a source of energy other than that inherent in the signal to perform its function.

adapter A fitting or electrical connector that links equipment that cannot be connected directly.

adaptive A device able to adjust or react to a condition or application, as an *adaptive circuit*. This term usually refers to filter circuits.

adaptive system A general name for a system that is capable of reconfiguring itself to meet new requirements.

adder A device whose output represents the sum of its inputs.

adjacent channel interference Interference to communications caused by a transmitter operating on an adjacent radio channel. The sidebands of the transmitter mix with the carrier being received on the desired channel, resulting in noise.

admittance A measure of how well alternating current flows in a conductor. It is the reciprocal of *impedance* and is expressed in *siemens*. The real part of admittance is *conductance*; the imaginary part is *susceptance*.

AFC (automatic frequency control) A circuit that automatically keeps an oscillator on frequency by comparing the output of the oscillator with a standard frequency source or signal.

air core An inductor with no magnetic material in its core.

algorithm A prescribed finite set of well-defined rules or processes for the solution of a problem in a finite number of steps.

alignment The adjustment of circuit components so that an entire system meets minimum performance values. For example, the stages in a radio are aligned to ensure proper reception.

allocation The planned use of certain facilities and equipment to meet current, pending, and/or forecasted circuit- and carrier-system requirements.

alternating current (ac) A continuously variable current, rising to a maximum in one direction, falling to zero, then reversing direction and rising to a maximum in the other direction, then falling to zero and repeating the cycle. Alternating current usually follows a sinusoidal growth and decay curve. Note that the correct usage of the term *ac* is lower case.

alternator A generator that produces alternating current electric power.

ambient electromagnetic environment The radiated or conducted electromagnetic signals and noise at a specific location and time.

ambient level The magnitude of radiated or conducted electromagnetic signals and noise at a specific test location when equipment-under-test is not powered.

ambient temperature The temperature of the surrounding medium, typically air, that comes into contact with an apparatus. Ambient temperature may also refer simply to room temperature.

American National Standards Institute (ANSI) A nonprofit organization that coordinates voluntary standards activities in the U.S.

American Wire Gauge (AWG) The standard American method of classifying wire diameter.

ammeter An instrument that measures and records the amount of current in amperes flowing in a circuit.

amp (A) An abbreviation of the term *ampere*.

ampacity A measure of the current carrying capacity of a power cable. *Ampacity* is determined by the maximum continuous-performance temperature of the insulation, by the heat generated in the cable (as a result of conductor and insulation losses), and by the heat-dissipating properties of the cable and its environment.

ampere (amp) The standard unit of electric current.

ampere per meter The standard unit of magnetic field strength.

ampere-hour The energy that is consumed when a current of one ampere flows for a period of one hour.

ampere-turns The product of the number of turns of a coil and the current in amperes flowing through the coil.

amplification The process that results when the output of a circuit is an enlarged reproduction of the input signal. Amplifiers may be designed to provide amplification of voltage, current, or power, or a combination of these quantities.

amplification factor In a vacuum tube, the ratio of the change in plate voltage to the change in grid voltage that causes a corresponding change in plate current. Amplification factor is expressed by the Greek letter μ (*mu*).

amplifier (1—general) A device that receives an input signal and provides as an output a magnified replica of the input waveform. **(2—audio)** An amplifier designed to cover the normal audio frequency range (20 Hz to 20 kHz). **(3—balanced)** A circuit with two identical connected signal branches that operate in phase opposition, with input and output connections each balanced to ground. **(4—bridging)** An amplifying circuit featuring high input impedance to prevent loading of the source. **(5—broadband)** An amplifier capable of operating over a specified broad band of frequencies with acceptably small amplitude variations as a function of frequency. **(6—buffer)** An amplifier stage used to isolate a frequency-sensitive circuit from variations in the load presented by following stages. **(7—linear)** An amplifier in which the instantaneous output signal is a linear function of the corresponding input signal. **(8—magnetic)** An amplifier incorporating a control device dependent on magnetic saturation. A small dc signal applied to a control circuit triggers a large change in operating impedance and, hence, in the output of the circuit. **(9—microphone)** A circuit that amplifies the low level output from a microphone to make it sufficient to be used as an input signal to a power amplifier or another stage in a modulation circuit. Such a circuit is commonly known as a *preamplifier*. **(10—push-pull)** A balanced amplifier with two similar amplifying units connected in phase opposition in order to cancel undesired harmonics and minimize distortion. **(11—tuned radio frequency)** An amplifier tuned to a particular radio frequency or band so that only selected frequencies are amplified.

amplifier operating class (1—general) The operating point of an amplifying stage. The operating point, termed the operating *class*, determines the period during which current flows in the output. **(2—class A)** An amplifier in which output current flows during the whole of the input current cycle. **(3—class AB)** An amplifier in which the output current flows for more than half but less than the whole of the input cycle. **(4—class B)** An amplifier in which output current is cut off at zero input signal; a half-wave rectified output is produced. **(5—class C)** An amplifier in which output current flows for less than half the input cycle. **(6—class D)** An amplifier operating in a pulse-only mode.

amplitude The magnitude of a signal in voltage or current, frequently expressed in terms of *peak*, *peak-to-peak*, or *root-mean-square* (RMS). The actual amplitude of a quantity at a particular instant often varies in a sinusoidal manner.

amplitude distortion A distortion mechanism occurring in an amplifier or other device when the output amplitude is not a linear function of the input amplitude under specified conditions.

amplitude equalizer A corrective network that is designed to modify the amplitude characteristics of a circuit or system over a desired frequency range.

amplitude-versus-frequency distortion The distortion in a transmission system caused by the nonuniform attenuation or gain of the system with respect to frequency under specified conditions.

analog carrier system A carrier system whose signal amplitude, frequency, or phase is varied continuously as a function of a modulating input.

anode (1 — general) A positive pole or element. **(2—vacuum tube)** The outermost positive element in a vacuum tube, also called the *plate*. **(3—battery)** The positive element of a battery or cell.

anodize The formation of a thin film of oxide on a metallic surface, usually to produce an insulating layer.

antenna (1—general) A device used to transmit or receive a radio signal. An antenna is usually designed for a specified frequency range and serves to couple electromagnetic energy from a transmission line to and/or from the free space through which it travels. Directional antennas concentrate the energy in a particular horizontal or vertical direction. **(2—aperiodic)** An antenna that is not periodic or resonant at particular frequencies, and so can be used over a wide band of frequencies. **(3—artificial)** A device that behaves, so far as the transmitter is concerned, like a proper antenna, but does not radiate any power at radio frequencies. **(4—broadband)** An antenna that operates within specified performance limits over a wide band of frequencies, without requiring retuning for each individual frequency. **(5—Cassegrain)** A double reflecting antenna, often used for ground stations in satellite systems. **(6—coaxial)** A dipole antenna made by folding back on itself a quarter wavelength of the outer conductor of a coaxial line, leaving a quarter wavelength of the inner conductor exposed. **(7—corner)** An antenna within the angle formed by two plane-reflecting surfaces. **(8—dipole)** A center-fed antenna, one half-wavelength long. **(9—directional)** An antenna designed to receive or emit radiation more efficiently in a particular direction. **(10—dummy)** An artificial antenna, designed to accept power from the transmitter but not to radiate it. **(11—ferrite)** A common AM broadcast receive antenna that uses a small coil mounted on a short rod of ferrite material. **(12—flat top)** An antenna in which all the horizontal components are in the same horizontal plane. **(13—folded dipole)** A radiating device consisting of two ordinary half-wave dipoles joined at their outer ends and fed at the center of one of the dipoles. **(14—horn reflector)** A radiator in which the feed horn extends into a parabolic reflector, and the power is radiated through a window in the horn. **(15—isotropic)** A theoretical antenna in free space that transmits or receives with the same efficiency in all directions. **(16—log-periodic)** A broadband directional antenna incorporating an array of dipoles of different lengths, the length and spacing between dipoles increasing logarithmically away from the feeder element. **(17—long wire)** An antenna made up of one or more conductors in a straight line pointing in the required direction with a total length of several wavelengths at the operating frequency. **(18—loop)** An antenna consisting of one or more turns of wire in the same or parallel planes. **(19—nested rhombic)** An assembly of two rhombic antennas, one smaller than the other, so that the complete diamond-shaped antenna fits inside the area occupied by the larger unit. (**20—omnidirectional**) An antenna whose radiating or receiving properties are the same in all horizontal plane directions. (**21—periodic**) A resonant antenna designed for use at a particular frequency. (**22—quarter-wave**) A dipole antenna whose length is equal to one quarter of a wavelength at the operating frequency. (**23—rhombic**) A large diamond-shaped antenna, with sides of the diamond several wavelengths long. The rhombic antenna is fed at one of the corners, with directional efficiency in the direction of the diagonal. (**24—series fed**) A vertical antenna that is fed at its lower end. (**25—shunt fed**) A vertical antenna whose base is grounded, and is fed at a specified point above ground. The point at which the antenna is fed above ground determines the operating impedance. (**26—steerable**) An antenna so constructed that its major lobe may readily be changed in direction. (**27—top-loaded**) A vertical antenna capacitively loaded at its upper end, often by simple enlargement or the attachment of a disc or plate. (**28—turnstile**) An antenna with one or more tiers of horizontal dipoles, crossed at right angles to each other and with excitation of the dipoles in phase quadrature. (**29—whip**) An antenna constructed of a thin semiflexible metal rod or tube, fed at its base. (**30—Yagi**) A directional antenna constructed of a series of dipoles cut to specific lengths. *Director* elements are placed in front of the active dipole and *reflector* elements are placed behind the active element.

antenna array A group of several antennas coupled together to yield a required degree of directivity.

antenna beamwidth The angle between the *half-power* points (3 dB points) of the main lobe of the antenna pattern when referenced to the peak power point of the antenna pattern. *Antenna beamwidth* is measured in degrees and normally refers to the horizontal radiation pattern.

antenna directivity factor The ratio of the power flux density in the desired direction to the average value of power flux density at crests in the antenna directivity pattern in the interference section.

antenna factor A factor that, when applied to the voltage appearing at the terminals of measurement equipment, yields the electrical field strength at an antenna. The unit of antenna factor is volts per meter per measured volt.

antenna gain The ratio of the power required at the input of a theoretically perfect omnidirectional reference antenna to the power supplied to the input of the given antenna to produce the same field at the same distance. When not specified otherwise, the figure expressing the gain of an antenna refers to the gain in the direction of the radiation main lobe. In services using *scattering* modes of propagation, the full gain of an antenna may not be realizable in practice and the apparent gain may vary with time.

antenna gain-to-noise temperature For a satellite earth terminal receiving system, a figure of merit that equals G/T, where G is the gain in dB of the earth terminal antenna at the receive frequency, and T is the equivalent noise temperature of the receiving system in Kelvins.

antenna matching The process of adjusting an antenna matching circuit (or the antenna itself) so that the input impedance of the antenna is equal to the characteristic impedance of the transmission line.

antenna monitor A device used to measure the ratio and phase between the currents flowing in the towers of a directional AM broadcast station.

antenna noise temperature The temperature of a resistor having an available noise power per unit bandwidth equal to that at the antenna output at a specified frequency.

antenna pattern A diagram showing the efficiency of radiation in all directions from the antenna.

antenna power rating The maximum continuous-wave power that can be applied to an antenna without degrading its performance.

antenna preamplifier A small amplifier, usually mast-mounted, for amplifying weak signals to a level sufficient to compensate for down-lead losses.

apparent power The product of the root-mean-square values of the voltage and current in an alternating-current circuit without a correction for the phase difference between the voltage and current.

arc A sustained luminous discharge between two or more electrodes.

arithmetic mean The sum of the values of several quantities divided by the number of quantities, also referred to as the *average*.

armature winding The winding of an electrical machine, either a motor or generator, in which current is induced.

array (1—antenna) An assembly of several directional antennas so placed and interconnected that directivity may be enhanced. **(2—broadside)** An antenna array whose elements are all in the same plane, producing a major lobe perpendicular to the plane. **(3—colinear)** An antenna array whose elements are in the same line, either horizontal or vertical. **(4—end-fire)** An antenna array whose elements are in parallel rows, one behind the other, producing a major lobe perpendicular to the plane in which individual elements are placed. **(5—linear)** An antenna array whose elements are arranged end-to-end. **(6—stacked)** An antenna array whose elements are stacked, one above the other.

artificial line An assembly of resistors, inductors, and capacitors that simulates the electrical characteristics of a transmission line.

assembly A manufactured part made by combining several other parts or subassemblies.

assumed values A range of values, parameters, levels, and other elements assumed for a mathematical model, hypothetical circuit, or network, from which analysis, additional estimates, or calculations will be made. The range of values, while not measured, represents the best engineering judgment and is generally derived from values found or measured in real circuits or networks of the same generic type, and includes projected improvements.

atmosphere The gaseous envelope surrounding the earth, composed largely of oxygen, carbon dioxide, and water vapor. The atmosphere is divided into four primary layers: *troposphere, stratosphere, ionosphere,* and *exosphere.*

atmospheric noise Radio noise caused by natural atmospheric processes, such as lightning.

attack time The time interval in seconds required for a device to respond to a control stimulus.

attenuation The decrease in amplitude of an electrical signal traveling through a transmission medium caused by dielectric and conductor losses.

attenuation coefficient The rate of decrease in the amplitude of an electrical signal caused by attenuation. The *attenuation coefficient* can be expressed in decibels or nepers per unit length. It may also be referred to as the *attenuation constant.*

attenuation distortion The distortion caused by attenuation that varies over the frequency range of a signal.

attenuation-limited operation The condition prevailing when the received signal amplitude (rather than distortion) limits overall system performance.

attenuator A fixed or adjustable component that reduces the amplitude of an electrical signal without causing distortion.

atto A prefix meaning one *quintillionth*.

attraction The attractive force between two unlike magnetic poles (N/S) or electrically charged bodies (+/–).

attributes The characteristics of equipment that aid planning and circuit design.

automatic frequency control (AFC) A system designed to maintain the correct operating frequency of a receiver. Any drift in tuning results in the production of a control voltage, which is used to adjust the frequency of a local oscillator so as to minimize the tuning error.

automatic gain control (AGC) An electronic circuit that compares the level of an incoming signal with a previously defined standard and automatically amplifies or attenuates the signal so it arrives at its destination at the correct level.

autotransformer A transformer in which both the primary and secondary currents flow through one common part of the coil.

auxiliary power An alternate source of electric power, serving as a back-up for the primary utility company ac power.

availability A measure of the degree to which a system, subsystem, or equipment is operable and not in a stage of congestion or failure at any given point in time.

avalanche effect The effect obtained when the electric field across a barrier region is sufficiently strong for electrons to collide with *valence electrons*, thereby releasing more electrons and giving a cumulative multiplication effect in a semiconductor.

average life The mean value for a normal distribution of product or component lives, generally applied to mechanical failures resulting from "wear-out."

B

back emf A voltage induced in the reverse direction when current flows through an inductance. *Back emf* is also known as *counter-emf*.

back scattering A form of wave scattering in which at least one component of the scattered wave is deflected opposite to the direction of propagation of the incident wave.

background noise The total system noise in the absence of information transmission, independent of the presence or absence of a signal.

backscatter The deflection or reflection of radiant energy through angles greater than 90° with respect to the original angle of travel.

backscatter range The maximum distance from which backscattered radiant energy can be measured.

backup A circuit element or facility used to replace an element that has failed.

backup supply A redundant power supply that takes over if the primary power supply fails.

balance The process of equalizing the voltage, current, or other parameter between two or more circuits or systems.

balanced A circuit having two sides (conductors) carrying voltages that are symmetrical about a common reference point, typically ground.

balanced circuit A circuit whose two sides are electrically equal in all transmission respects.

balanced line A transmission line consisting of two conductors in the presence of ground capable of being operated in such a way that when the voltages of the two conductors at all transverse planes are equal in magnitude and opposite in polarity with respect to ground, the currents in the two conductors are equal in magnitude and opposite in direction.

balanced modulator A modulator that combines the information signal and the carrier so that the output contains the two sidebands without the carrier.

balanced three-wire system A power distribution system using three conductors, one of which is balanced to have a potential midway between the potentials of the other two.

balanced-to-ground The condition when the impedance to ground on one wire of a two-wire circuit is equal to the impedance to ground on the other wire.

balun (balanced/unbalanced) A device used to connect balanced circuits with unbalanced circuits.

band A range of frequencies between a specified upper and lower limit.

band elimination filter A filter having a single continuous attenuation band, with neither the upper nor lower cut-off frequencies being zero or infinite. A *band elimination filter* may also be referred to as a *band-stop, notch,* or *band reject* filter.

bandpass filter A filter having a single continuous transmission band with neither the upper nor the lower cut-off frequencies being zero or infinite. A bandpass filter permits only a specific band of frequencies to pass; frequencies above or below are attenuated.

bandwidth The range of signal frequencies that can be transmitted by a communications channel with a defined maximum loss or distortion. Bandwidth indicates the information-carrying capacity of a channel.

bandwidth expansion ratio The ratio of the necessary bandwidth to the baseband bandwidth.

bandwidth-limited operation The condition prevailing when the frequency spectrum or bandwidth, rather than the amplitude (or power) of the signal, is the limiting factor in communication capability. This condition is reached when the system distorts the shape of the waveform beyond tolerable limits.

bank A group of similar items connected together in a specified manner and used in conjunction with one another.

bare A wire conductor that is not enameled or enclosed in an insulating sheath.

baseband The band of frequencies occupied by a signal before it modulates a carrier wave to form a transmitted radio or line signal.

baseband channel A channel that carries a signal without modulation, in contrast to a *passband* channel.

baseband signal The original form of a signal, unchanged by modulation.

bath tub The shape of a typical graph of component failure rates: high during an initial period of operation, falling to an acceptable low level during the normal usage period, and then rising again as the components become time-expired.

battery A group of several cells connected together to furnish current by conversion of chemical, thermal, solar, or nuclear energy into electrical energy. A single cell is itself sometimes also called a battery.

bay A row or suite of racks on which transmission, switching, and/or processing equipment is mounted.

Bel A unit of power measurement, named in honor of Alexander Graham Bell. The commonly used unit is one tenth of a Bel, or a decibel (dB). One Bel is defined as a tenfold increase in power. If an amplifier increases the power of a signal by 10 times, the power gain of the amplifier is equal to 1 Bel or 10 *decibels* (dB). If power is increased by 100 times, the power gain is 2 Bels or 20 decibels.

bend A transition component between two elements of a transmission waveguide.

bending radius The smallest bend that may be put into a cable under a stated pulling force. The bending radius is typically expressed in inches.

bias A dc voltage difference applied between two elements of an active electronic device, such as a vacuum tube, transistor, or integrated circuit. Bias currents may or may not be drawn, depending on the device and circuit type.

bidirectional An operational qualification which implies that the transmission of information occurs in both directions.

bifilar winding A type of winding in which two insulated wires are placed side by side. In some components, bifilar winding is used to produce balanced circuits.

bipolar A signal that contains both positive-going and negative-going amplitude components. A bipolar signal may also contain a zero amplitude state.

bleeder A high resistance connected in parallel with one or more filter capacitors in a high voltage dc system. If the power supply load is disconnected, the capacitors discharge through the bleeder.

block diagram An overview diagram that uses geometric figures to represent the principal divisions or sections of a circuit, and lines and arrows to show the path of a signal, or to show program functionalities. It is not a *schematic*, which provides greater detail.

blocking capacitor A capacitor included in a circuit to stop the passage of direct current.

BNC An abbreviation for *bayonet Neill-Concelman*, a type of cable connector used extensively in RF applications (named for its inventor).

Boltzmann's constant 1.38×10^{-23} joules.

bridge A type of network circuit used to match different circuits to each other, ensuring minimum transmission impairment.

bridging The shunting or paralleling of one circuit with another.

broadband The quality of a communications link having essentially uniform response over a given range of frequencies. A communications link is said to be *broadband* if it offers no perceptible degradation to the signal being transported.

buffer A circuit or component that isolates one electrical circuit from another.

burn-in The operation of a device, sometimes under extreme conditions, to stabilize its characteristics and identify latent component failures before bringing the device into normal service.

bus A central conductor for the primary signal path. The term bus may also refer to a signal path to which a number of inputs may be connected for feed to one or more outputs.

busbar A main dc power bus.

bypass capacitor A capacitor that provides a signal path that effectively shunts or bypasses other components.

bypass relay A switch used to bypass the normal electrical route of a signal or current in the event of power, signal, or equipment failure.

C

cable An electrically and/or optically conductive interconnecting device.

cable loss Signal loss caused by passing a signal through a coaxial cable. Losses are the result of resistance, capacitance, and inductance in the cable.

cable splice The connection of two pieces of cable by joining them mechanically and closing the joint with a weather-tight case or sleeve.

cabling The wiring used to interconnect electronic equipment.

calibrate The process of checking, and adjusting if necessary, a test instrument against one known to be set correctly.

calibration The process of identifying and measuring errors in instruments and/or procedures.

capacitance The property of a device or component that enables it to store energy in an electrostatic field and to release it later. A capacitor consists of two conductors separated by an insulating material. When the conductors have a voltage difference between them, a charge will be stored in the electrostatic field between the conductors.

capacitor A device that stores electrical energy. A capacitor allows the apparent flow of alternating current, while blocking the flow of direct current. The degree to which the device permits ac current flow depends on the frequency of the signal and the size of the capacitor. Capacitors are used in filters, delay-line components, couplers, frequency selectors, timing elements, voltage transient suppression, and other applications.

carrier A single frequency wave that, prior to transmission, is modulated by another wave containing information. A carrier may be modulated by manipulating its amplitude and/or frequency in direct relation to one or more applied signals.

carrier frequency The frequency of an unmodulated oscillator or transmitter. Also, the average frequency of a transmitter when a signal is frequency modulated by a symmetrical signal.

cascade connection A tandem arrangement of two or more similar component devices or circuits, with the output of one connected to the input of the next.

cascaded An arrangement of two or more circuits in which the output of one circuit is connected to the input of the next circuit.

cathode ray tube (CRT) A vacuum tube device, usually glass, that is narrow at one end and widens at the other to create a surface onto which images can be projected. The narrow end contains the necessary circuits to generate and focus an electron beam on the luminescent screen at the other end. CRTs are used to display pictures in TV receivers, video monitors, oscilloscopes, computers, and other systems.

cell An elementary unit of communication, of power supply, or of equipment.

Celsius A temperature measurement scale, expressed in degrees C, in which water freezes at 0°C and boils at 100°C. To convert to degrees Fahrenheit, multiply by 0.555 and add 32. To convert to Kelvins add 273 (approximately).

center frequency In frequency modulation, the resting frequency or initial frequency of the carrier before modulation.

center tap A connection made at the electrical center of a coil.

channel The smallest subdivision of a circuit that provides a single type of communication service.

channel decoder A device that converts an incoming modulated signal on a given channel back into the source-encoded signal.

channel encoder A device that takes a given signal and converts it into a form suitable for transmission over the communications channel.

channel noise level The ratio of the channel noise at any point in a transmission system to some arbitrary amount of circuit noise chosen as a reference. This ratio is usually expressed in *decibels above reference noise*, abbreviated *dBrn*.

channel reliability The percent of time a channel is available for use in a specific direction during a specified period.

channelization The allocation of communication circuits to channels and the forming of these channels into groups for higher order multiplexing.

characteristic The property of a circuit or component.

characteristic impedance The impedance of a transmission line, as measured at the driving point, if the line were of infinite length. In such a line, there would be no standing waves. The *characteristic impedance* may also be referred to as the *surge impedance*.

charge The process of replenishing or replacing the electrical charge in a secondary cell or storage battery.

charger A device used to recharge a battery. Types of charging include: (1) *constant voltage charge*, (2) *equalizing charge*, and (3) *trickle charge*.

chassis ground A connection to the metal frame of an electronic system that holds the components in a place. The chassis ground connection serves as the ground return or electrical common for the system.

circuit Any closed path through which an electrical current can flow. In a *parallel circuit*, components are connected between common inputs and outputs such that all paths are parallel to each other. The same voltage appears across all paths. In a *series circuit*, the same current flows through all components.

circuit noise level The ratio of the circuit noise at some given point in a transmission system to an established reference, usually expressed in decibels above the reference.

circuit reliability The percentage of time a circuit is available to the user during a specified period of scheduled availability.

circular mil The measurement unit of the cross-sectional area of a circular conductor. A *circular mil* is the area of a circle whose diameter is one mil, or 0.001 inch.

clear channel A transmission path wherein the full bandwidth is available to the user, with no portions of the channel used for control, framing, or signaling. Can also refer to a classification of AM broadcast station.

clipper A limiting circuit which ensures that a specified output level is not exceeded by restricting the output waveform to a maximum peak amplitude.

clipping The distortion of a signal caused by removing a portion of the waveform through restriction of the amplitude of the signal by a circuit or device.

coax A short-hand expression for *coaxial cable*, which is used to transport high-frequency signals.

coaxial cable A transmission line consisting of an inner conductor surrounded first by an insulating material and then by an outer conductor, either solid or braided. The mechanical dimensions of the cable determine its *characteristic impedance*.

coherence The correlation between the phases of two or more waves.

coherent The condition characterized by a fixed phase relationship among points on an electromagnetic wave.

coherent pulse The condition in which a fixed phase relationship is maintained between consecutive pulses during pulse transmission.

cold joint A soldered connection that was inadequately heated, with the result that the wire is held in place by rosin flux, not solder. A cold joint is sometimes referred to as a *dry joint*.

comb filter An electrical filter circuit that passes a series of frequencies and rejects the frequencies in between, producing a frequency response similar to the teeth of a comb.

common A point that acts as a reference for circuits, often equal in potential to the local ground.

common mode Signals identical with respect to amplitude, frequency, and phase that are applied to both terminals of a cable and/or both the input and reference of an amplifier.

common return A return path that is common to two or more circuits, and returns currents to their source or to ground.

common return offset The dc common return potential difference of a line.

communications system A collection of individual communications networks, transmission systems, relay stations, tributary stations, and terminal equipment capable of interconnection and interoperation to form an integral whole. The individual components must serve a common purpose, be technically compatible, employ common procedures, respond to some form of control, and, in general, operate in unison.

commutation A successive switching process carried out by a commutator.

commutator A circular assembly of contacts, insulated one from another, each leading to a different portion of the circuit or machine.

compatibility The ability of diverse systems to exchange necessary information at appropriate levels of command directly and in usable form. Communications equipment items are compatible if signals can be exchanged between them without the addition of buffering or translation for the specific purpose of achieving workable interface connections, and if the equipment or systems being interconnected possess comparable performance characteristics, including the suppression of undesired radiation.

complex wave A waveform consisting of two or more sinewave components. At any instant of time, a complex wave is the algebraic sum of all its sinewave components.

compliance For mechanical systems, a property which is the reciprocal of stiffness.

component An assembly, or part thereof, that is essential to the operation of some larger circuit or system. A *component* is an immediate subdivision of the assembly to which it belongs.

COMSAT The *Communications Satellite Corporation*, an organization established by an act of Congress in 1962. COMSAT launches and operates the international satellites for the INTELSAT consortium of countries.

concentricity A measure of the deviation of the center conductor position relative to its ideal location in the exact center of the dielectric cross-section of a coaxial cable.

conditioning The adjustment of a channel in order to provide the appropriate transmission characteristics needed for data or other special services.

conditioning equipment The equipment used to match transmission levels and impedances, and to provide equalization between facilities.

conductance A measure of the capability of a material to conduct electricity. It is the reciprocal of *resistance* (ohm) and is expressed in *siemens*. (Formerly expressed as *mho*.)

conducted emission An electromagnetic energy propagated along a conductor.

conduction The transfer of energy through a medium, such as the conduction of electricity by a wire, or of heat by a metallic frame.

conduction band A partially filled or empty atomic energy band in which electrons are free to move easily, allowing the material to carry an electric current.

conductivity The conductance per unit length.

conductor Any material that is capable of carrying an electric current.

configuration A relative arrangement of parts.

connection A point at which a junction of two or more conductors is made.

connector A device mounted on the end of a wire or fiber optic cable that mates to a similar device on a specific piece of equipment or another cable.

constant-current source A source with infinitely high output impedance so that output current is independent of voltage, for a specified range of output voltages.

constant-voltage charge A method of charging a secondary cell or storage battery during which the terminal voltage is kept at a constant value.

constant-voltage source A source with low, ideally zero, internal impedance, so that voltage will remain constant, independent of current supplied.

contact The points that are brought together or separated to complete or break an electrical circuit.

contact bounce The rebound of a contact, which temporarily opens the circuit after its initial *make*.

contact form The configuration of a contact assembly on a relay. Many different configurations are possible from simple *single-make* contacts to complex arrangements involving *breaks* and *makes*.

contact noise A noise resulting from current flow through an electrical contact that has a rapidly varying resistance, as when the contacts are corroded or dirty.

contact resistance The resistance at the surface when two conductors make contact.

continuity A continuous path for the flow of current in an electrical circuit.

continuous wave An electromagnetic signal in which successive oscillations of the waves are identical.

control The supervision that an operator or device exercises over a circuit or system.

control grid The grid in an electron tube that controls the flow of current from the cathode to the anode.

convention A generally acceptable symbol, sign, or practice in a given industry.

Coordinated Universal Time (UTC) The time scale, maintained by the BIH (Bureau International de l'Heure), that forms the basis of a coordinated dissemination of standard frequencies and time signals.

copper loss The loss resulting from the heating effect of current.

corona A bluish luminous discharge resulting from ionization of the air near a conductor carrying a voltage gradient above a certain *critical level*.

corrective maintenance The necessary tests, measurements, and adjustments required to remove or correct a fault.

cosmic noise The random noise originating outside the earth's atmosphere.

coulomb The standard unit of electric quantity or charge. One *coulomb* is equal to the quantity of electricity transported in 1 second by a current of 1 ampere.

Coulomb's Law The attraction and repulsion of electric charges act on a line between them. The charges are inversely proportional to the square of the distance between them, and proportional to the product of their magnitudes. (Named for the French physicist Charles-Augustine de Coulomb, 1736–1806.)

counter-electromotive force The effective electromotive force within a system that opposes the passage of current in a specified direction.

couple The process of linking two circuits by inductance, so that energy is transferred from one circuit to another.

coupled mode The selection of either ac or dc coupling.

coupling The relationship between two components that enables the transfer of energy between them. Included are *direct coupling* through a direct electrical connection, such as a wire; *capacitive coupling* through the capacitance formed by two adjacent conductors; and *inductive coupling* in which energy is transferred through a magnetic field. Capacitive coupling is also called *electrostatic coupling*. Inductive coupling is often referred to as *electromagnetic coupling*.

coupling coefficient A measure of the electrical coupling that exists between two circuits. The *coupling coefficient* is equal to the ratio of the mutual impedance to the square root of the product of the self impedances of the coupled circuits.

cross coupling The coupling of a signal from one channel, circuit, or conductor to another, where it becomes an undesired signal.

crossover distortion A distortion that results in an amplifier when an irregularity is introduced into the signal as it crosses through a zero reference point. If an amplifier is properly designed and biased, the upper half cycle and lower half cycle of the signal coincide at the zero crossover reference.

crossover frequency The frequency at which output signals pass from one channel to the other in a *crossover network*. At the *crossover frequency* itself, the outputs to each side are equal.

crossover network A type of filter that divides an incoming signal into two or more outputs, with higher frequencies directed to one output, and lower frequencies to another.

crosstalk Undesired transmission of signals from one circuit into another circuit in the same system. Crosstalk is usually caused by unintentional capacitive (ac) coupling.

crosstalk coupling The ratio of the power in a disturbing circuit to the induced power in the disturbed circuit, observed at a particular point under specified conditions. Crosstalk coupling is typically expressed in dB.

crowbar A short-circuit or low resistance path placed across the input to a circuit, usually for protective purposes.

CRT (cathode ray tube) A vacuum tube device that produces light when energized by the electron beam generated inside the tube. A CRT includes an electron gun, deflection mechanism, and phosphor-covered faceplate.

crystal A solidified form of a substance that has atoms and molecules arranged in a symmetrical pattern.

crystal filter A filter that uses piezoelectric crystals to create resonant or antiresonant circuits.

crystal oscillator An oscillator using a piezoelectric crystal as the tuned circuit that controls the resonant frequency.

crystal-controlled oscillator An oscillator in which a piezoelectric-effect crystal is coupled to a tuned oscillator circuit in such a way that the crystal pulls the oscillator frequency to its own natural frequency and does not allow frequency drift.

current (1—general) A general term for the transfer of electricity, or the movement of electrons or *holes*. **(2—alternating)** An electric current that is constantly varying in amplitude and periodically reversing direction. **(3—average)** The arithmetic mean of the instantaneous values of current, averaged over one complete half cycle. **(4—charging)** The current that flows in to charge a capacitor when it is first connected to a source of electric potential. **(5—direct)** Electric current that flows in one direction only. **(6—eddy)** A wasteful current that flows in the core of a transformer and produces heat. *Eddy currents* are largely eliminated through the use of laminated cores. **(7—effective)** The ac current that will produce the same effective heat in a resistor as is produced by dc. If the ac is sinusoidal, the *effective current* value is 0.707 times the peak ac value. **(8—fault)** The current that flows between conductors or to ground during a fault condition. **(9—ground fault)** A fault current that flows to ground. **(10—ground return)** A current that returns through the earth. **(11—lagging)** A phenomenon observed in an inductive circuit where alternating current lags behind the voltage that produces it. **(12—leading)** A phenomenon observed in a capacitive circuit where alternating current leads the voltage that produces it. **(13—magnetizing)** The current in a transformer primary winding that is just sufficient to magnetize the core and offset iron losses. **(14—neutral)** The current that flows in the neutral conductor of an unbalanced polyphase power circuit. If correctly balanced, the neutral would carry no net current. **(15—peak)** The maximum value reached by a varying current during one cycle. **(16—pick-up)** The minimum current at which a relay just begins to operate. **(17—plate)** The anode current of an electron tube. **(18—residual)** The vector sum of the currents in the phase wires of an unbalanced polyphase power circuit. **(19—space)** The total current flowing through an electron tube.

current amplifier A low output impedance amplifier capable of providing high current output.

current probe A sensor, clamped around an electrical conductor, in which an induced current is developed from the magnetic field surrounding the conductor. For measurements, the current probe is connected to a suitable test instrument.

current transformer A transformer-type of instrument in which the primary carries the current to be measured and the secondary is in series with a low current ammeter. A current transformer is used to measure high values of alternating current.

current-carrying capacity A measure of the maximum current that can be carried continuously without damage to components or devices in a circuit.

cut-off frequency The frequency above or below which the output current in a circuit is reduced to a specified level.

cycle The interval of time or space required for a periodic signal to complete one period.

cycles per second The standard unit of frequency, expressed in Hertz (one cycle per second).

D

damped oscillation An oscillation exhibiting a progressive diminution of amplitude with time.

damping The dissipation and resultant reduction of any type of energy, such as electromagnetic waves.

dB (decibel) A measure of voltage, current, or power gain equal to 0.1 Bel. Decibels are given by the equations

$$20 \log \frac{V_{out}}{V_{in}},\ 20 \log \frac{I_{out}}{I_{in}},\ \text{or } 10 \log \frac{P_{out}}{P_{in}}.$$

dBk A measure of power relative to 1 kilowatt. 0 dBk equals 1 kW.

dBm (decibels above 1 milliwatt) A logarithmic measure of power with respect to a reference power of one milliwatt.

dBmv A measure of voltage gain relative to 1 millivolt at 75 ohms.

dBr The power difference expressed in dB between any point and a reference point selected as the *zero relative transmission level* point. A power expressed in *dBr* does not specify the absolute power; it is a relative measurement only.

dBu A term that reflects comparison between a measured value of voltage and a reference value of 0.775 V, expressed under conditions in which the impedance at the point of measurement (and of the reference source) are not considered.

dbV A measure of voltage gain relative to 1 V.

dBW A measure of power relative to 1 watt. 0 dBW equals 1 W.

dc An abbreviation for *direct current*. Note that the preferred usage of the term *dc* is lower case.

dc amplifier A circuit capable of amplifying dc and slowly varying alternating current signals.

dc component The portion of a signal that consists of direct current. This term may also refer to the average value of a signal.

dc coupled A connection configured so that both the signal (ac component) and the constant voltage on which it is riding (dc component) are passed from one stage to the next.

dc coupling A method of coupling one circuit to another so as to transmit the static (dc) characteristics of the signal as well as the varying (ac) characteristics. Any dc offset present on the input signal is maintained and will be present in the output.

dc offset The amount that the dc component of a given signal has shifted from its correct level.

dc signal bounce Overshoot of the proper dc voltage level resulting from multiple ac couplings in a signal path.

de-energized A system from which sources of power have been disconnected.

deca A prefix meaning *ten*.

decay The reduction in amplitude of a signal on an exponential basis.

decay time The time required for a signal to fall to a certain fraction of its original value.

decibel (dB) One tenth of a Bel. The decibel is a logarithmic measure of the ratio between two powers.

decode The process of recovering information from a signal into which the information has been encoded.

decoder A device capable of deciphering encoded signals. A decoder interprets input instructions and initiates the appropriate control operations as a result.

decoupling The reduction or removal of undesired coupling between two circuits or stages.

deemphasis The reduction of the high-frequency components of a received signal to reverse the preemphasis that was placed on them to overcome attenuation and noise in the transmission process.

defect An error made during initial planning that is normally detected and corrected during the development phase. Note that a *fault* is an error that occurs in an in-service system.

deflection The control placed on electron direction and motion in CRTs and other vacuum tube devices by varying the strengths of electrostatic (electrical) or electromagnetic fields.

degradation In susceptibility testing, any undesirable change in the operational performance of a test specimen. This term does not necessarily mean malfunction or catastrophic failure.

degradation failure A failure that results from a gradual change in performance characteristics of a system or part with time.

delay The amount of time by which a signal is delayed or an event is retarded.

delay circuit A circuit designed to delay a signal passing through it by a specified amount.

delay distortion The distortion resulting from the difference in phase delays at two frequencies of interest.

delay equalizer A network that adjusts the velocity of propagation of the frequency components of a complex signal to counteract the delay distortion characteristics of a transmission channel.

delay line A transmission network that increases the propagation time of a signal traveling through it.

delta connection A common method of joining together a three-phase power supply, with each phase across a different pair of the three wires used.

delta-connected system A 3-phase power distribution system where a single-phase output can be derived from each of the adjacent pairs of an equilateral triangle formed by the service drop transformer secondary windings.

demodulator Any device that recovers the original signal after it has modulated a high-frequency carrier. The output from the unit may be in baseband composite form.

demultiplexer (demux) A device used to separate two or more signals that were previously combined by a compatible multiplexer and are transmitted over a single channel.

derating factor An operating safety margin provided for a component or system to ensure reliable performance. A *derating allowance* also is typically provided for operation under extreme environmental conditions, or under stringent reliability requirements.

desiccant A drying agent used for drying out cable splices or sensitive equipment.

design A layout of all the necessary equipment and facilities required to make a special circuit, piece of equipment, or system work.

design objective The desired electrical or mechanical performance characteristic for electronic circuits and equipment.

detection The rectification process that results in the modulating signal being separated from a modulated wave.

detectivity The reciprocal of *noise equivalent power*.

detector A device that converts one type of energy into another.

device A functional circuit, component, or network unit, such as a vacuum tube or transistor.

dewpoint The temperature at which moisture will condense out.

diagnosis The process of locating errors in software, or equipment faults in hardware.

diagnostic routine A software program designed to trace errors in software, locate hardware faults, or identify the cause of a breakdown.

dielectric An insulating material that separates the elements of various components, including capacitors and transmission lines. Dielectric materials include air, plastic, mica, ceramic, and Teflon. A dielectric material must be an insulator. (*Teflon* is a registered trademark of Du Pont.)

dielectric constant The ratio of the capacitance of a capacitor with a certain dielectric material to the capacitance with a vacuum as the dielectric. The *dielectric constant* is considered a measure of the capability of a dielectric material to store an electrostatic charge.

dielectric strength The potential gradient at which electrical breakdown occurs.

differential amplifier An input circuit that rejects voltages that are the same at both input terminals but amplifies any voltage difference between the inputs. Use of a differential amplifier causes any signal present on both terminals, such as common mode hum, to cancel itself.

differential dc The maximum dc voltage that can be applied between the differential inputs of an amplifier while maintaining linear operation.

differential gain The difference in output amplitude (expressed in percent or dB) of a small high frequency sinewave signal at two stated levels of a low frequency signal on which it is superimposed.

differential phase The difference in output phase of a small high frequency sinewave signal at two stated levels of a low frequency signal on which it is superimposed.

differential-mode interference An interference source that causes a change in potential of one side of a signal transmission path relative to the other side.

diffuse reflection The scattering effect that occurs when light, radio, or sound waves strike a rough surface.

diffusion The spreading or scattering of a wave, such as a radio wave.

diode A semiconductor or vacuum tube with two electrodes that passes electric current in one direction only. Diodes are used in rectifiers, gates, modulators, and detectors.

direct coupling A coupling method between stages that permits dc current to flow between the stages.

direct current An electrical signal in which the direction of current flow remains constant.

discharge The conversion of stored energy, as in a battery or capacitor, into an electric current.

discontinuity An abrupt nonuniform point of change in a transmission circuit that causes a disruption of normal operation.

discrete An individual circuit component.

discrete component A separately contained circuit element with its own external connections.

discriminator A device or circuit whose output amplitude and polarity vary according to how much the input signal varies from a standard or from another signal. A discriminator can be used to recover the modulating waveform in a frequency modulated signal.

dish An antenna system consisting of a parabolic shaped reflector with a signal feed element at the focal point. Dish antennas commonly are used for transmission and reception from microwave stations and communications satellites.

dispersion The wavelength dependence of a parameter.

display The representation of text and images on a cathode-ray tube, an array of light-emitting diodes, a liquid-crystal readout, or another similar device.

display device An output unit that provides a visual representation of data.

distortion The difference between the wave shape of an original signal and the signal after it has traversed a transmission circuit.

distortion-limited operation The condition prevailing when the shape of the signal, rather than the amplitude (or power), is the limiting factor in communication capability. This condition is reached when the system distorts the shape of the waveform beyond tolerable limits. For linear systems, *distortion-limited* operation is equivalent to *bandwidth-limited* operation.

disturbance The interference with normal conditions and communications by some external energy source.

disturbance current The unwanted current of any irregular phenomenon associated with transmission that tends to limit or interfere with the interchange of information.

disturbance power The unwanted power of any irregular phenomenon associated with transmission that tends to limit or interfere with the interchange of information.

disturbance voltage The unwanted voltage of any irregular phenomenon associated with transmission that tends to limit or interfere with the interchange of information.

diversity receiver A receiver using two antennas connected through circuitry that senses which antenna is receiving the stronger signal. Electronic gating permits the stronger source to be routed to the receiving system.

documentation A written description of a program. *Documentation* can be considered as any record that has permanence and can be read by humans or machines.

down-lead A lead-in wire from an antenna to a receiver.

downlink The portion of a communication link used for transmission of signals from a satellite or airborne platform to a surface terminal.

downstream A specified signal modification occurring after other given devices in a signal path.

downtime The time during which equipment is not capable of doing useful work because of malfunction. This does not include preventive maintenance time. In other words, *downtime* is measured from the occurrence of a malfunction to the correction of that malfunction.

drift A slow change in a nominally constant signal characteristic, such as frequency.

drift-space The area in a klystron tube in which electrons drift at their entering velocities and form electron *bunches*.

drive The input signal to a circuit, particularly to an amplifier.

driver An electronic circuit that supplies an isolated output to drive the input of another circuit.

drop-out value The value of current or voltage at which a relay will cease to be operated.

dropout The momentary loss of a signal.

dropping resistor A resistor designed to carry current that will make a required voltage available.

duplex separation The frequency spacing required in a communications system between the *forward* and *return* channels to maintain interference at an acceptably low level.

duplex signaling A configuration permitting signaling in both transmission directions simultaneously.

duty cycle The ratio of operating time to total elapsed time of a device that operates intermittently, expressed in percent.

dynamic A situation in which the operating parameters and/or requirements of a given system are continually changing.

dynamic range The maximum range or extremes in amplitude, from the lowest to the highest (noise floor to system clipping), that a system is capable of reproducing. The dynamic range is expressed in dB against a reference level.

dynamo A rotating machine, normally a dc generator.

dynamotor A rotating machine used to convert dc into ac.

E

earth A large conducting body with no electrical potential, also called *ground*.

earth capacitance The capacitance between a given circuit or component and a point at ground potential.

earth current A current that flows to earth/ground, especially one that follows from a fault in the system. *Earth current* may also refer to a current that flows in the earth, resulting from ionospheric disturbances, lightning, or faults on power lines.

earth fault A fault that occurs when a conductor is accidentally grounded/earthed, or when the resistance to earth of an insulator falls below a specified value.

earth ground A large conducting body that represents *zero level* in the scale of electrical potential. An *earth ground* is a connection made either accidentally or by design between a conductor and earth.

earth potential The potential taken to be the arbitrary zero in a scale of electric potential.

effective ground A connection to ground through a medium of sufficiently low impedance and adequate current-carrying capacity to prevent the buildup of voltages that might be hazardous to equipment or personnel.

effective resistance The increased resistance of a conductor to an alternating current resulting from the *skin effect*, relative to the direct-current resistance of the conductor. Higher frequencies tend to travel only on the outer skin of the conductor, whereas dc flows uniformly through the entire area.

efficiency The useful power output of an electrical device or circuit divided by the total power input, expressed in percent.

electric Any device or circuit that produces, operates on, transmits, or uses electricity.

electric charge An excess of either electrons or protons within a given space or material.

electric field strength The magnitude, measured in volts per meter, of the electric field in an electromagnetic wave.

electric flux The amount of electric charge, measured in coulombs, across a dielectric of specified area. *Electric flux* may also refer simply to electric lines of force.

electricity An energy force derived from the movement of negative and positive electric charges.

electrode An electrical terminal that emits, collects, or controls an electric current.

electrolysis A chemical change induced in a substance resulting from the passage of electric current through an electrolyte.

electrolyte A nonmetallic conductor of electricity in which current is carried by the physical movement of ions.

electromagnet An iron or steel core surrounded by a wire coil. The core becomes magnetized when current flows through the coil but loses its magnetism when the current flow is stopped.

electromagnetic compatibility The capability of electronic equipment or systems to operate in a specific electromagnetic environment, at designated levels of efficiency and within a defined margin of safety, without interfering with itself or other systems.

electromagnetic field The electric and magnetic fields associated with radio and light waves.

electromagnetic induction An electromotive force created with a conductor by the relative motion between the conductor and a nearby magnetic field.

electromagnetism The study of phenomena associated with varying magnetic fields, electromagnetic radiation, and moving electric charges.

electromotive force (EMF) An electrical potential, measured in volts, that can produce the movement of electrical charges.

electron A stable elementary particle with a negative charge that is mainly responsible for electrical conduction. Electrons move when under the influence of an electric field. This movement constitutes an *electric current.*

electron beam A stream of emitted electrons, usually in a vacuum.

electron gun A hot cathode that produces a finely focused stream of fast electrons, which are necessary for the operation of a vacuum tube, such as a cathode ray tube. The gun is made up of a hot cathode electron source, a control grid, accelerating anodes, and (usually) focusing electrodes.

electron lens A device used for focusing an electron beam in a cathode ray tube. Such focusing can be accomplished by either magnetic forces, in which external coils are used to create the proper magnetic field within the tube, or electrostatic forces, where metallic plates within the tube are charged electrically in such a way as to control the movement of electrons in the beam.

electron volt The energy acquired by an electron in passing through a potential difference of one volt in a vacuum.

electronic A description of devices (or systems) that are dependent on the flow of electrons in electron tubes, semiconductors, and other devices, and not solely on electron flow in ordinary wires, inductors, capacitors, and similar passive components.

Electronic Industries Association (EIA) A trade organization, based in Washington, DC, representing the manufacturers of electronic systems and parts, including communications systems. The association develops standards for electronic components and systems.

electronic switch A transistor, semiconductor diode, or a vacuum tube used as an on/off switch in an electrical circuit. Electronic switches can be controlled manually, by other circuits, or by computers.

electronics The field of science and engineering that deals with electron devices and their utilization.

electroplate The process of coating a given material with a deposit of metal by electrolytic action.

electrostatic The condition pertaining to electric charges that are at rest.

electrostatic field The space in which there is electric stress produced by static electric charges.

electrostatic induction The process of inducing static electric charges on a body by bringing it near other bodies that carry high electrostatic charges.

element A substance that consists of atoms of the same atomic number. Elements are the basic units in all chemical changes other than those in which *atomic changes*, such as fusion and fission, are involved.

EMI (electromagnetic interference) Undesirable electromagnetic waves that are radiated unintentionally from an electronic circuit or device into other circuits or devices, disrupting their operation.

emission (1—radiation) The radiation produced, or the production of radiation by a radio transmitting system. The emission is considered to be a *single emission* if the modulating signal and other characteristics are the same for every transmitter of the radio transmitting system and the spacing between antennas is not more than a few wavelengths. **(2—cathode)** The release of electrons from the cathode of a vacuum tube. **(3—parasitic)** A spurious radio frequency emission unintentionally generated at frequencies that are independent of the carrier frequency being amplified or modulated. **(4—secondary)** In an electron tube, emission of electrons by a plate or grid because of bombardment by *primary emission* electrons from the cathode of the tube. **(5—spurious)** An emission outside the radio frequency band authorized for a transmitter. **(6—thermonic)** An emission from a cathode resulting from high temperature.

emphasis The intentional alteration of the frequency-amplitude characteristics of a signal to reduce the adverse effects of noise in a communication system.

empirical A conclusion not based on pure theory, but on practical and experimental work.

emulation The use of one system to imitate the capabilities of another system.

enable To prepare a circuit for operation or to allow an item to function.

enabling signal A signal that permits the occurrence of a specified event.

encode The conversion of information from one form into another to obtain characteristics required by a transmission or storage system.

encoder A device that processes one or more input signals into a specified form for transmission and/or storage.

energized The condition when a circuit is switched on, or powered up.

energy spectral density A frequency-domain description of the energy in each of the frequency components of a pulse.

envelope The boundary of the family of curves obtained by varying a parameter of a wave.

envelope delay The difference in absolute delay between the fastest and slowest propagating frequencies within a specified bandwidth.

envelope delay distortion The maximum difference or deviation of the envelope-delay characteristic between any two specified frequencies.

envelope detection A demodulation process that senses the shape of the modulated RF envelope. A diode detector is one type of envelope detection device.

environmental An equipment specification category relating to temperature and humidity.

EQ (equalization) network A network connected to a circuit to correct or control its transmission frequency characteristics.

equalization (EQ) The reduction of frequency distortion and/or phase distortion of a circuit through the introduction of one or more networks to compensate for the difference in attenuation, time delay, or both, at the various frequencies in the transmission band.

equalize The process of inserting in a line a network with complementary transmission characteristics to those of the line, so that when the loss or delay in the line and that in the equalizer are combined, the overall loss or delay is approximately equal at all frequencies.

equalizer A network that corrects the transmission-frequency characteristics of a circuit to allow it to transmit selected frequencies in a uniform manner.

equatorial orbit The plane of a satellite orbit which coincides with that of the equator of the primary body.

equipment A general term for electrical apparatus and hardware, switching systems, and transmission components.

equipment failure The condition when a hardware fault stops the successful completion of a task.

equipment ground A protective ground consisting of a conducting path to ground of noncurrent carrying metal parts.

equivalent circuit A simplified network that emulates the characteristics of the real circuit it replaces. An equivalent circuit is typically used for mathematical analysis.

equivalent noise resistance A quantitative representation in resistance units of the spectral density of a noise voltage generator at a specified frequency.

error A collective term that includes all types of inconsistencies, transmission deviations, and control failures.

excitation The current that energizes field coils in a generator.

expandor A device with a nonlinear gain characteristic that acts to increase the gain more on larger input signals than it does on smaller input signals.

extremely high frequency (EHF) The band of microwave frequencies between the limits of 30 GHz and 300 GHz (wavelengths between 1 cm and 1 mm).

extremely low frequency The radio signals with operating frequencies below 300 Hz (wavelengths longer than 1000 km).

F

fail-safe operation A type of control architecture for a system that prevents improper functioning in the event of circuit or operator failure.

failure A detected cessation of ability to perform a specified function or functions within previously established limits. A *failure* is beyond adjustment by the operator by means of controls normally accessible during routine operation of the system. (This requires that measurable limits be established to define "satisfactory performance".)

failure effect The result of the malfunction or failure of a device or component.

failure in time (FIT) A unit value that indicates the reliability of a component or device. One failure in time corresponds to a failure rate of 10^{-9} per hour.

failure mode and effects analysis (FMEA) An iterative documented process performed to identify basic faults at the component level and determine their effects at higher levels of assembly.

failure rate The ratio of the number of actual failures to the number of times each item has been subjected to a set of specified stress conditions.

fall time The length of time during which a pulse decreases from 90 percent to 10 percent of its maximum amplitude.

farad The standard unit of capacitance equal to the value of a capacitor with a potential of one volt between its plates when the charge on one plate is one coulomb and there is an equal and opposite charge on the other plate. The farad is a large value and is more commonly expressed in *microfarads* or *picofarads*. The *farad* is named for the English chemist and physicist Michael Faraday (1791–1867).

fast frequency shift keying (FFSK) A system of digital modulation where the digits are represented by different frequencies that are related to the baud rate, and where transitions occur at the zero crossings.

fatigue The reduction in strength of a metal caused by the formation of crystals resulting from repeated flexing of the part in question.

fault A condition that causes a device, a component, or an element to fail to perform in a required manner. Examples include a short-circuit, broken wire, or intermittent connection.

fault to ground A fault caused by the failure of insulation and the consequent establishment of a direct path to ground from a part of the circuit that should not normally be grounded.

fault tree analysis (FTA) An iterative documented process of a systematic nature performed to identify basic faults, determine their causes and effects, and establish their probabilities of occurrence.

feature A distinctive characteristic or part of a system or piece of equipment, usually visible to end users and designed for their convenience.

Federal Communications Commission (FCC) The federal agency empowered by law to regulate all interstate radio and wireline communications services originating in the United States, including radio, television, facsimile, telegraph, data transmission, and telephone systems. The agency was established by the Communications Act of 1934.

feedback The return of a portion of the output of a device to the input. *Positive feedback* adds to the input; *negative feedback* subtracts from the input.

feedback amplifier An amplifier with the components required to feed a portion of the output back into the input to alter the characteristics of the output signal.

feedline A transmission line, typically coaxial cable, that connects a high frequency energy source to its load.

femto A prefix meaning *one quadrillionth* (10^{-15}).

ferrite A ceramic material made of powdered and compressed ferric oxide, plus other oxides (mainly cobalt, nickel, zinc, yttrium-iron, and manganese). These materials have low eddy current losses at high frequencies.

ferromagnetic material A material with low relative permeability and high coercive force so that it is difficult to magnetize and demagnetize. Hard ferromagnetic materials retain magnetism well, and are commonly used in permanent magnets.

fidelity The degree to which a system, or a portion of a system, accurately reproduces at its output the essential characteristics of the signal impressed upon its input.

field strength The strength of an electric, magnetic, or electromagnetic field.

filament A wire that becomes hot when current is passed through it, used either to emit light (for a light bulb) or to heat a cathode to enable it to emit electrons (for an electron tube).

film resistor A type of resistor made by depositing a thin layer of resistive material on an insulating core.

filter A network that passes desired frequencies but greatly attenuates other frequencies.

filtered noise White noise that has been passed through a filter. The power spectral density of filtered white noise has the same shape as the transfer function of the filter.

fitting A coupling or other mechanical device that joins one component with another.

fixed A system or device that is not changeable or movable.

flashover An arc or spark between two conductors.

flashover voltage The voltage between conductors at which flashover just occurs.

flat face tube The design of CRT tube with almost a flat face, giving improved legibility of text and reduced reflection of ambient light.

flat level A signal that has an equal amplitude response for all frequencies within a stated range.

flat loss A circuit, device, or channel that attenuates all frequencies of interest by the same amount, also called *flat slope*.

flat noise A noise whose power per unit of frequency is essentially independent of frequency over a specified frequency range.

flat response The performance parameter of a system in which the output signal amplitude of the system is a faithful reproduction of the input amplitude over some range of specified input frequencies.

floating A circuit or device that is not connected to any source of potential or to ground.

fluorescence The characteristic of a material to produce light when excited by an external energy source. Minimal or no heat results from the process.

flux The electric or magnetic lines of force resulting from an applied energy source.

flywheel effect The characteristic of an oscillator that enables it to sustain oscillations after removal of the control stimulus. This characteristic may be desirable, as in the case of a phase-locked loop employed in a synchronous system, or undesirable, as in the case of a voltage-controlled oscillator.

focusing A method of making beams of radiation converge on a target, such as the face of a CRT.

Fourier analysis A mathematical process for transforming values between the frequency domain and the time domain. This term also refers to the decomposition of a time-domain signal into its frequency components.

Fourier transform An integral that performs an actual transformation between the frequency domain and the time domain in Fourier analysis.

frame A segment of an analog or digital signal that has a repetitive characteristic, in that corresponding elements of successive *frames* represent the same things.

free electron An electron that is not attached to an atom and is, thus, mobile when an electromotive force is applied.

free running An oscillator that is not controlled by an external synchronizing signal.

free-running oscillator An oscillator that is not synchronized with an external timing source.

frequency The number of complete cycles of a periodic waveform that occur within a given length of time. Frequency is usually specified in cycles per second (*Hertz*). Frequency is the reciprocal of wavelength. The higher the frequency, the shorter the wavelength. In general, the higher the frequency of a signal, the more capacity it has to carry information, the smaller an antenna is required, and the more susceptible the signal is to absorption by the atmosphere and by physical structures. At microwave frequencies, radio signals take on a *line-of-sight* characteristic and require highly directional and focused antennas to be used successfully.

frequency accuracy The degree of conformity of a given signal to the specified value of a frequency.

frequency allocation The designation of radio-frequency bands for use by specific radio services.

frequency content The band of frequencies or specific frequency components contained in a signal.

frequency converter A circuit or device used to change a signal of one frequency into another of a different frequency.

frequency coordination The process of analyzing frequencies in use in various bands of the spectrum to achieve reliable performance for current and new services.

frequency counter An instrument or test set used to measure the frequency of a radio signal or any other alternating waveform.

frequency departure An unintentional deviation from the nominal frequency value.

frequency difference The algebraic difference between two frequencies. The two frequencies can be of identical or different nominal values.

frequency displacement The end-to-end shift in frequency that may result from independent frequency translation errors in a circuit.

frequency distortion The distortion of a multifrequency signal caused by unequal attenuation or amplification at the different frequencies of the signal. This term may also be referred to as *amplitude distortion*.

frequency domain A representation of signals as a function of frequency, rather than of time.

frequency modulation (FM) The modulation of a carrier signal so that its instantaneous frequency is proportional to the instantaneous value of the modulating wave.

frequency multiplier A circuit that provides as an output an exact multiple of the input frequency.

frequency offset A frequency shift that occurs when a signal is sent over an analog transmission facility in which the modulating and demodulating frequencies are not identical. A channel with frequency offset does not preserve the waveform of a transmitted signal.

frequency response The measure of system linearity in reproducing signals across a specified bandwidth. Frequency response is expressed as a frequency range with a specified amplitude tolerance in dB.

frequency response characteristic The variation in the transmission performance (gain or loss) of a system with respect to variations in frequency.

frequency reuse A technique used to expand the capacity of a given set of frequencies or channels by separating the signals either geographically or through the use of different polarization techniques. Frequency reuse is a common element of the *frequency coordination* process.

frequency selectivity The ability of equipment to separate or differentiate between signals at different frequencies.

frequency shift The difference between the frequency of a signal applied at the input of a circuit and the frequency of that signal at the output.

frequency shift keying (FSK) A commonly used method of digital modulation in which a one and a zero (the two possible states) are each transmitted as separate frequencies.

frequency stability A measure of the variations of the frequency of an oscillator from its mean frequency over a specified period of time.

frequency standard An oscillator with an output frequency sufficiently stable and accurate that it is used as a reference.

frequency-division multiple access (FDMA) The provision of multiple access to a transmission facility, such as an earth satellite, by assigning each transmitter its own frequency band.

frequency-division multiplexing (FDM) The process of transmitting multiple analog signals by an orderly assignment of frequency slots, that is, by dividing transmission bandwidth into several narrow bands, each of which carries a single communication and is sent simultaneously with others over a common transmission path.

full duplex A communications system capable of transmission simultaneously in two directions.

full-wave rectifier A circuit configuration in which both positive and negative half-cycles of the incoming ac signal are rectified to produce a unidirectional (dc) current through the load.

functional block diagram A diagram illustrating the definition of a device, system, or problem on a logical and functional basis.

functional unit An entity of hardware and/or software capable of accomplishing a given purpose.

fundamental frequency The lowest frequency component of a complex signal.

fuse A protective device used to limit current flow in a circuit to a specified level. The fuse consists of a metallic link that melts and opens the circuit at a specified current level.

fuse wire A fine-gauge wire made of an alloy that overheats and melts at the relatively low temperatures produced when the wire carries overload currents. When used in a fuse, the wire is called a fuse (or fusible) link.

G

gain An increase or decrease in the level of an electrical signal. Gain is measured in terms of decibels or number-of-times of magnification. Strictly speaking, *gain* refers to an increase in level. Negative numbers, however, are commonly used to denote a decrease in level.

gain-bandwidth The gain times the frequency of measurement when a device is biased for maximum obtainable gain.

gain/frequency characteristic The gain-versus-frequency characteristic of a channel over the bandwidth provided, also referred to as *frequency response*.

gain/frequency distortion A circuit defect in which a change in frequency causes a change in signal amplitude.

galvanic A device that produces direct current by chemical action.

gang The mechanical connection of two or more circuit devices so that they can all be adjusted simultaneously.

gang capacitor A variable capacitor with more than one set of moving plates linked together.

gang tuning The simultaneous tuning of several different circuits by turning a single shaft on which ganged capacitors are mounted.

ganged One or more devices that are mechanically coupled, normally through the use of a shared shaft.

gas breakdown The ionization of a gas between two electrodes caused by the application of a voltage that exceeds a threshold value. The ionized path has a low impedance. Certain types of circuit and line protectors rely on gas breakdown to divert hazardous currents away from protected equipment.

gas tube A protection device in which a sufficient voltage across two electrodes causes a gas to ionize, creating a low impedance path for the discharge of dangerous voltages.

gas-discharge tube A gas-filled tube designed to carry current during gas breakdown. The gas-discharge tube is commonly used as a protective device, preventing high voltages from damaging sensitive equipment.

gauge A measure of wire diameter. In measuring wire gauge, the lower the number, the thicker the wire.

Gaussian distribution A statistical distribution, also called the *normal* distribution. The graph of a Gaussian distribution is a bell-shaped curve.

Gaussian noise Noise in which the distribution of amplitude follows a Gaussian model; that is, the noise is random but distributed about a reference voltage of zero.

Gaussian pulse A pulse that has the same form as its own Fourier transform.

generator A machine that converts mechanical energy into electrical energy, or one form of electrical energy into another form.

geosynchronous The attribute of a satellite in which the relative position of the satellite as viewed from the surface of a given planet is stationary. For earth, the geosynchronous position is 22,300 miles above the planet.

getter A metal used in vaporized form to remove residual gas from inside an electron tube during manufacture.

giga A prefix meaning one billion.

gigahertz (GHz) A measure of frequency equal to one billion cycles per second. Signals operating above 1 gigahertz are commonly known as *microwaves*, and begin to take on the characteristics of visible light.

glitch A general term used to describe a wide variety of momentary signal discontinuities.

graceful degradation An equipment failure mode in which the system suffers reduced capability, but does not fail altogether.

graticule A fixed pattern of reference markings used with oscilloscope CRTs to simplify measurements. The graticule may be etched on a transparent plate covering the front of the CRT or, for greater accuracy in readings, may be electrically generated within the CRT itself.

grid (1—general) A mesh electrode within an electron tube that controls the flow of electrons between the cathode and plate of the tube. **(2—bias)** The potential applied to a grid in an electron tube to control its center operating point. **(3—control)** The grid in an electron tube to which the input signal is usually applied. **(4—screen)** The grid in an electron tube, typically held at a steady potential, that screens the control grid from changes in anode potential. **(5—suppressor)** The grid in an electron tube near the anode (plate) that suppresses the emission of secondary electrons from the plate.

ground An electrical connection to earth or to a common conductor usually connected to earth.

ground clamp A clamp used to connect a ground wire to a ground rod or system.

ground loop An undesirable circulating ground current in a circuit grounded via multiple connections or at multiple points.

ground plane A conducting material at ground potential, physically close to other equipment, so that connections may be made readily to ground the equipment at the required points.

ground potential The point at zero electric potential.

ground return A conductor used as a path for one or more circuits back to the ground plane or central facility ground point.

ground rod A metal rod driven into the earth and connected into a mesh of interconnected rods so as to provide a low resistance link to ground.

ground window A single-point interface between the integrated ground plane of a building and an isolated ground plane.

ground wire A copper conductor used to extend a good low-resistance earth ground to protective devices in a facility.

grounded The connection of a piece of equipment to earth via a low resistance path.

grounding The act of connecting a device or circuit to ground or to a conductor that is grounded.

group delay A condition where the different frequency elements of a given signal suffer differing propagation delays through a circuit or a system. The delay at a lower frequency is different from the delay at a higher frequency, resulting in a time-related distortion of the signal at the receiving point.

group delay time The rate of change of the total phase shift of a waveform with angular frequency through a device or transmission facility.

group velocity The speed of a pulse on a transmission line.

guard band A narrow bandwidth between adjacent channels intended to reduce interference or crosstalk.

H

half-wave rectifier A circuit or device that changes only positive or negative half-cycle inputs of alternating current into direct current.

Hall effect The phenomenon by which a voltage develops between the edges of a current-carrying metal strip whose faces are perpendicular to an external magnetic field.

hard-wired Electrical devices connected through physical wiring.

harden The process of constructing military telecommunications facilities so as to protect them from damage by enemy action, especially *electromagnetic pulse* (EMP) radiation.

hardware Physical equipment, such as mechanical, magnetic, electrical, or electronic devices or components.

harmonic A periodic wave having a frequency that is an integral multiple of the fundamental frequency. For example, a wave with twice the frequency of the fundamental is called the *second harmonic*.

harmonic analyzer A test set capable of identifying the frequencies of the individual signals that make up a complex wave.

harmonic distortion The production of harmonics at the output of a circuit when a periodic wave is applied to its input. The level of the distortion is usually expressed as a percentage of the level of the input.

hazard A condition that could lead to danger for operating personnel.

headroom The difference, in decibels, between the typical operating signal level and a peak overload level.

heat loss The loss of useful electrical energy resulting from conversion into unwanted heat.

heat sink A device that conducts heat away from a heat-producing component so that it stays within a safe working temperature range.

heater In an electron tube, the filament that heats the cathode to enable it to emit electrons.

hecto A prefix meaning 100.

henry The standard unit of electrical inductance, equal to the self-inductance of a circuit or the mutual inductance of two circuits when there is an induced electromotive force of one volt and a current change of one ampere per second. The symbol for inductance is H, named for the American physicist Joseph Henry (1797–1878).

hertz (Hz) The unit of frequency that is equal to one cycle per second. Hertz is the reciprocal of the *period*, the interval after which the same portion of a periodic waveform recurs. Hertz was named for the German physicist Heinrich R. Hertz (1857–1894).

heterodyne The mixing of two signals in a nonlinear device in order to produce two additional signals at frequencies that are the sum and difference of the original frequencies.

heterodyne frequency The sum of, or the difference between, two frequencies, produced by combining the two signals together in a modulator or similar device.

heterodyne wavemeter A test set that uses the heterodyne principle to measure the frequencies of incoming signals.

high-frequency loss Loss of signal amplitude at higher frequencies through a given circuit or medium. For example, high frequency loss could be caused by passing a signal through a coaxial cable.

high Q An inductance or capacitance whose ratio of reactance to resistance is high.

high tension A high voltage circuit.

high-pass filter A network that passes signals of higher than a specified frequency but attenuates signals of all lower frequencies.

homochronous Signals whose corresponding significant instants have a constant but uncontrolled phase relationship with each other.

horn gap A lightning arrester utilizing a gap between two horns. When lightning causes a discharge between the horns, the heat produced lengthens the arc and breaks it.

horsepower The basic unit of mechanical power. One horsepower (hp) equals 550 foot-pounds per second or 746 watts.

hot A charged electrical circuit or device.

hot dip galvanized The process of galvanizing steel by dipping it into a bath of molten zinc.

hot standby System equipment that is fully powered but not in service. A *hot standby* can rapidly replace a primary system in the event of a failure.

hum Undesirable coupling of the 60 Hz power sine wave into other electrical signals and/or circuits.

HVAC An abbreviation for *heating, ventilation, and air conditioning* system.

hybrid system A communication system that accommodates both digital and analog signals.

hydrometer A testing device used to measure specific gravity, particularly the specific gravity of the dilute sulphuric acid in a lead-acid storage battery, to learn the state of charge of the battery.

hygrometer An instrument that measures the relative humidity of the atmosphere.

hygroscopic The ability of a substance to absorb moisture from the air.

hysteresis The property of an element evidenced by the dependence of the value of the output, for a given excursion of the input, upon the history of prior excursions and direction of the input. Originally, *hysteresis* was the name for magnetic phenomena only—the lagging of flux density behind the change in value of the magnetizing flux—but now, the term is also used to describe other inelastic behavior.

hysteresis loop The plot of magnetizing current against magnetic flux density (or of other similarly related pairs of parameters), which appears as a loop. The area within the loop is proportional to the power loss resulting from hysteresis.

hysteresis loss The loss in a magnetic core resulting from hysteresis.

I

I^2R **loss** The power lost as a result of the heating effect of current passing through resistance.

idling current The current drawn by a circuit, such as an amplifier, when no signal is present at its input.

image frequency A frequency on which a carrier signal, when heterodyned with the local oscillator in a superheterodyne receiver, will cause a sum or difference frequency that is the same as the intermediate frequency of the receiver. Thus, a signal on an *image frequency* will be demodulated along with the desired signal and will interfere with it.

impact ionization The ionization of an atom or molecule as a result of a high energy collision.

impedance The total passive opposition offered to the flow of an alternating current. *Impedance* consists of a combination of resistance, inductive reactance, and capacitive reactance. It is the vector sum of resistance and reactance $(R + jX)$ or the vector of magnitude Z at an angle θ.

impedance characteristic A graph of the impedance of a circuit showing how it varies with frequency.

impedance irregularity A discontinuity in an impedance characteristic caused, for example, by the use of different coaxial cable types.

impedance matching The adjustment of the impedances of adjoining circuit components to a common value so as to minimize reflected energy from the junction and to maximize energy transfer across it. Incorrect adjustment results in an *impedance mismatch*.

impedance matching transformer A transformer used between two circuits of different impedances with a turns ratio that provides for maximum power transfer and minimum loss by reflection.

impulse A short high energy surge of electrical current in a circuit or on a line.

impulse current A current that rises rapidly to a peak then decays to zero without oscillating.

impulse excitation The production of an oscillatory current in a circuit by impressing a voltage for a relatively short period compared with the duration of the current produced.

impulse noise A noise signal consisting of random occurrences of energy spikes, having random amplitude and bandwidth.

impulse response The amplitude-versus-time output of a transmission facility or device in response to an impulse.

impulse voltage A unidirectional voltage that rises rapidly to a peak and then falls to zero, without any appreciable oscillation.

in-phase The property of alternating current signals of the same frequency that achieve their peak positive, peak negative, and zero amplitude values simultaneously.

incidence angle The angle between the perpendicular to a surface and the direction of arrival of a signal.

increment A small change in the value of a quantity.

induce To produce an electrical or magnetic effect in one conductor by changing the condition or position of another conductor.

induced current The current that flows in a conductor because a voltage has been induced across two points in, or connected to, the conductor.

induced voltage A voltage developed in a conductor when the conductor passes through magnetic lines of force.

inductance The property of an inductor that opposes any change in a current that flows through it. The standard unit of inductance is the *Henry*.

induction The electrical and magnetic interaction process by which a changing current in one circuit produces a voltage change not only in its own circuit (*self inductance*) but also in other circuits to which it is linked magnetically.

inductive A circuit element exhibiting inductive reactance.

inductive kick A voltage surge produced when a current flowing through an inductance is interrupted.

inductive load A load that possesses a net inductive reactance.

inductive reactance The reactance of a circuit resulting from the presence of inductance and the phenomenon of induction.

inductor A coil of wire, usually wound on a core of high permeability, that provides high inductance without necessarily exhibiting high resistance.

inert An inactive unit, or a unit that has no power requirements.

infinite line A transmission line that appears to be of infinite length. There are no reflections back from the far end because it is terminated in its characteristic impedance.

infra low frequency (ILF) The frequency band from 300 Hz to 3000 Hz.

inhibit A control signal that prevents a device or circuit from operating.

injection The application of a signal to an electronic device.

input The waveform fed into a circuit, or the terminals that receive the input waveform.

insertion gain The gain resulting from the insertion of a transducer in a transmission system, expressed as the ratio of the power delivered to that part of the system following the transducer to the power delivered to that same part before insertion. If more than one component is involved in the input or output, the particular component used must be specified. This ratio is usually expressed in decibels. If the resulting number is negative, an *insertion loss* is indicated.

insertion loss The signal loss within a circuit, usually expressed in decibels as the ratio of input power to output power.

insertion loss-vs.-frequency characteristic The amplitude transfer characteristic of a system or component as a function of frequency. The amplitude response may be stated as actual gain, loss, amplification, or attenuation, or as a ratio of any one of these quantities at a particular frequency, with respect to that at a specified reference frequency.

inspection lot A collection of units of product from which a sample is drawn and inspected to determine conformance with acceptability criteria.

instantaneous value The value of a varying waveform at a given instant of time. The value can be in volts, amperes, or phase angle.

Institute of Electrical and Electronics Engineers (IEEE) The organization of electrical and electronics scientists and engineers formed in 1963 by the merger of the Institute of Radio Engineers (IRE) and the American Institute of Electrical Engineers (AIEE).

instrument multiplier A measuring device that enables a high voltage to be measured using a meter with only a low voltage range.

instrument rating The range within which an instrument has been designed to operate without damage.

insulate The process of separating one conducting body from another conductor.

insulation The material that surrounds and insulates an electrical wire from other wires or circuits. *Insulation* may also refer to any material that does not ionize easily and thus presents a large impedance to the flow of electrical current.

insulator A material or device used to separate one conducting body from another.

intelligence signal A signal containing information.

intensity The strength of a given signal under specified conditions.

interconnect cable A short distance cable intended for use between equipment (generally less than 3 m in length).

interface A device or circuit used to interconnect two pieces of electronic equipment.

interface device A unit that joins two interconnecting systems.

interference emission An emission that results in an electrical signal being propagated into and interfering with the proper operation of electrical or electronic equipment.

interlock A protection device or system designed to remove all dangerous voltages from a machine or piece of equipment when access doors or panels are opened or removed.

intermediate frequency A frequency that results from combining a signal of interest with a signal generated within a radio receiver. In superheterodyne receivers, all incoming signals are converted to a single intermediate frequency for which the amplifiers and filters of the receiver have been optimized.

intermittent A noncontinuous recurring event, often used to denote a problem that is difficult to find because of its unpredictable nature.

intermodulation The production, in a nonlinear transducer element, of frequencies corresponding to the sums and differences of the fundamentals and harmonics of two or more frequencies that are transmitted through the transducer.

intermodulation distortion (IMD) The distortion that results from the mixing of two input signals in a nonlinear system. The resulting output contains new frequencies that represent the sum and difference of the input signals and the sums and differences of their harmonics. IMD is also called *intermodulation noise.*

intermodulation noise In a transmission path or device, the noise signal that is contingent upon modulation and demodulation, resulting from nonlinear characteristics in the path or device.

internal resistance The actual resistance of a source of electric power. The total electromotive force produced by a power source is not available for external use; some of the energy is used in driving current through the source itself.

International Standards Organization (ISO) An international body concerned with worldwide standardization for a broad range of industrial products, including telecommunications equipment. Members are represented by national standards organizations, such as ANSI (American National Standards Institute) in the United States. ISO was established in 1947 as a specialized agency of the United Nations.

International Telecommunications Satellite Consortium (Intelsat) A nonprofit cooperative of member nations that owns and operates a satellite system for international and, in many instances, domestic communications.

International Telecommunications Union (ITU) A specialized agency of the United Nations established to maintain and extend international cooperation for the maintenance, development, and efficient use of telecommunications. The union does this through standards and recommended regulations, and through technical and telecommunications studies.

interoperability The condition achieved among communications and electronics systems or equipment when information or services can be exchanged directly between them or their users, or both.

interpolate The process of estimating unknown values based on a knowledge of comparable data that falls on both sides of the point in question.

interrupting capacity The rating of a circuit breaker or fuse that specifies the maximum current the device is designed to interrupt at its rated voltage.

interval The points or numbers lying between two specified endpoints.

inverse voltage The effective value of voltage across a rectifying device, which conducts a current in one direction during one half cycle of the alternating input, during the half cycle when current is not flowing.

inversion The change in the polarity of a pulse, such as from positive to negative.

inverter A circuit or device that converts a direct current into an alternating current.

ionizing radiation The form of electromagnetic radiation that can turn an atom into an ion by knocking one or more of its electrons loose. Examples of ionizing radiation include X rays, gamma rays, and cosmic rays

IR **drop** A drop in voltage because of the flow of current (I) through a resistance (R), also called *resistance drop.*

IR **loss** The conversion of electrical power to heat caused by the flow of electrical current through a resistance.

isochronous A signal in which the time interval separating any two significant instants is theoretically equal to a specified unit interval or to an integral multiple of the unit interval.

isolated ground A ground circuit that is isolated from all equipment framework and any other grounds, except for a single-point external connection.

isolated ground plane A set of connected frames that are grounded through a single connection to a ground reference point. That point and all parts of the frames are insulated from any other ground system in a building.

isolated pulse A pulse uninfluenced by other pulses in the same signal.

isophasing amplifier A timing device that corrects for small timing errors.

isotropic A quantity exhibiting the same properties in all planes and directions.

J

jack A receptacle or connector that makes electrical contact with the mating contacts of a plug. In combination, the plug and jack provide a ready means for making connections in electrical circuits.

jacket An insulating layer of material surrounding a wire in a cable.

jitter Small, rapid variations in a waveform resulting from fluctuations in a supply voltage or other causes.

joule The standard unit of work that is equal to the work done by one newton of force when the point at which the force is applied is displaced a distance of one meter in the direction of the force. The *joule* is named for the English physicist James Prescott Joule (1818–1889).

Julian date A chronological date in which days of the year are numbered in sequence. For example, the first day is 001, the second is 002, and the last is 365 (or 366 in a leap year).

K

Kelvin (K) The standard unit of thermodynamic temperature. Zero degrees Kelvin represents *absolute zero*. Water freezes at 273 K and water boils at 373 K under standard pressure conditions.

kilo A prefix meaning one thousand.

kilohertz (kHz) A unit of measure of frequency equal to 1,000 Hz.

kilovar A unit equal to one thousand volt-amperes.

kilovolt (kV) A unit of measure of electrical voltage equal to 1,000 V.

kilowatt A unit equal to one thousand watts.

Kirchoff's Law At any point in a circuit, there is as much current flowing into the point as there is flowing away from it.

klystron (1—general) A family of electron tubes that function as microwave amplifiers and oscillators. Simplest in form are two-cavity klystrons in which an electron beam passes through a cavity that is excited by a microwave input, producing a velocity-modulated beam which passes through a second cavity a precise distance away that is coupled to a tuned circuit, thereby producing an amplified output of the original input signal frequency. If part of the output is fed back to the input, an oscillator can be the result. **(2—multi-cavity)** An amplifier device for UHF and microwave signals based on velocity modulation of an electron beam. The beam is directed through an input cavity, where the input RF signal polarity initializes a *bunching effect* on electrons in the beam. The bunching effect excites subsequent cavities, which increase the bunching through an energy flywheel concept. Finally, the beam passes to an output cavity that couples the amplified signal to the load (antenna system). The beam falls onto a collector element that forms the return path for the current and dissipates the heat resulting from electron beam bombardment. **(3—reflex)** A klystron with only one cavity. The action is the same as in a two-cavity klystron but the beam is reflected back into the cavity in which it was first excited, after being sent out to a reflector. The one cavity, therefore, acts both as the original exciter (or buncher) and as the collector from which the output is taken.

knee In a response curve, the region of maximum curvature.

ku band Radio frequencies in the range of 15.35 GHz to 17.25 GHz, typically used for satellite telecommunications.

L

ladder network A type of filter with components alternately across the line and in the line.

lag The difference in phase between a current and the voltage that produced it, expressed in electrical degrees.

lagging current A current that lags behind the alternating electromotive force that produced it. A circuit that produces a *lagging current* is one containing inductance alone, or whose effective impedance is inductive.

lagging load A load whose combined inductive reactance exceeds its capacitive reactance. When an alternating voltage is applied, the current lags behind the voltage.

laminate A material consisting of layers of the same or different materials bonded together and built up to the required thickness.

latitude An angular measurement of a point on the earth above or below the equator. The equator represents 0°, the north pole +90°, and the south pole –90°.

layout A proposed or actual arrangement or allocation of equipment.

LC circuit An electrical circuit with both inductance (L) and capacitance (C) that is resonant at a particular frequency.

LC ratio The ratio of inductance to capacitance in a given circuit.

lead An electrical wire, usually insulated.

leading edge The initial portion of a pulse or wave in which voltage or current rise rapidly from zero to a final value.

leading load A reactive load in which the reactance of capacitance is greater than that of inductance. Current through such a load *leads* the applied voltage causing the current.

leakage The loss of energy resulting from the flow of electricity past an insulating material, the escape of electromagnetic radiation beyond its shielding, or the extension of magnetic lines of force beyond their intended working area.

leakage resistance The resistance of a path through which leakage current flows.

level The strength or intensity of a given signal.

level alignment The adjustment of transmission levels of single links and links in tandem to prevent overloading of transmission subsystems.

life cycle The predicted useful life of a class of equipment, operating under normal (specified) working conditions.

life safety system A system designed to protect life and property, such as emergency lighting, fire alarms, smoke exhaust and ventilating fans, and site security.

life test A test in which random samples of a product are checked to see how long they can continue to perform their functions satisfactorily. A form of *stress testing* is used, including temperature, current, voltage, and/or vibration effects, cycled at many times the rate that would apply in normal usage.

limiter An electronic device in which some characteristic of the output is automatically prevented from exceeding a predetermined value.

limiter circuit A circuit of nonlinear elements that restricts the electrical excursion of a variable in accordance with some specified criteria.

limiting A process by which some characteristic at the output of a device is prevented from exceeding a predetermined value.

line loss The total end-to-end loss in decibels in a transmission line.

line-up The process of adjusting transmission parameters to bring a circuit to its specified values.

linear A circuit, device, or channel whose output is directly proportional to its input.

linear distortion A distortion mechanism that is independent of signal amplitude.

linearity A constant relationship, over a designated range, between the input and output characteristics of a circuit or device.

lines of force A group of imaginary lines indicating the direction of the electric or magnetic field at all points along it.

lissajous pattern The looping patterns generated by a CRT spot when the horizontal (X) and vertical (Y) deflection signals are sinusoids. The lissajous pattern is useful for evaluating the delay or phase of two sinusoids of the same frequency.

live A device or system connected to a source of electric potential.

load The work required of an electrical or mechanical system.

load factor The ratio of the average load over a designated period of time to the peak load occurring during the same period.

load line A straight line drawn across a grouping of plate current/plate voltage characteristic curves showing the relationship between grid voltage and plate current for a particular plate load resistance of an electron tube.

logarithm The power to which a base must be raised to produce a given number. Common logarithms are to base 10.

logarithmic scale A meter scale with displacement proportional to the logarithm of the quantity represented.

long persistence The quality of a cathode ray tube that has phosphorescent compounds on its screen (in addition to fluorescent compounds) so that the image continues to glow after the original electron beam has ceased to create it by producing the usual fluorescence effect. Long persistence is often used in radar screens or where photographic evidence is needed of a display. Most such applications, however, have been superseded through the use of digital storage techniques.

longitude The angular measurement of a point on the surface of the earth in relation to the meridian of Greenwich (London). The earth is divided into 360° of longitude, beginning at the Greenwich mean. As one travels west around the globe, the longitude increases.

longitudinal current A current that travels in the same direction on both wires of a pair. The return current either flows in another pair or via a ground return path.

loss The power dissipated in a circuit, usually expressed in decibels, that performs no useful work.

loss deviation The change of actual loss in a circuit or system from a designed value.

loss variation The change in actual measured loss over time.

lossy The condition when the line loss per unit length is significantly greater than some defined normal parameter.

lossy cable A coaxial cable constructed to have high transmission loss so it can be used as an artificial load or as an attenuator.

lot size A specific quantity of similar material or a collection of similar units from a common source; in inspection work, the quantity offered for inspection and acceptance at any one time. The **lot size** may be a collection of raw material, parts, subassemblies inspected during production, or a consignment of finished products to be sent out for service.

low tension A low voltage circuit.

low-pass filter A filter network that passes all frequencies below a specified frequency with little or no loss, but that significantly attenuates higher frequencies.

lug A tag or projecting terminal onto which a wire may be connected by wrapping, soldering, or crimping.

lumped constant A resistance, inductance, or capacitance connected at a point, and not distributed uniformly throughout the length of a route or circuit.

M

mA An abbreviation for *milliamperes* (0.001 A).

magnet A device that produces a magnetic field and can attract iron, and attract or repel other magnets.

magnetic field An energy field that exists around magnetic materials and current-carrying conductors. Magnetic fields combine with electric fields in light and radio waves.

magnetic flux The field produced in the area surrounding a magnet or electric current. The standard unit of flux is the *Weber*.

magnetic flux density A vector quantity measured by a standard unit called the *Tesla*. The *magnetic flux density* is the number of magnetic lines of force per unit area, at right angles to the lines.

magnetic leakage The magnetic flux that does not follow a useful path.

magnetic pole A point that appears from the outside to be the center of magnetic attraction or repulsion at or near one end of a magnet.

magnetic storm A violent local variation in the earth's magnetic field, usually the result of sunspot activity.

magnetism A property of iron and some other materials by which external magnetic fields are maintained, other magnets being thereby attracted or repelled.

magnetization The exposure of a magnetic material to a magnetizing current, field, or force.

magnetizing force The force producing magnetization.

magnetomotive force The force that tends to produce lines of force in a magnetic circuit. The *magnetomotive force* bears the same relationship to a magnetic circuit that voltage does to an electrical circuit.

magnetron A high-power, ultra high frequency electron tube oscillator that employs the interaction of a strong electric field between an anode and cathode with the field of a strong permanent magnet to cause oscillatory electron flow through multiple internal cavity resonators. The magnetron may operate in a continuous or pulsed mode.

maintainability The probability that a failure will be repaired within a specified time after the failure occurs.

maintenance Any activity intended to keep a functional unit in satisfactory working condition. The term includes the tests, measurements, replacements, adjustments, and repairs necessary to keep a device or system operating properly.

malfunction An equipment failure or a fault.

manometer A test device for measuring gas pressure.

margin The difference between the value of an operating parameter and the value that would result in unsatisfactory operation. Typical *margin* parameters include signal level, signal-to-noise ratio, distortion, crosstalk coupling, and/or undesired emission level.

Markov model A statistical model of the behavior of a complex system over time in which the probabilities of the occurrence of various future states depend only on the present state of the system, and not on the path by which the present state was achieved. This term was named for the Russian mathematician Andrei Andreevich Markov (1856-1922).

master clock An accurate timing device that generates a synchronous signal to control other clocks or equipment.

master oscillator A stable oscillator that provides a standard frequency signal for other hardware and/or systems.

matched termination A termination that absorbs all the incident power and so produces no reflected waves or mismatch loss.

matching The connection of channels, circuits, or devices in a manner that results in minimal reflected energy.

matrix A logical network configured in a rectangular array of intersections of input/output signals.

Maxwell's equations Four differential equations that relate electric and magnetic fields to electromagnetic waves. The equations are a basis of electrical and electronic engineering.

mean An arithmetic average in which values are added and divided by the number of such values.

mean time between failures (MTBF) For a particular interval, the total functioning life of a population of an item divided by the total number of failures within the population during the measurement interval.

mean time to failure (MTTF) The measured operating time of a single piece of equipment divided by the total number of failures during the measured period of time. This measurement is normally made during that period between early life and wear-out failures.

mean time to repair (MTTR) The total corrective maintenance time on a component or system divided by the total number of corrective maintenance actions during a given period of time.

measurement A procedure for determining the amount of a quantity.

median A value in a series that has as many readings or values above it as below.

medium An electronic pathway or mechanism for passing information from one point to another.

mega A prefix meaning one million.

megahertz (MHz) A quantity equal to one million Hertz (cycles per second).

megohm A quantity equal to one million ohms.

metric system A decimal system of measurement based on the meter, the kilogram, and the second.

micro A prefix meaning one millionth.

micron A unit of length equal to one millionth of a meter (1/25,000 of an inch).

microphonic(s) Unintended noise introduced into an electronic system by mechanical vibration of electrical components.

microsecond One millionth of a second (0.000001 s).

microvolt A quantity equal to one-millionth of a volt.

milli A prefix meaning one thousandth.

milliammeter A test instrument for measuring electrical current, often part of a *multimeter*.

millihenry A quantity equal to one-thousandth of a henry.

milliwatt A quantity equal to one thousandth of a watt.

minimum discernible signal The smallest input that will produce a discernible change in the output of a circuit or device.

mixer A circuit used to combine two or more signals to produce a third signal that is a function of the input waveforms.

mixing ratio The ratio of the mass of water vapor to the mass of dry air in a given volume of air. The *mixing ratio* affects radio propagation.

mode An electromagnetic field distribution that satisfies theoretical requirements for propagation in a waveguide or oscillation in a cavity.

modified refractive index The sum of the refractive index of the air at a given height above sea level, and the ratio of this height to the radius of the earth.

modular An equipment design in which major elements are readily separable, and which the user may replace, reducing the mean-time-to-repair.

modulation The process whereby the amplitude, frequency, or phase of a single-frequency wave (the *carrier*) is varied in step with the instantaneous value of, or samples of, a complex wave (the *modulating wave*).

modulator A device that enables the intelligence in an information-carrying modulating wave to be conveyed by a signal at a higher frequency. A *modulator* modifies a carrier wave by amplitude, phase, and/or frequency as a function of a control signal that carries intelligence. Signals are *modulated* in this way to permit more efficient and/or reliable transmission over any of several media.

module An assembly replaceable as an entity, often as an interchangeable plug-in item. A *module* is not normally capable of being disassembled.

monostable A device that is stable in one state only. An input pulse causes the device to change state, but it reverts immediately to its stable state.

motor A machine that converts electrical energy into mechanical energy.

motor effect The repulsion force exerted between adjacent conductors carrying currents in opposite directions.

moving coil Any device that utilizes a coil of wire in a magnetic field in such a way that the coil is made to move by varying the applied current, or itself produces a varying voltage because of its movement.

ms An abbreviation for *millisecond* (0.001 s).

multimeter A test instrument fitted with several ranges for measuring voltage, resistance, and current, and equipped with an analog meter or digital display readout. The *multimeter* is also known as a *volt-ohm-milliammeter*, or *VOM*.

multiplex (MUX) The use of a common channel to convey two or more channels. This is done either by splitting of the common channel frequency band into narrower bands, each of which is used to constitute a distinct channel (*frequency division multiplex*), or by allotting this common channel to multiple users in turn to constitute different intermittent channels (*time division multiplex*).

multiplexer A device or circuit that combines several signals onto a single signal.

multiplexing A technique that uses a single transmission path to carry multiple channels. In *time division multiplexing* (TDM), path time is shared. For *frequency division multiplexing* (FDM) or *wavelength division multiplexing* (WDM), signals are divided into individual channels sent along the same path but at different frequencies.

multiplication Signal mixing that occurs within a multiplier circuit.

multiplier A circuit in which one or more input signals are mixed under the direction of one or more control signals. The resulting output is a composite of the input signals, the characteristics of which are determined by the scaling specified for the circuit.

mutual induction The property of the magnetic flux around a conductor that induces a voltage in a nearby conductor. The voltage generated in the secondary conductor in turn induces a voltage in the primary conductor. The inductance of two conductors so coupled is referred to as *mutual inductance*.

mV An abbreviation for *millivolt* (0.001 V).

mW An abbreviation for *milliwatt* (0.001 W).

N

nano A prefix meaning one billionth.

nanometer 1×10^{-9} meter.

nanosecond (ns) One billionth of a second (1×10^{-9} s).

narrowband A communications channel of restricted bandwidth, often resulting in degradation of the transmitted signal.

narrowband emission An emission having a spectrum exhibiting one or more sharp peaks that are narrow in width compared to the nominal bandwidth of the measuring instrument, and are far enough apart in frequency to be resolvable by the instrument.

National Electrical Code (NEC) A document providing rules for the installation of electric wiring and equipment in public and private buildings, published by the National Fire Protection Association. The NEC has been adopted as law by many states and municipalities in the U.S.

National Institute of Standards and Technology (NIST) A nonregulatory agency of the Department of Commerce that serves as a national reference and measurement laboratory for the physical and engineering sciences. Formerly called the *National Bureau of Standards*, the agency was renamed in 1988 and given the additional responsibility of aiding U.S. companies in adopting new technologies to increase their international competitiveness.

negative In a conductor or semiconductor material, an excess of electrons or a deficiency of positive charge.

negative feedback The return of a portion of the output signal from a circuit to the input but 180° out of phase. This type of feedback decreases signal amplitude but stabilizes the amplifier and reduces distortion and noise.

negative impedance An impedance characterized by a decrease in voltage drop across a device as the current through the device is increased, or a decrease in current through the device as the voltage across it is increased.

neutral A device or object having no electrical charge.

neutral conductor A conductor in a power distribution system connected to a point in the system that is designed to be at neutral potential. In a balanced system, the neutral conductor carries no current.

neutral ground An intentional ground applied to the neutral conductor or neutral point of a circuit, transformer, machine, apparatus, or system.

newton The standard unit of force. One *newton* is the force that, when applied to a body having a mass of 1 kg, gives it an acceleration of 1 m/s^2.

nitrogen A gas widely used to pressurize radio frequency transmission lines. If a small puncture occurs in the cable sheath, the nitrogen keeps moisture out so that service is not adversely affected.

node The points at which the current is at minimum in a transmission system in which standing waves are present.

noise Any random disturbance or unwanted signal in a communication system that tends to obscure the clarity or usefulness of a signal in relation to its intended use.

noise factor (NF) The ratio of the noise power measured at the output of a receiver to the noise power that would be present at the output if the thermal noise resulting from the resistive component of the source impedance were the only source of noise in the system.

noise figure A measure of the noise in dB generated at the input of an amplifier, compared with the noise generated by an impedance-method resistor at a specified temperature.

noise filter A network that attenuates noise frequencies.

noise generator A generator of wideband random noise.

noise power ratio (NPR) The ratio, expressed in decibels, of signal power to intermodulation product power plus residual noise power, measured at the baseband level.

noise suppressor A filter or digital signal processing circuit in a receiver or transmitter that automatically reduces or eliminates noise.

noise temperature The temperature, expressed in Kelvin, at which a resistor will develop a particular noise voltage. The noise temperature of a radio receiver is the value by which the temperature of the resistive component of the source impedance should be increased—if it were the only source of noise in the system—to cause the noise power at the output of the receiver to be the same as in the real system.

nominal The most common value for a component or parameter that falls between the maximum and minimum limits of a tolerance range.

nominal value A specified or intended value independent of any uncertainty in its realization.

nomogram A chart showing three or more scales across which a straight edge may be held in order to read off a graphical solution to a three-variable equation.

nonconductor A material that does not conduct energy, such as electricity, heat, or sound.

noncritical technical load That part of the technical power load for a facility not required for minimum acceptable operation.

noninductive A device or circuit without significant inductance.

nonionizing radiation Electromagnetic radiation that does not turn an atom into an ion. Examples of nonionizing radiation include visible light and radio waves.

nonlinearity A distortion in which the output of a circuit or system does not rise or fall in direct proportion to the input.

nontechnical load The part of the total operational load of a facility used for such purposes as general lighting, air conditioning, and ventilating equipment during normal operation.

normal A line perpendicular to another line or to a surface.

normal-mode noise Unwanted signals in the form of voltages appearing in line-to-line and line-to-neutral signals.

normalized frequency The ratio between the actual frequency and its nominal value.

normalized frequency departure The frequency departure divided by the nominal frequency value.

normalized frequency difference The algebraic difference between two normalized frequencies.

normalized frequency drift The frequency drift divided by the nominal frequency value.

normally closed Switch contacts that are closed in their nonoperated state, or relay contacts that are closed when the relay is de-energized.

normally open Switch contacts that are open in their nonoperated state, or relay contacts that are open when the relay is de-energized.

north pole The pole of a magnet that seeks the north magnetic pole of the earth.

notch filter A circuit designed to attenuate a specific frequency band; also known as a *band stop filter*.

notched noise A noise signal in which a narrow band of frequencies has been removed.

ns An abbreviation for *nanosecond*.

null A zero or minimum amount or position.

O

octave Any frequency band in which the highest frequency is twice the lowest frequency.

off-line A condition wherein devices or subsystems are not connected into, do not form a part of, and are not subject to the same controls as an operational system.

offset An intentional difference between the realized value and the nominal value.

ohm The unit of electric resistance through which one ampere of current will flow when there is a difference of one volt. The quantity is named for the German physicist Georg Simon Ohm (1787-1854).

Ohm's law A law that sets forth the relationship between voltage (E), current (I), and resistance (R). The law states that $E = I \times R$, $I = \dfrac{E}{R}$, and $R = \dfrac{E}{I}$. *Ohm's Law* is named for the German physicist Georg Simon Ohm (1787–1854).

ohmic loss The power dissipation in a line or circuit caused by electrical resistance.

ohmmeter A test instrument used for measuring resistance, often part of a *multimeter*.

ohms-per-volt A measure of the sensitivity of a voltmeter.

on-line A device or system that is energized and operational, and ready to perform useful work.

open An interruption in the flow of electrical current, as caused by a broken wire or connection.

open-circuit A defined loop or path that closes on itself and contains an infinite impedance.

open-circuit impedance The input impedance of a circuit when its output terminals are open, that is, not terminated.

open-circuit voltage The voltage measured at the terminals of a circuit when there is no load and, hence, no current flowing.

operating lifetime The period of time during which the principal parameters of a component or system remain within a prescribed range.

optimize The process of adjusting for the best output or maximum response from a circuit or system.

orbit The path, relative to a specified frame of reference, described by the center of mass of a satellite or other object in space, subjected solely to natural forces (mainly gravitational attraction).

order of diversity The number of independently fading propagation paths or frequencies, or both, used in a diversity reception system.

original equipment manufacturer (OEM) A manufacturer of equipment that is used in systems assembled and sold by others.

oscillation A variation with time of the magnitude of a quantity with respect to a specified reference when the magnitude is alternately greater than and smaller than the reference.

oscillator A nonrotating device for producing alternating current, the output frequency of which is determined by the characteristics of the circuit.

oscilloscope A test instrument that uses a display, usually a cathode-ray tube, to show the instantaneous values and waveforms of a signal that varies with time or some other parameter.

out-of-band energy Energy emitted by a transmission system that falls outside the frequency spectrum of the intended transmission.

outage duration The average elapsed time between the start and the end of an outage period.

outage probability The probability that an outage state will occur within a specified time period. In the absence of specific known causes of outages, the *outage probability* is the sum of all outage durations divided by the time period of measurement.

outage threshold A defined value for a supported performance parameter that establishes the minimum operational service performance level for that parameter.

output impedance The impedance presented at the output terminals of a circuit, device, or channel.

output stage The final driving circuit in a piece of electronic equipment.

ovenized crystal oscillator (OXO) A crystal oscillator enclosed within a temperature regulated heater (oven) to maintain a stable frequency despite external temperature variations.

overcoupling A degree of coupling greater than the *critical coupling* between two resonant circuits. *Overcoupling* results in a wide bandwidth circuit with two peaks in the response curve.

overload In a transmission system, a power greater than the amount the system was designed to carry. In a power system, an overload could cause excessive heating. In a communications system, distortion of a signal could result.

overshoot The first maximum excursion of a pulse beyond the 100% level. Overshoot is the portion of the pulse that exceeds its defined level temporarily before settling to the correct level. Overshoot amplitude is expressed as a percentage of the defined level.

P

pentode An electron tube with five electrodes, the cathode, control grid, screen grid, suppressor grid, and plate.

photocathode An electrode in an electron tube that will emit electrons when bombarded by photons of light.

picture tube A cathode-ray tube used to produce an image by variation of the intensity of a scanning beam on a phosphor screen.

pin A terminal on the base of a component, such as an electron tube.

plasma (1—arc) An ionized gas in an arc-discharge tube that provides a conducting path for the discharge. **(2—solar)** The ionized gas at extremely high temperature found in the sun.

plate (1—electron tube) The anode of an electron tube. **(2—battery)** An electrode in a storage battery. **(3—capacitor)** One of the surfaces in a capacitor. **(4—chassis)** A mounting surface to which equipment may be fastened.

propagation time delay The time required for a signal to travel from one point to another.

protector A device or circuit that prevents damage to lines or equipment by conducting dangerously high voltages or currents to ground. Protector types include spark gaps, semiconductors, varistors, and gas tubes.

proximity effect A nonuniform current distribution in a conductor, caused by current flow in a nearby conductor.

pseudonoise In a spread-spectrum system, a seemingly random series of pulses whose frequency spectrum resembles that of continuous noise.

pseudorandom A sequence of signals that appears to be completely random but have, in fact, been carefully drawn up and repeat after a significant time interval.

pseudorandom noise A noise signal that satisfies one or more of the standard tests for statistical randomness. Although it seems to lack any definite pattern, there is a sequence of pulses that repeats after a long time interval.

pseudorandom number sequence A sequence of numbers that satisfies one or more of the standard tests for statistical randomness. Although it seems to lack any definite pattern, there is a sequence that repeats after a long time interval.

pulsating direct current A current changing in value at regular or irregular intervals but which has the same direction at all times.

pulse One of the elements of a repetitive signal characterized by the rise and decay in time of its magnitude. A *pulse* is usually short in relation to the time span of interest.

pulse decay time The time required for the trailing edge of a pulse to decrease from 90 percent to 10 percent of its peak amplitude.

pulse duration The time interval between the points on the leading and trailing edges of a pulse at which the instantaneous value bears a specified relation to the peak pulse amplitude.

pulse duration modulation (PDM) The modulation of a pulse carrier by varying the width of the pulses according to the instantaneous values of the voltage samples of the modulating signal (also called *pulse width modulation*).

pulse edge The leading or trailing edge of a pulse, defined as the 50 percent point of the pulse rise or fall time.

pulse fall time The interval of time required for the edge of a pulse to fall from 90 percent to 10 percent of its peak amplitude.

pulse interval The time between the start of one pulse and the start of the next.

pulse length The duration of a pulse (also called *pulse width*).

pulse level The voltage amplitude of a pulse.

pulse period The time between the start of one pulse and the start of the next.

pulse ratio The ratio of the length of any pulse to the total pulse period.

pulse repetition period The time interval from the beginning of one pulse to the beginning of the next pulse.

pulse repetition rate The number of times each second that pulses are transmitted.

pulse rise time The time required for the leading edge of a pulse to rise from 10 percent to 90 percent of its peak amplitude.

pulse train A series of pulses having similar characteristics.

pulse width The measured interval between the 50 percent amplitude points of the leading and trailing edges of a pulse.

puncture A breakdown of insulation or of a dielectric, such as in a cable sheath or in the insulant around a conductor.

pW An abbreviation for picowatt, a unit of power equal to 10^{-12} W (–90 dBm).

Q

Q (quality factor) A figure of merit that defines how close a coil comes to functioning as a pure inductor. *High Q* describes an inductor with little energy loss resulting from resistance. *Q* is found by dividing the inductive reactance of a device by its resistance.

quadrature A state of alternating current signals separated by one quarter of a cycle (90°).

quadrature amplitude modulation (QAM) A process that allows two different signals to modulate a single carrier frequency. The two signals of interest amplitude modulate two samples of the carrier that are of the same frequency, but differ in phase by 90°. The two resultant signals can be added and transmitted. Both signals may be recovered at a decoder when they are demodulated 90° apart.

quadrature component The component of a voltage or current at an angle of 90° to a reference signal, resulting from inductive or capacitive reactance.

quadrature phase shift keying (QPSK) A type of phase shift keying using four phase states.

quality The absence of objectionable distortion.

quality assurance (QA) All those activities, including surveillance, inspection, control, and documentation, aimed at ensuring that a given product will meet its performance specifications.

quality control (QC) A function whereby management exercises control over the quality of raw material or intermediate products in order to prevent the production of defective devices or systems.

quantum noise Any noise attributable to the discrete nature of electromagnetic radiation. Examples include shot noise, photon noise, and recombination noise.

quantum-limited operation An operation wherein the minimum detectable signal is limited by quantum noise.

quartz A crystalline mineral that when electrically excited vibrates with a stable period. Quartz is typically used as the frequency-determining element in oscillators and filters.

quasi-peak detector A detector that delivers an output voltage that is some fraction of the peak value of the regularly repeated pulses applied to it. The fraction increases toward unity as the pulse repetition rate increases.

quick-break fuse A fuse in which the fusible link is under tension, providing for rapid operation.

quiescent An inactive device, signal, or system.

quiescent current The current that flows in a device in the absence of an applied signal.

R

rack An equipment rack, usually measuring 19 in (48.26 cm) wide at the front mounting rails.

rack unit (RU) A unit of measure of vertical space in an equipment enclosure. One rack unit is equal to 1.75 in (4.45 cm).

radiate The process of emitting electromagnetic energy.

radiation The emission and propagation of electromagnetic energy in the form of waves. *Radiation* is also called *radiant energy*.

radiation scattering The diversion of thermal, electromagnetic, or nuclear radiation from its original path as a result of interactions or collisions with atoms, molecules, or large particles in the atmosphere or other media between the source of radiation and a point some distance away. As a result of scattering, radiation (especially gamma rays and neutrons) will be received at such a point from many directions, rather than only from the direction of the source.

radio The transmission of signals over a distance by means of electromagnetic waves in the approximate frequency range of 150 kHz to 300 GHz. The term may also be used to describe the equipment used to transmit or receive electromagnetic waves.

radio detection The detection of the presence of an object by radio location without precise determination of its position.

radio frequency interference (RFI) The intrusion of unwanted signals or electromagnetic noise into various types of equipment resulting from radio frequency transmission equipment or other devices using radio frequencies.

radio frequency spectrum Those frequency bands in the electromagnetic spectrum that range from several hundred thousand cycles per second (*very low frequency*) to several billion cycles per second (*microwave frequencies*).

radio recognition In military communications, the determination by radio means of the "friendly" or "unfriendly" character of an aircraft or ship.

random noise Electromagnetic signals that originate in transient electrical disturbances and have random time and amplitude patterns. Random noise is generally undesirable; however, it may also be generated for testing purposes.

rated output power The power available from an amplifier or other device under specified conditions of operation.

RC constant The time constant of a resistor-capacitor circuit. The *RC constant* is the time in seconds required for current in an RC circuit to rise to 63 percent of its final steady value or fall to 37 percent of its original steady value, obtained by multiplying the resistance value in ohms by the capacitance value in farads.

RC network A circuit that contains resistors and capacitors, normally connected in series.

reactance The part of the impedance of a network resulting from inductance or capacitance. The *reactance* of a component varies with the frequency of the applied signal.

reactive power The power circulating in an ac circuit. It is delivered to the circuit during part of the cycle and is returned during the other half of the cycle. The *reactive power* is obtained by multiplying the voltage, current, and the sine of the phase angle between them.

reactor A component with inductive reactance.

received signal level (RSL) The value of a specified bandwidth of signals at the receiver input terminals relative to an established reference.

receiver Any device for receiving electrical signals and converting them to audible sound, visible light, data, or some combination of these elements.

receptacle An electrical socket designed to receive a mating plug.

reception The act of receiving, listening to, or watching information-carrying signals.

rectification The conversion of alternating current into direct current.

rectifier A device for converting alternating current into direct current. A *rectifier* normally includes filters so that the output is, within specified limits, smooth and free of ac components.

rectify The process of converting alternating current into direct current.

redundancy A system design that provides a back-up for key circuits or components in the event of a failure. Redundancy improves the overall reliability of a system.

redundant A configuration when two complete systems are available at one time. If the online system fails, the backup will take over with no loss of service.

reference voltage A voltage used for control or comparison purposes.

reflectance The ratio of reflected power to incident power.

reflection An abrupt change, resulting from an impedance mismatch, in the direction of propagation of an electromagnetic wave. For light, at the interface of two dissimilar materials, the incident wave is returned to its medium of origin.

reflection coefficient The ratio between the amplitude of a reflected wave and the amplitude of the incident wave. For large smooth surfaces, the reflection coefficient may be near unity.

reflection gain The increase in signal strength that results when a reflected wave combines, in phase, with an incident wave.

reflection loss The apparent loss of signal strength caused by an impedance mismatch in a transmission line or circuit. The loss results from the reflection of part of the signal back toward the source from the point of the impedance discontinuity. The greater the mismatch, the greater the loss.

reflectometer A device that measures energy traveling in each direction in a waveguide, used in determining the standing wave ratio.

refraction The bending of a sound, radio, or light wave as it passes obliquely from a medium of one density to a medium of another density that varies its speed.

regulation The process of adjusting the level of some quantity, such as circuit gain, by means of an electronic system that monitors an output and feeds back a controlling signal to constantly maintain a desired level.

regulator A device that maintains its output voltage at a constant level.

relative envelope delay The difference in envelope delay at various frequencies when compared with a reference frequency that is chosen as having zero delay.

relative humidity The ratio of the quantity of water vapor in the atmosphere to the quantity that would cause saturation at the ambient temperature.

relative transmission level The ratio of the signal power in a transmission system to the signal power at some point chosen as a reference. The ratio is usually determined by applying a standard test signal at the input to the system and measuring the gain or loss at the location of interest.

relay A device by which current flowing in one circuit causes contacts to operate that control the flow of current in another circuit.

relay armature The movable part of an electromechanical relay, usually coupled to spring sets on which contacts are mounted.

relay bypass A device that, in the event of a loss of power or other failure, routes a critical signal around the equipment that has failed.

release time The time required for a pulse to drop from steady-state level to zero, also referred to as the *decay time*.

reliability The ability of a system or subsystem to perform within the prescribed parameters of quality of service. *Reliability* is often expressed as the probability that a system or subsystem will perform its intended function for a specified interval under stated conditions.

reliability growth The action taken to move a hardware item toward its reliability potential, during development or subsequent manufacturing or operation.

reliability predictions The compiled failure rates for parts, components, subassemblies, assemblies, and systems. These generic failure rates are used as basic data to predict the reliability of a given device or system.

remote control A system used to control a device from a distance.

remote station A station or terminal that is physically remote from a main station or center but can gain access through a communication channel.

repeater The equipment between two circuits that receives a signal degraded by normal factors during transmission and amplifies the signal to its original level for re-transmission.

repetition rate The rate at which regularly recurring pulses are repeated.

reply A transmitted message that is a direct response to an original message.

repulsion The mechanical force that tends to separate like magnetic poles, like electric charges, or conductors carrying currents in opposite directions.

reset The act of restoring a device to its default or original state.

residual flux The magnetic flux that remains after a magnetomotive force has been removed.

residual magnetism The magnetism or flux that remains in a core after current ceases to flow in the coil producing the magnetomotive force.

residual voltage The vector sum of the voltages in all the phase wires of an unbalanced polyphase power system.

resistance The opposition of a material to the flow of electrical current. Resistance is equal to the voltage drop through a given material divided by the current flow through it. The standard unit of resistance is the *ohm*, named for the German physicist Georg Simon Ohm (1787–1854).

resistance drop The fall in potential (volts) between two points, the product of the current and resistance.

resistance-grounded A circuit or system grounded for safety through a resistance, which limits the value of the current flowing through the circuit in the event of a fault.

resistive load A load in which the voltage is in phase with the current.

resistivity The resistance per unit volume or per unit area.

resistor A device the primary function of which is to introduce a specified resistance into an electrical circuit.

resonance A tuned condition conducive to oscillation, when the reactance resulting from capacitance in a circuit is equal in value to the reactance resulting from inductance.

resonant frequency The frequency at which the inductive reactance and capacitive reactance of a circuit are equal.

resonator A resonant cavity.

return A return path for current, sometimes through ground.

reversal A change in magnetic polarity, in the direction of current flow.

reverse current A small current that flows through a diode when the voltage across it is such that normal forward current does not flow.

reverse voltage A voltage in the reverse direction from that normally applied.

rheostat A two-terminal variable resistor, usually constructed with a sliding or rotating shaft that can be used to vary the resistance value of the device.

ripple An ac voltage superimposed on the output of a dc power supply, usually resulting from imperfect filtering.

rise time The time required for a pulse to rise from 10 percent to 90 percent of its peak value.

roll-off A gradual attenuation of gain-frequency response at either or both ends of a transmission pass band.

root-mean-square (RMS) The square root of the average value of the squares of all the instantaneous values of current or voltage during one half-cycle of an alternating current. For an alternating current, the RMS voltage or current is equal to the amount of direct current or voltage that would produce the same heating effect in a purely resistive circuit. For a sinewave, the root-mean-square value is equal to 0.707 times the peak value. RMS is also called the *effective value*.

rotor The rotating part of an electric generator or motor.

RU An abbreviation for *rack unit*.

S

scan One sweep of the target area in a camera tube, or of the screen in a picture tube.

screen grid A grid in an electron tube that improves performance of the device by shielding the control grid from the plate.

self-bias The provision of bias in an electron tube through a voltage drop in the cathode circuit.

shot noise The noise developed in a vacuum tube or photoconductor resulting from the random number and velocity of emitted charge carriers.

slope The rate of change, with respect to frequency, of transmission line attenuation over a given frequency spectrum.

slope equalizer A device or circuit used to achieve a specified slope in a transmission line.

smoothing circuit A filter designed to reduce the amount of ripple in a circuit, usually a dc power supply.

snubber An electronic circuit used to suppress high frequency noise.

solar wind Charged particles from the sun that continuously bombard the surface of the earth.

solid A single wire conductor, as contrasted with a stranded, braided, or rope-type wire.

solid-state The use of semiconductors rather than electron tubes in a circuit or system.

source The part of a system from which signals or messages are considered to originate.

source terminated A circuit whose output is terminated for correct impedance matching with standard cable.

spare A system that is available but not presently in use.

spark gap A gap between two electrodes designed to produce a spark under given conditions.

specific gravity The ratio of the weight of a volume, liquid, or solid to the weight of the same volume of water at a specified temperature.

spectrum A continuous band of frequencies within which waves have some common characteristics.

spectrum analyzer A test instrument that presents a graphic display of signals over a selected frequency bandwidth. A cathode-ray tube is often used for the display.

spectrum designation of frequency A method of referring to a range of communication frequencies. In American practice, the designation is a two or three letter acronym for the name. The ranges are: below 300 Hz, ELF (extremely low frequency); 300 Hz–3000 Hz, ILF (infra low frequency); 3 kHz–30 kHz, VLF (very low frequency); 30 kHz–300 kHz, LF (low frequency); 300 kHz–3000 kHz, MF (medium frequency); 3 MHz–30 MHz, HF (high frequency); 30 MHz–300 MHz, VHF (very high frequency); 300 MHz–3000 MHz, UHF (ultra high frequency); 3 GHz–30 GHz, SHF (super high frequency); 30 GHz–300 GHz, EHF (extremely high frequency); 300 GHz–3000 GHz, THF (tremendously high frequency).

spherical antenna A type of satellite receiving antenna that permits more than one satellite to be accessed at any given time. A spherical antenna has a broader angle of acceptance than a parabolic antenna.

spike A high amplitude, short duration pulse superimposed on an otherwise regular waveform.

split-phase A device that derives a second phase from a single phase power supply by passing it through a capacitive or inductive reactor.

splitter A circuit or device that accepts one input signal and distributes it to several outputs.

splitting ratio The ratio of the power emerging from the output ports of a coupler.

sporadic An event occurring at random and infrequent intervals.

spread spectrum A communications technique in which the frequency components of a narrowband signal are spread over a wide band. The resulting signal resembles white noise. The technique is used to achieve signal security and privacy, and to enable the use of a common band by many users.

spurious signal Any portion of a given signal that is not part of the fundamental wave-form. Spurious signals include transients, noise, and hum.

square wave A square or rectangular-shaped periodic wave that alternately assumes two fixed values for equal lengths of time, the transition being negligible in comparison with the duration of each fixed value.

square wave testing The use of a square wave containing many odd harmonics of the fundamental frequency as an input signal to a device. Visual examination of the output signal on an oscilloscope indicates the amount of distortion introduced.

stability The ability of a device or circuit to remain stable in frequency, power level, and/or other specified parameters.

standard The specific signal configuration, reference pulses, voltage levels, and other parameters that describe the input/output requirements for a particular type of equipment.

standard time and frequency signal A time-controlled radio signal broadcast at scheduled intervals on a number of different frequencies by government-operated radio stations to provide a method for calibrating instruments.

standing wave ratio (SWR) The ratio of the maximum to the minimum value of a component of a wave in a transmission line or waveguide, such as the maximum voltage to the minimum voltage.

static charge An electric charge on the surface of an object, particularly a dielectric.

station One of the input or output points in a communications system.

stator The stationary part of a rotating electric machine.

status The present condition of a device.

statute mile A unit of distance equal to 1,609 km or 5,280 ft.

steady-state A condition in which circuit values remain essentially constant, occurring after all initial transients or fluctuating conditions have passed.

steady-state condition A condition occurring after all initial transient or fluctuating conditions have damped out in which currents, voltages, or fields remain essentially constant or oscillate uniformly without changes in characteristics such as amplitude, frequency, or wave shape.

steep wavefront A rapid rise in voltage of a given signal, indicating the presence of high frequency odd harmonics of a fundamental wave frequency.

step up (or down) The process of increasing (or decreasing) the voltage of an electrical signal, as in a step-up (or step-down) transformer.

straight-line capacitance A capacitance employing a variable capacitor with plates so shaped that capacitance varies directly with the angle of rotation.

stray capacitance An unintended—and usually undesired—capacitance between wires and components in a circuit or system.

stray current A current through a path other than the intended one.

stress The force per unit of cross-sectional area on a given object or structure.

subassembly A functional unit of a system.

subcarrier (SC) A carrier applied as modulation on another carrier, or on an intermediate subcarrier.

subharmonic A frequency equal to the fundamental frequency of a given signal divided by a whole number.

submodule A small circuit board or device that mounts on a larger module or device.

subrefraction A refraction for which the refractivity gradient is greater than standard.

subsystem A functional unit of a system.

superheterodyne receiver A radio receiver in which all signals are first converted to a common frequency for which the intermediate stages of the receiver have been optimized, both for tuning and filtering. Signals are converted by mixing them with the output of a local oscillator whose output is varied in accordance with the frequency of the received signals so as to maintain the desired *intermediate frequency*.

suppressor grid The fifth grid of a pentode electron tube, which provides screening between plate and screen grid.

surface leakage A leakage current from line to ground over the face of an insulator supporting an open wire route.

surface refractivity The refractive index, calculated from observations of pressure, temperature, and humidity at the surface of the earth.

surge A rapid rise in current or voltage, usually followed by a fall back to the normal value.

susceptance The reciprocal of reactance, and the imaginary component of admittance, expressed in siemens.

sweep The process of varying the frequency of a signal over a specified bandwidth.

sweep generator A test oscillator, the frequency of which is constantly varied over a specified bandwidth.

switching The process of making and breaking (connecting and disconnecting) two or more electrical circuits.

synchronization The process of adjusting the corresponding significant instants of signals—for example, the zero-crossings—to make them synchronous. The term *synchronization* is often abbreviated as *sync*.

synchronize The process of causing two systems to operate at the same speed.

synchronous In step or in phase, as applied to two or more devices; a system in which all events occur in a predetermined timed sequence.

synchronous detection A demodulation process in which the original signal is recovered by multiplying the modulated signal by the output of a synchronous oscillator locked to the carrier.

synchronous system A system in which the transmitter and receiver are operating in a fixed time relationship.

system standards The minimum required electrical performance characteristics of a specific collection of hardware and/or software.

systems analysis An analysis of a given activity to determine precisely what must be accomplished and how it is to be done.

T

tetrode A four element electron tube consisting of a cathode, control grid, screen grid, and plate.

thyratron A gas-filled electron tube in which plate current flows when the grid voltage reaches a predetermined level. At that point, the grid has no further control over the current, which continues to flow until it is interrupted or reversed.

tolerance The permissible variation from a standard.

torque A moment of force acting on a body and tending to produce rotation about an axis.

total harmonic distortion (THD) The ratio of the sum of the amplitudes of all signals harmonically related to the fundamental versus the amplitude of the fundamental signal. THD is expressed in percent.

trace The pattern on an oscilloscope screen when displaying a signal.

tracking The locking of tuned stages in a radio receiver so that all stages are changed appropriately as the receiver tuning is changed.

trade-off The process of weighing conflicting requirements and reaching a compromise decision in the design of a component or a subsystem.

transceiver Any circuit or device that receives and transmits signals.

transconductance The mutual conductance of an electron tube expressed as the change in plate current divided by the change in control grid voltage that produced it.

transducer A device that converts energy from one form to another.

transfer characteristics The intrinsic parameters of a system, subsystem, or unit of equipment which, when applied to the input of the system, subsystem, or unit of equipment, will fully describe its output.

transformer A device consisting of two or more windings wrapped around a single core or linked by a common magnetic circuit.

transformer ratio The ratio of the number of turns in the secondary winding of a transformer to the number of turns in the primary winding, also known as the *turns ratio*.

transient A sudden variance of current or voltage from a steady-state value. A transient normally results from changes in load or effects related to switching action.

transient disturbance A voltage pulse of high energy and short duration impressed upon the ac waveform. The overvoltage pulse can be one to 100 times the normal ac potential (or more) and can last up to 15 ms. Rise times measure in the nanosecond range.

transient response The time response of a system under test to a stated input stimulus.

transition A sequence of actions that occurs when a process changes from one state to another in response to an input.

transmission The transfer of electrical power, signals, or an intelligence from one location to another by wire, fiber optic, or radio means.

transmission facility A transmission medium and all the associated equipment required to transmit information.

transmission loss The ratio, in decibels, of the power of a signal at a point along a transmission path to the power of the same signal at a more distant point along the same path. This value is often used as a measure of the quality of the transmission medium for conveying signals. Changes in power level are normally expressed in decibels by calculating ten times the logarithm (base 10) of the ratio of the two powers.

transmission mode One of the field patterns in a waveguide in a plane transverse to the direction of propagation.

transmission system The set of equipment that provides single or multichannel communications facilities capable of carrying audio, video, or data signals.

transmitter The device or circuit that launches a signal into a passive medium, such as the atmosphere.

transparency The property of a communications system that enables it to carry a signal without altering or otherwise affecting the electrical characteristics of the signal.

tray The metal cabinet that holds circuit boards.

tremendously high frequency (THF) The frequency band from 300 GHz to 3000 GHz.

triangular wave An oscillation, the values of which rise and fall linearly, and immediately change upon reaching their peak maximum and minimum. A graphical representation of a triangular wave resembles a triangle.

trim The process of making fine adjustments to a circuit or a circuit element.

trimmer A small mechanically-adjustable component connected in parallel or series with a major component so that the net value of the two can be finely adjusted for tuning purposes.

triode A three-element electron tube, consisting of a cathode, control grid, and plate.

triple beat A third-order beat whose three beating carriers all have different frequencies, but are spaced at equal frequency separations.

troposphere The layer of the earth's atmosphere, between the surface and the stratosphere, in which about 80 percent of the total mass of atmospheric air is concentrated and in which temperature normally decreases with altitude.

trouble A failure or fault affecting the service provided by a system.

troubleshoot The process of investigating, localizing, and (if possible) correcting a fault.

tube (1—electron) An evacuated or gas-filled tube enclosed in a glass or metal case in which the electrodes are maintained at different voltages, giving rise to a controlled flow of electrons from the cathode to the anode. **(2—cathode ray, CRT)** An electron beam tube used for the display of changing electrical phenomena, generally similar to a television picture tube. **(3—cold-cathode)** An electron tube whose cathode emits electrons without the need of a heating filament. **(4—gas)** A gas-filled electron tube in which the gas plays an essential role in operation of the device. **(5—mercury-vapor)** A tube filled with mercury vapor at low pressure, used as a rectifying device. **(6—metal)** An electron tube enclosed in a metal case. **(7—traveling wave, TWT)** A wide band microwave amplifier in which a stream of electrons interacts with a guided electromagnetic wave moving substantially in synchronism with the electron stream, resulting in a net transfer of energy from the electron stream to the wave. **(8—velocity-modulated)** An electron tube in which the velocity of the electron stream is continually changing, as in a klystron.

tune The process of adjusting the frequency of a device or circuit, such as for resonance or for maximum response to an input signal.

tuned trap A series resonant network bridged across a circuit that eliminates ("traps") the frequency of the resonant network.

tuner The radio frequency and intermediate frequency parts of a radio receiver that produce a low level output signal.

tuning The process of adjusting a given frequency; in particular, to adjust for resonance or for maximum response to a particular incoming signal.

turns ratio In a transformer, the ratio of the number of turns on the secondary to the number of turns on the primary.

tweaking The process of adjusting an electronic circuit to optimize its performance.

twin-line A feeder cable with two parallel, insulated conductors.

two-phase A source of alternating current circuit with two sinusoidal voltages that are 90° apart.

U

ultra high frequency (UHF) The frequency range from 300 MHz to 3000 MHz.

ultraviolet radiation Electromagnetic radiation in a frequency range between visible light and high-frequency X-rays.

unattended A device or system designed to operate without a human attendant.

unattended operation A system that permits a station to receive and transmit messages without the presence of an attendant or operator.

unavailability A measure of the degree to which a system, subsystem, or piece of equipment is not operable and not in a committable state at the start of a mission, when the mission is called for at a random point in time.

unbalanced circuit A two-wire circuit with legs that differ from one another in resistance, capacity to earth or to other conductors, leakage, or inductance.

unbalanced line A transmission line in which the magnitudes of the voltages on the two conductors are not equal with respect to ground. A coaxial cable is an example of an unbalanced line.

unbalanced modulator A modulator whose output includes the carrier signal.

unbalanced output An output with one leg at ground potential.

unbalanced wire circuit A circuit whose two sides are inherently electrically unlike.

uncertainty An expression of the magnitude of a possible deviation of a measured value from the true value. Frequently, it is possible to distinguish two components: the *systematic uncertainty* and the *random uncertainty*. The random uncertainty is expressed by the standard deviation or by a multiple of the standard deviation. The systematic uncertainty is generally estimated on the basis of the parameter characteristics.

undamped wave A signal with constant amplitude.

underbunching A condition in a traveling wave tube wherein the tube is not operating at its optimum bunching rate.

Underwriters Laboratories, Inc. A laboratory established by the National Board of Fire Underwriters which tests equipment, materials, and systems that may affect insurance risks, with special attention to fire dangers and other hazards to life.

ungrounded A circuit or line not connected to ground.

unicoupler A device used to couple a balanced circuit to an unbalanced circuit.

unidirectional A signal or current flowing in one direction only.

uniform transmission line A transmission line with electrical characteristics that are identical, per unit length, over its entire length.

unit An assembly of equipment and associated wiring that together forms a complete system or independent subsystem.

unity coupling In a theoretically perfect transformer, complete electromagnetic coupling between the primary and secondary windings with no loss of power.

unity gain An amplifier or active circuit in which the output amplitude is the same as the input amplitude.

unity power factor A power factor of 1.0, which means that the load is—in effect—a pure resistance, with ac voltage and current completely in phase.

unterminated A device or system that is not terminated.

up-converter A frequency translation device in which the frequency of the output signal is greater than that of the input signal. Such devices are commonly found in microwave radio and satellite systems.

uplink A transmission system for sending radio signals from the ground to a satellite or aircraft.

upstream A device or system placed ahead of other devices or systems in a signal path.

useful life The period during which a low, constant failure rate can be expected for a given device or system. The *useful life* is the portion of a product life cycle between break-in and wear out.

user A person, organization, or group that employs the services of a system for the transfer of information or other purposes.

V

VA An abbreviation for *volt-amperes*, volts times amperes.

vacuum relay A relay whose contacts are enclosed in an evacuated space, usually to provide reliable long-term operation.

vacuum switch A switch whose contacts are enclosed in an evacuated container so that spark formation is discouraged.

vacuum tube An electron tube. The most common vacuum tubes include the diode, triode, tetrode, and pentode.

validity check A test designed to ensure that the quality of transmission is maintained over a given system.

varactor A semiconductor that behaves like a capacitor under the influence of an external control voltage.

varactor diode A semiconductor device whose capacitance is a function of the applied voltage. A varactor diode, also called a *variable reactance diode* or simply a *varactor*, is often used to tune the operating frequency of a radio circuit.

variable frequency oscillator (VFO) An oscillator whose frequency can be set to any required value in a given range of frequencies.

variable impedance A capacitor, inductor, or resistor that is adjustable in value.

variable-gain amplifier An amplifier whose gain can be controlled by an external signal source.

variable-reluctance A transducer in which the input (usually a mechanical movement) varies the magnetic reluctance of a device.

variation monitor A device used for sensing a deviation in voltage, current, or frequency, which is capable of providing an alarm and/or initiating transfer to another power source when programmed limits of voltage, frequency, current, or time are exceeded.

varicap A diode used as a variable capacitor.

VCXO (voltage controlled crystal oscillator) A device whose output frequency is determined by an input control voltage.

vector A quantity having both magnitude and direction.

vector diagram A diagram using vectors to indicate the relationship between voltage and current in a circuit.

vector sum The sum of two vectors which, when they are at right angles to each other, equal the length of the hypotenuse of the right triangle so formed. In the general case, the vector sum of the two vectors equals the diagonal of the parallelogram formed on the two vectors.

velocity of light The speed of propagation of electromagnetic waves in a vacuum, equal to 299,792,458 m/s, or approximately 186,000 mi/s. For rough calculations, the figure of 300,000 km/s is used.

velocity of propagation The velocity of signal transmission. In free space, electromagnetic waves travel at the speed of light. In a cable, the velocity is substantially lower.

vernier A device that enables precision reading of a measuring set or gauge, or the setting of a dial with precision.

very low frequency (VLF) A radio frequency in the band 3 kHz to 30 kHz.

vestigial sideband A form of transmission in which one sideband is significantly attenuated. The carrier and the other sideband are transmitted without attenuation.

vibration testing A testing procedure whereby subsystems are mounted on a test base that vibrates, thereby revealing any faults resulting from badly soldered joints or other poor mechanical design features.

volt The standard unit of electromotive force, equal to the potential difference between two points on a conductor that is carrying a constant current of one ampere when the power dissipated between the two points is equal to one watt. One *volt* is equivalent to the potential difference across a resistance of one ohm when one ampere is flowing through it. The volt is named for the Italian physicist Alessandro Volta (1745–1827).

volt-ampere (VA) The apparent power in an ac circuit (volts times amperes).

volt-ohm-milliammeter (VOM) A general purpose multirange test meter used to measure voltage, resistance, and current.

voltage The potential difference between two points.

voltage drop A decrease in electrical potential resulting from current flow through a resistance.

voltage gradient The continuous drop in electrical potential, per unit length, along a uniform conductor or thickness of a uniform dielectric.

voltage level The ratio of the voltage at a given point to the voltage at an arbitrary reference point.

voltage reference circuit A stable voltage reference source.

voltage regulation The deviation from a nominal voltage, expressed as a percentage of the nominal voltage.

voltage regulator A circuit used for controlling and maintaining a voltage at a constant level.

voltage stabilizer A device that produces a constant or substantially constant output voltage despite variations in input voltage or output load current.

voltage to ground The voltage between any given portion of a piece of equipment and the ground potential.

voltmeter An instrument used to measure differences in electrical potential.

vox A voice-operated relay circuit that permits the equivalent of push-to-talk operation of a transmitter by the operator.

VSAT (very small aperture terminal) A satellite Ku-band earth station intended for fixed or portable use. The antenna diameter of a VSAT is on the order of 1.5 m or less.

W

watt The unit of power equal to the work done at one joule per second, or the rate of work measured as a current of one ampere under an electric potential of one volt. Designated by the symbol *W*, the watt is named after the Scottish inventor James Watt (1736–1819).

watt meter A meter indicating in watts the rate of consumption of electrical energy.

watt-hour The work performed by one watt over a one hour period.

wave A disturbance that is a function of time or space, or both, and is propagated in a medium or through space.

wave number The reciprocal of wavelength; the number of wave lengths per unit distance in the direction of propagation of a wave.

waveband A band of wavelengths defined for some given purpose.

waveform The characteristic shape of a periodic wave, determined by the frequencies present and their amplitudes and relative phases.

wavefront A continuous surface that is a locus of points having the same phase at a given instant. A *wavefront* is a surface at right angles to rays that proceed from the wave source. The surface passes through those parts of the wave that are in the same phase and travel in the same direction. For parallel rays the wavefront is a plane; for rays that radiate from a point, the wavefront is spherical.

waveguide Generally, a rectangular or circular pipe that constrains the propagation of an acoustic or electromagnetic wave along a path between two locations. The dimensions of a waveguide determine the frequencies for optimum transmission.

wavelength For a sinusoidal wave, the distance between points of corresponding phase of two consecutive cycles.

weber The unit of magnetic flux equal to the flux that, when linked to a circuit of one turn, produces an electromotive force of one volt as the flux is reduced at a uniform rate to zero in one second. The *weber* is named for the German physicist Wilhelm Eduard Weber (1804–1891).

weighted The condition when a correction factor is applied to a measurement.

weighting The adjustment of a measured value to account for conditions that would otherwise be different or appropriate during a measurement.

weighting network A circuit, used with a test instrument, that has a specified amplitude-versus-frequency characteristic.

wideband The passing or processing of a wide range of frequencies. The meaning varies with the context.

Wien bridge An ac bridge used to measure capacitance or inductance.

winding A coil of wire used to form an inductor.

wire A single metallic conductor, usually solid-drawn and circular in cross section.

working range The permitted range of values of an analog signal over which transmitting or other processing equipment can operate.

working voltage The rated voltage that may safely be applied continuously to a given circuit or device.

X

x-band A microwave frequency band from 5.2 GHz to 10.9 GHz.

x-cut A method of cutting a quartz plate for an oscillator, with the x-axis of the crystal perpendicular to the faces of the plate.

X ray An electromagnetic radiation of approximately 100 nm to 0.1 nm, capable of penetrating nonmetallic materials.

Y

y-cut A method of cutting a quartz plate for an oscillator, with the y-axis of the crystal perpendicular to the faces of the plate.

yield strength The magnitude of mechanical stress at which a material will begin to deform. Beyond the *yield strength* point, extension is no longer proportional to stress and rupture is possible.

yoke A material that interconnects magnetic cores. *Yoke* can also refer to the deflection windings of a CRT.

yttrium-iron garnet (YIG) A crystalline material used in microwave devices.

22

Abbreviations and Acronyms

A

a	atto (10^{-18})
Å	angstrom
A	ampere
AAR	automatic alternate routing
AARTS	automatic audio remote test set
ac	alternating current
ACA	automatic circuit assurance
ACC	automatic callback calling
ACD	automatic call distributor
ac-dc	alternating current - direct current
ACK	acknowledge character
ACTS	Advanced Communications Technology Satellite
ACU	automatic calling unit
A-D	analog-to-digital
ADC	analog-to-digital converter; analog-to-digital conversion
ADCCP	Advanced Data Communication Control Procedures
ADH	automatic data handling
ADP	automatic data processing
ADPCM	adaptive differential pulse-code modulation
ADPE	automatic data processing equipment
ADU	automatic dialing unit
ADX	automatic data exchange
AECS	Aeronautical Emergency Communications System [Plan]
AF	audio frequency
AFC	area frequency coordinator; automatic frequency control
AFRS	Armed Forces Radio Service

This chapter adapted from: General Services Administration, Information Technology Service, *National Communications System: Glossary of Telecommunication Terms*, Technology and Standards Division, Federal Standard 1037C, General Services Administration, Washington, D.C., August, 7, 1996.

AGC	automatic gain control
AGE	aerospace ground equipment
AI	artificial intelligence
AIG	address indicator group; address indicating group
AIM	amplitude intensity modulation
AIN	advanced intelligent network
AIOD	automatic identified outward dialing
AIS	automated information system
AJ	anti-jamming
ALC	automatic level control; automatic load control
ALE	automatic link establishment
ALU	arithmetic and logic unit
AM	amplitude modulation
AMA	automatic message accounting
AMC	administrative management complex
AME	amplitude modulation equivalent; automatic message exchange
AMI	alternate mark inversion [signal]
AM/PM/VSB	amplitude modulation/phase modulation/vestigial sideband
AMPS	automatic message processing system
AMPSSO	automated message processing system security officer
AMSC	American Mobile Satellite Corporation
AMTS	automated maritime telecommunications system
ANI	automatic number identification
ANL	automatic noise limiter
ANMCC	Alternate National Military Command Center
ANS	American National Standard
ANSI	American National Standards Institute
AP	anomalous propagation
APC	adaptive predictive coding
API	application program interface
APK	amplitude phase-shift keying
APL	average picture level
ARP	address resolution protocol
ARPA	Advanced Research Projects Agency [now DARPA]
ARPANET	Advanced Research Projects Agency Network
ARQ	automatic repeat-request
ARS	automatic route selection
ARSR	air route surveillance radar
ARU	audio response unit
ASCII	American Standard Code for Information Interchange
ASIC	application-specific integrated circuits
ASP	Aggregated Switch Procurement; adjunct service point
ASR	automatic send and receive; airport surveillance radar
AT	access tandem
ATACS	Army Tactical Communications System
ATB	all trunks busy
ATCRBS	air traffic control radar beacon system
ATDM	asynchronous time-division multiplexing

ATE	automatic test equipment
ATM	asynchronous transfer mode
ATV	advanced television
au	astronomical unit
AUI	attachment unit interface
AUTODIN	Automatic Digital Network
AUTOVON	Automatic Voice Network
AVD	alternate voice/data
AWG	American wire gauge
AWGN	additive white Gaussian noise
AZ	azimuth

B

b	bit
B	bel; byte
balun	balanced to unbalanced
basecom	base communications
BASIC	beginners' all-purpose symbolic instruction code
BCC	block check character
BCD	binary coded decimal; binary-coded decimal notation
BCI	bit-count integrity
Bd	baud
B8ZS	bipolar with eight-zero substitution
bell	BEL character
BER	bit error ratio
BERT	bit error ratio tester
BETRS	basic exchange telecommunications radio service
BEX	broadband exchange
BIH	International Time Bureau
B-ISDN	broadband ISDN
bi-sync	binary synchronous [communication]
bit	binary digit
BIT	built-in test
BITE	built-in test equipment
BIU	bus interface unit
BNF	Backus Naur form
BOC	Bell Operating Company
bpi	bits per inch; bytes per inch
BPSK	binary phase-shift keying
b/s	bits per second
b/in	bits per inch
BPSK	binary phase-shift keying
BR	bit rate
BRI	basic rate interface
BSA	basic serving arrangement
BSE	basic service element
BSI	British Standards Institution
B6ZS	bipolar with six-zero substitution

B3ZS	bipolar with three-zero substitution
BTN	billing telephone number
BW	bandwidth

C

c	centi (10^{-2})
CACS	centralized alarm control system
CAM	computer-aided manufacturing
CAMA	centralized automatic message accounting
CAN	cancel character
CAP	competitive access provider; customer administration panel
CARS	cable television relay service [station]
CAS	centralized attendant services
CASE	computer-aided software engineering; computer aided system engineering; computer-assisted software engineering
CATV	cable TV; cable television; community antenna television
CBX	computer branch exchange
C^2	command and control
C^3	command, control, and communications
C^3CM	C^3 countermeasures
C^3I	command, control, communications and intelligence
CCA	carrier-controlled approach
CCD	charge-coupled device
CCH	connections per circuit hour
CCIF	International Telephone Consultative Committee
CCIR	International Radio Consultative Committee
CCIS	common-channel interoffice signaling
CCIT	International Telegraph Consultative Committee
CCITT	International Telegraph and Telephone Consultative Committee
CCL	continuous communications link
CCS	hundred call-seconds
CCSA	common control switching arrangement
CCTV	closed-circuit television
CCW	cable cutoff wavelength
cd	candela
CD	collision detection; compact disk
CDF	combined distribution frame; cumulative distribution function
CDMA	code-division multiple access
CDPSK	coherent differential phase-shift keying
CDR	call detail recording
CD ROM	compact disk read-only memory
CDT	control data terminal
CDU	central display unit
C-E	communications-electronics
CEI	comparably efficient interconnection
CELP	code-excited linear prediction
CEP	circular error probable
CFE	contractor-furnished equipment

cgs	centimeter-gram-second
ChR	channel reliability
CIAS	circuit inventory and analysis system
CIC	content indicator code
CIF	common intermediate format
CIFAX	ciphered facsimile
CiR	circuit reliability
C/kT	carrier-to-receiver noise density
CLASS	custom local area signaling service
cm	centimeter
CMI	coded mark inversion
CMIP	Common Management Information Protocol
CMIS	common management information service
CMOS	complementary metal oxide substrate
CMRR	common-mode rejection ratio
CNR	carrier-to-noise ratio; combat-net radio
CNS	complementary network service
C.O.	central office
COAM	customer owned and maintained equipment
COBOL	common business oriented language
codec	coder-decoder
COG	centralized ordering group
COMINT	communications intelligence
COMJAM	communications jamming
compandor	compressor-expander
COMPUSEC	computer security
COMSAT	Communications Satellite Corporation
COMSEC	communications security
CONEX	connectivity exchange
CONUS	Continental United States
COP	Committee of Principals
COR	Council of Representatives
COT	customer office terminal
CPAS	cellular priority access services
CPE	customer premises equipment
cpi	characters per inch
cpm	counts per minute
cps	characters per second
CPU	central processing unit; communications processor unit
CR	channel reliability; circuit reliability
CRC	cyclic redundancy check
CRITICOM	Critical Intelligence Communications
CROM	control read-only memory
CRT	cathode ray tube
c/s	cycles per second
CSA	Canadian Standards Association
CSC	circuit-switching center; common signaling channel
CSMA	carrier sense multiple access

CSMA/CA	carrier sense multiple access with collision avoidance
CSMA/CD	carrier sense multiple access with collision detection
CSU	channel service unit; circuit switching unit; customer service unit
CTS	clear to send
CTX	Centrex® [service]; clear to transmit
CVD	chemical vapor deposition
CVSD	continuously variable slope delta [modulation]
cw	carrier wave; composite wave; continuous wave
CX	composite signaling
cxr	carrier

D

d	deci (10^{-1})
da	deka (10)
D-A	digital-to-analog; digital-to-analog converter
D/L	downlink
DACS	digital access and cross-connect system
DAMA	demand assignment multiple access
DARPA	Defense Advanced Research Projects Agency
dB	decibel
dBa	decibels adjusted
dBa0	noise power measured at zero transmission level point
dBc	dB relative to carrier power
dBm	dB referred to 1 milliwatt
dBm(psoph)	noise power in dBm measured by a set with psophometric weighting
DBMS	database management system
dBmV	dB referred to 1 millivolt across 75 ohms
dBm0	noise power in dBm referred to or measured at 0TLP
dBm0p	noise power in dBm0 measured by a psophometric or noise measuring set having psophometric weighting
dBr	power difference in dB between any point and a reference point
dBrn	dB above reference noise
dBrnC	noise power in dBrn measured by a set with C-message weighting
dBrnC0	noise power in dBrnC referred to or measured at 0TLP
dBrn(f_1-f_2)	flat noise power in dBrn
dBrn(144)	noise power in dBrn measured by a set with 144-line weighting
dBv	dB relative to 1 V (volt) peak-to-peak
dBW	dB referred to 1 W (watt)
dBx	dB above reference coupling
dc	direct current
DCA	Defense Communications Agency
DCE	data circuit-terminating equipment
DCL	direct communications link
DCPSK	differentially coherent phase-shift keying
DCS	Defense Communications System
DCTN	Defense Commercial Telecommunications Network
DCWV	direct-current working volts
DDD	direct distance dialing

DDN	Defense Data Network
DDS	digital data service
DEL	delete character
demarc	demarcation point
demux	demultiplex; demultiplexer; demultiplexing
dequeue	double-ended queue
DES	Data Encryption Standard
detem	detector/emitter
DFSK	double-frequency shift keying
DIA	Defense Intelligence Agency
DID	direct inward dialing
DIN	Deutsches Institut für Normung
DIP	dual in-line package
DISA	Defense Information Systems Agency
DISC	disconnect command
DISN	Defense Information System Network
DISNET	Defense Integrated Secure Network
DLA	Defense Logistic Agency
DLC	digital loop carrier
DLE	data link escape character
DM	delta modulation
DMA	Defense Mapping Agency; direct memory access
DME	distance measuring equipment
DMS	Defense Message System
DNA	Defense Nuclear Agency
DNIC	data network identification code
DNPA	data numbering plan area
DNS	Domain Name System
DO	design objective
DoC	Department of Commerce
DOD	Department of Defense; direct outward dialing
DODD	Department of Defense Directive
DODISS	Department of Defense Index of Specifications and Standards
DOD-STD	Department of Defense Standard
DOS	Department of State
DPCM	differential pulse-code modulation
DPSK	differential phase-shift keying
DQDB	distributed-queue dual-bus [network]
DRAM	dynamic random access memory
DRSI	destination station routing indicator
DS	digital signal; direct support
DS0	digital signal 0
DS1	digital signal 1
DS1C	digital signal 1C
DS2	digital signal 2
DS3	digital signal 3
DS4	digital signal 4
DSA	dial service assistance

DSB	double sideband (transmission); Defense Science Board
DSB-RC	double-sideband reduced carrier transmission
DSB-SC	double-sideband suppressed-carrier transmission
DSC	digital selective calling
DSCS	Defense Satellite Communications System
DSE	data switching exchange
DSI	digital speech interpolation
DSL	digital subscriber line
DSN	Defense Switched Network
DSR	data signaling rate
DSS	direct station selection
DSSCS	Defense Special Service Communications System
DSTE	data subscriber terminal equipment
DSU	data service unit
DTE	data terminal equipment
DTG	date-time group
DTMF	dual-tone multifrequency (signaling)
DTN	data transmission network
DTS	Diplomatic Telecommunications Service
DTU	data transfer unit; data tape unit; digital transmission unit; direct to user
DVL	direct voice link
DX signaling	direct current signaling; duplex signaling

E

E	exa (10^{18})
E-MAIL	electronic mail
EAS	extended area service
EBCDIC	extended binary coded decimal interchange code
E_b/N_0	signal energy per bit per hertz of thermal noise
EBO	embedded base organization
EBS	Emergency Broadcast System
EBX	electronic branch exchange
EC	Earth coverage; Earth curvature
ECC	electronically controlled coupling; enhance call completion
ECCM	electronic counter-countermeasures
ECM	electronic countermeasures
EDC	error detection and correction
EDI	electronic data interchange
EDTV	extended-definition television
EHF	extremely high frequency
EIA	Electronic Industries Association
eirp	effective isotropically radiated power; equivalent isotopically radiated power
EIS	Emergency Information System
el	elevation
ELF	extremely low frequency
ELINT	electronics intelligence; electromagnetic intelligence

ELSEC	electronics security
ELT	emergency locator transmitter
EMC	electromagnetic compatibility
EMCON	emission control
EMD	equilibrium mode distribution
EME	electromagnetic environment
emf	electromotive force
EMI	electromagnetic interference; electromagnetic interference control
EMP	electromagnetic pulse
EMR	electromagnetic radiation
EMR hazards	electromagnetic radiation hazards
e.m.r.p.	effective monopole radiated power; equivalent monopole radiated power
EMS	electronic message system
EMSEC	emanations security
emu	electromagnetic unit
EMV	electromagnetic vulnerability
EMW	electromagnetic warfare; electromagnetic wave
ENQ	enquiry character
EO	end office
E.O.	Executive Order
EOD	end of data
EOF	end of file
EOL	end of line
EOM	end of message
EOP	end of program; end output
EOS	end-of-selection character
EOT	end-of-transmission character; end of tape
EOW	engineering orderwire
EPROM	erasable programmable read-only memory
EPSCS	enhanced private switched communications system
ERL	echo return loss
ERLINK	emergency response link
ERP, e.r.p.	effective radiated power
ES	end system; expert system
ESC	escape character; enhanced satellite capability
ESF	extended superframe
ESM	electronic warfare support measures
ESP	enhanced service provider
ESS	electronic switching system
ETB	end-of-transmission-block character
ETX	end-of-text character
EW	electronic warfare
EXCSA	Exchange Carriers Standards Association

F

f	femto (10^{-15})
f	frequency

FAA	Federal Aviation Administration
FAQ file	Frequently Asked Questions file
FAX	facsimile
FC	functional component
FCC	Federal Communications Commission
FCS	frame check sequence
FDDI	fiber distributed data interface
FDDI-2	fiber distributed data interface-2
FDHM	full duration at half maximum
FDM	frequency-division multiplexing
FDMA	frequency-division multiple access
FDX	full duplex
FEC	forward error correction
FECC	Federal Emergency Communications Coordinators
FED-STD	Federal Standard
FEMA	Federal Emergency Management Agency
FEP	front-end processor
FET	field effect transistor
FIFO	first-in first-out
FIP	Federal Information Processing
FIPS	Federal Information Processing Standards
FIR	finite impulse response
FIRMR	Federal Information Resources Management Regulations
FISINT	foreign instrumentation signals intelligence
flops	floating-point operations per second
FM	frequency modulation
FO	fiber optics
FOC	final operational capability; full operational capability
FOT	frequency of optimum traffic; frequency of optimum transmission
FPIS	forward propagation ionospheric scatter
fps	foot-pound-second
FPS	frames per second; focus projection and scanning
FRP	Federal Response Plan
FSDPSK	filtered symmetric differential phase-shift keying
FSK	frequency-shift keying
FSS	fully separate subsidiary
FT	fiber optic T-carrier
FTAM	file transfer, access, and management
FTF	Federal Telecommunications Fund
ft/min	feet per minute
FTP	file transfer protocol
ft/s	feet per second
FTS	Federal Telecommunications System
FTS2000	Federal Telecommunications System 2000
FTSC	Federal Telecommunications Standards Committee
FWHM	full width at half maximum
FX	fixed service; foreign exchange service
FYDP	Five Year Defense Plan

G

g	profile parameter
G	giga (10^9)
GBH	group busy hour
GCT	Greenwich Civil Time
GDF	group distribution frame
GETS	Government Emergency Telecommunications Service
GFE	Government-furnished equipment
GGCL	government-to-government communications link
GHz	gigahertz
GII	Global Information Infrastructure
GMT	Greenwich Mean Time
GOS	grade of service
GOSIP	Government Open Systems Interconnection Profile
GSA	General Services Administration
GSTN	general switched telephone network
G/T	antenna gain-to-noise-temperature
GTP	Government Telecommunications Program
GTS	Government Telecommunications System
GUI	graphical user interface

H

h	hecto (10^2); hour; Planck's constant
HCS	hard clad silica (fiber)
HDLC	high-level data link control
HDTV	high-definition television
HDX	half-duplex (operation)
HE_{11} mode	the fundamental hybrid mode (of an optical fiber)
HEMP	high-altitude electromagnetic pulse
HERF	hazards of electromagnetic radiation to fuel
HERO	hazards of electromagnetic radiation to ordnance
HERP	hazards of electromagnetic radiation to personnel
HF	high frequency
HFDF	high-frequency distribution frame
HLL	high-level language
HPC	high probability of completion
HV	high voltage
Hz	hertz

I

IA	International Alphabet
I&C	installation and checkout
IC	integrated circuit
ICI	incoming call identification
ICNI	Integrated Communications, Navigation, and Identification
ICW	interrupted continuous wave
IDDD	International Direct Distance Dialing
IDF	intermediate distribution frame

IDN	integrated digital network
IDTV	improved-definition television
IEC	International Electrotechnical Commission
IEEE	Institute of Electrical and Electronics Engineers
IES	Industry Executive Subcommittee
IF	intermediate frequency
I/F	interface
IFF	identification, friend or foe
IFRB	International Frequency Registration Board
IFS	ionospheric forward scatter
IIR	infinite impulse response
IITF	Information Infrastructure Task Force
ILD	injection laser diode
ILS	instrument landing system
IM	intensity modulation; intermodulation
I&M	installation and maintenance
IMD	intermodulation distortion
IMP	interface message processor
IN	intelligent network
INFOSEC	information systems security
INS	inertial navigation system
INTELSAT	International Telecommunications Satellite Consortium
INWATS	Inward Wide-Area Telephone Service
I/O	input/output (device)
IOC	integrated optical circuit; initial operational capability; input-output controller
IP	Internet protocol; intelligent peripheral
IPA	intermediate power amplifier
IPC	information processing center
IPM	impulses per minute; interference prediction model; internal polarization modulation; interruptions per minute
in/s	inches per second
ips	interruptions per second
IPX	Internet Packet Exchange
IQF	intrinsic quality factor
IR	infrared
IRAC	Interdepartment Radio Advisory Committee
IRC	international record carrier; Interagency Radio Committee
ISB	independent-sideband (transmission)
ISDN	Integrated Services Digital Network
ISM	industrial, scientific, and medical (applications)
ISO	International Organization for Standardization
ITA	International Telegraph Alphabet
ITA-5	International Telegraph Alphabet Number 5
ITC	International Teletraffic Congress
ITS	Institute for Telecommunication Sciences
ITSO	International Telecommunications Satellite Organization
ITU	International Telecommunication Union

IVDT integrated voice data terminal
IXC interexchange carrier

J
JANAP Joint Army-Navy-Air Force Publication(s)
JCS Joint Chiefs of Staff
JPL Jet Propulsion Laboratory
JSC Joint Steering Committee; Joint Spectrum Center
JTC^{3}A Joint Tactical Command, Control and Communications Agency
JTIDS Joint Tactical Information Distribution System
JTRB Joint Telecommunications Resources Board
JTSSG Joint Telecommunications Standards Steering Group
JWID Joint Warrior Interoperability Demonstration

K
k kilo (10^3); Boltzmann's constant
K coefficient of absorption; kelvin
KDC key distribution center
KDR keyboard data recorder
KDT keyboard display terminal
kg kilogram
kg·m·s kilogram-meter-second
kHz kilohertz
km kilometer
kΩ, k kilohm
KSR keyboard send/receive device
kT noise power density
KTS key telephone system
KTU key telephone unit

L
LAN local area network
LAP-B Data Link Layer protocol (CCITT Recommendation X.25 [1989])
LAP-D link access procedure D
laser light amplification by stimulated emission of radiation
LASINT laser intelligence
LATA local access and transport area
LBO line buildout
LC limited capability
LCD liquid crystal display
LD long distance
LDM limited distance modem
LEC local exchange carrier
LED light-emitting diode
LF low frequency
LFB look-ahead-for-busy (information)
LIFO last-in first-out
LLC logical link control (sublayer)

l/m	lines per minute
LMF	language media format
LMR	land mobile radio
LNA	launch numerical aperture
LOF	lowest operating frequency
loran	long-range aid to navigation system; long-range radio navigation; long-range radio aid to navigation system
LOS	line of sight, loss of signal
LP	linearly polarized (mode); linear programming; linking protection; log-periodic (antenna); log-periodic (array)
LPA	linear power amplifier
LPC	linear predictive coding
LPD	low probability of detection
LPI	low probability of interception
lpi	lines per inch
lpm	lines per minute
LP_{01}	the fundamental mode (of an optical fiber)
LQA	link quality analysis
LRC	longitudinal redundancy check
LSB	lower sideband, least significant bit
LSI	large scale integrated (circuit); large scale integration; line status indication
LTC	line traffic coordinator
LUF	lowest usable high frequency
LULT	line-unit-line termination
LUNT	line-unit-network termination
LV	low voltage

M

m	meter
M	mega (10^6)
MAC	medium access control [sublayer]
MACOM	major command
MAN	metropolitan area network
MAP	manufacturers' automation protocol
maser	microwave amplification by the stimulated emission of radiation
MAU	medium access unit
MCC	maintenance control circuit
MCEB	Military Communications-Electronics Board
MCM	multicarrier modulation
MCS	Master Control System
MCW	modulated continuous wave
MCXO	microcomputer compensated crystal oscillator
MDF	main distribution frame
MDT	mean downtime
MEECN	Minimum Essential Emergency Communications Network
MERCAST	merchant-ship broadcast system
MF	medium frequency; multifrequency (signaling)

MFD	mode field diameter
MFJ	Modification of Final Judgment
MFSK	multiple frequency-shift keying
MHF	medium high frequency
MHS	message handling service; message handling system
MHz	megahertz
mi	mile
MIC	medium interface connector; microphone; microwave integrated circuit; minimum ignition current; monolithic integrated circuit; mutual interface chart
MILNET	military network
MIL-STD	Military Standard
min	minute
MIP	medium interface point
MIPS, mips	million instructions per second
MIS	management information system
MKS	meter-kilogram-second
MLPP	multilevel precedence and preemption
MMW	millimeter wave
modem	modulator-demodulator
mol	mole
ms	millisecond (10^{-3} second)
MSB	most significant bit
MSK	minimum-shift keying
MTBF	mean time between failures
MTBM	mean time between maintenance
MTBO	mean time between outages
MTBPM	mean time between preventive maintenance
MTF	modulation transfer function
MTSO	mobile telephone switching office
MTSR	mean time to service restoration
MTTR	mean time to repair
μ	micro (10^{-6})
μs	microsecond
MUF	maximum usable frequency
MUX	multiplex; multiplexer
MUXing	multiplexing
mw	microwave
MWI	message waiting indicator
MWV	maximum working voltage

N

n	nano (10^{-9}); refractive index
N_0	sea level refractivity; spectral noise density
NA	numerical aperture
NACSEM	National Communications Security Emanation Memorandum
NACSIM	National Communications Security Information Memorandum
NAK	negative-acknowledge character

NASA	National Aeronautics and Space Administration
NATA	North American Telecommunications Association
NATO	North Atlantic Treaty Organization
NAVSTAR	Navigational Satellite Timing and Ranging
NBFM	narrowband frequency modulation
NBH	network busy hour
NBRVF	narrowband radio voice frequency
NBS	National Bureau of Standards
NBSV	narrowband secure voice
NCA	National Command Authorities
NCC	National Coordinating Center for Telecommunications
NCS	National Communications System; net control station
NCSC	National Communications Security Committee
NDCS	network data control system
NDER	National Defense Executive Reserve
NEACP	National Emergency Airborne Command Post
NEC	National Electric Code®
NEP	noise equivalent power
NES	noise equivalent signal
NF	noise figure
NFS	Network File System
NIC	network interface card
NICS	NATO Integrated Communications System
NID	network interface device; network inward dialing; network information database
NII	National Information Infrastructure
NIOD	network inward/outward dialing
NIST	National Institute of Standards and Technology
NIU	network interface unit
NLP	National Level Program
nm	nanometer
NMCS	National Military Command System
nmi	nautical mile
NOD	network outward dialing
Np	neper
NPA	numbering plan area
NPR	noise power ratio
NRI	net radio interface
NRM	network resource manager
NRRC	Nuclear Risk Reduction Center
NRZ	non-return-to-zero
NRZI	non-return-to-zero inverted
NRZ-M	non-return-to-zero mark
NRZ-S	non-return-to-zero space
NRZ1	non-return-to-zero, change on ones
NRZ-1	non-return-to-zero mark
ns	nanosecond
NSA	National Security Agency

NSC	National Security Council
NS/EP	National Security or Emergency Preparedness telecommunications
NSTAC	National Security Telecommunications Advisory Committee
NTCN	National Telecommunications Coordinating Network
NTDS	Naval Tactical Data System
NT1	Network termination 1
NT2	Network termination 2
NTI	network terminating interface
NTIA	National Telecommunications and Information Administration
NTMS	National Telecommunications Management Structure
NTN	network terminal number
NTSC	National Television Standards Committee; National Television Standards Committee (standard)
NUL	null character
NVIS	near vertical incidence skywave

O

O&M	operations and maintenance
OC	operations center
OCC	other common carrier
OCR	optical character reader; optical character recognition
OCU	orderwire control unit
OCVCXO	oven controlled-voltage controlled crystal oscillator
OCXO	oven controlled crystal oscillator
OD	optical density; outside diameter
OFC	optical fiber, conductive
OFCP	optical fiber, conductive, plenum
OFCR	optical fiber, conductive, riser
OFN	optical fiber, nonconductive
OFNP	optical fiber, nonconductive, plenum
OFNR	optical fiber, nonconductive, riser
OMB	Office of Management and Budget
ONA	open network architecture
opm	operations per minute
OPMODEL	operations model
OPSEC	operations security
OPX	off-premises extension
OR	off-route service; off-route aeronautical mobile service
OSHA	Occupational Safety and Health Administration
OSI	open switching interval; Open Systems Interconnection
OSI-RM	Open Systems Interconnection—Reference Model
OSRI	originating stations routing indicator
OSSN	originating stations serial number
OTAM	over-the-air management of automated HF network nodes
OTAR	over-the-air rekeying
OTDR	optical time domain reflectometer; optical time domain reflectometry
OW	orderwire [circuit]

P

p	pico (10^{-12})
P	peta (10^{15})
PABX	private automatic branch exchange
PAD	packet assembler/disassembler
PAL	phase alternation by line
PAL-M	phase alternation by line—modified
PAM	pulse-amplitude modulation
PAMA	pulse-address multiple access
p/a r	peak-to-average ratio
PAR	performance analysis and review
PARAMP	parametric amplifier
par meter	peak-to-average ratio meter
PAX	private automatic exchange
PBER	pseudo-bit-error-ratio
PBX	private branch exchange
PC	carrier power (of a radio transmitter); personal computer
PCB	power circuit breaker; printed circuit board
PCM	pulse-code modulation; plug compatible module; process control module
PCS	Personal Communications Services; personal communications system; plastic-clad silica (fiber)
PCSR	parallel channels signaling rate
PD	photodetector
PDM	pulse delta modulation; pulse-duration modulation
PDN	public data network
PDS	protected distribution system; power distribution system; program data source
PDT	programmable data terminal
PDU	protocol data unit
PE	phase-encoded (recording)
PEP	peak envelope power (of a radio transmitter)
pF	picofarad
PF	power factor
PFM	pulse-frequency modulation
PI	protection interval
PIC	plastic insulated cable
ping	packet Internet groper
PIV	peak inverse voltage
PLA	programmable logic array
PL/I	programming language 1
PLL	phase-locked loop
PLN	private line network
PLR	pulse link repeater
PLS	physical signaling sublayer
pm	phase modulation
PM	mean power; polarization-maintaining (optical fiber); preventive maintenance; pulse modulation

PMB	pilot-make-busy (circuit)
PMO	program management office
POI	point of interface
POP	point of presence
POSIX	portable operating system interface for computer environments
POTS	plain old telephone service
PP	polarization-preserving (optical fiber)
P-P	peak-to-peak (value)
P/P	point-to-point
PPM	pulse-position modulation
pps	pulses per second
PR	pulse rate
PRF	pulse-repetition frequency
PRI	primary rate interface
PRM	pulse-rate modulation
PROM	programmable read-only memory
PRR	pulse repetition rate
PRSL	primary area switch locator
PS	permanent signal
psi	pounds (force) per square inch
PSK	phase-shift keying
PSN	public switched network
p-static	precipitation static
PSTN	public switched telephone network
PTF	patch and test facility
PTM	pulse-time modulation
PTT	postal, telephone, and telegraph; push-to-talk (operation)
PTTC	paper tape transmission code
PTTI	precise time and time interval
PU	power unit
PUC	public utility commission; public utilities commission
PVC	permanent virtual circuit; polyvinyl chloride (insulation)
pW	picowatt
PWM	pulse-width modulation
PX	private exchange

Q

QA	quality assurance
QAM	quadrature amplitude modulation
QC	quality control
QCIF	quarter common intermediate format
QMR	qualitative material requirement
QOS	quality of service
QPSK	quadrature phase-shift keying
QRC	quick reaction capability

R

racon	radar beacon

rad	radian; radiation absorbed dose
radar	radio detection and ranging
RADHAZ	electromagnetic radiation hazards
RADINT	radar intelligence
RAM	random access memory; reliability, availability, and maintainability
R&D	research and development
RATT	radio teletypewriter system
RBOC	Regional Bell Operating Company
RbXO	rubidium-crystal oscillator
RC	reflection coefficient; resource controller
RCC	radio common carrier
RCVR	receiver
RDF	radio-direction finding
REA	Rural Electrification Administration
REN	ringer equivalency number
RF	radio frequency; range finder
RFI	radio frequency interference
RFP	request for proposal
RFQ	request for quotation
RGB	red-green-blue
RH	relative humidity
RHR	radio horizon range
RI	routing indicator
RISC	reduced instruction set chip
RJ	registered jack
RJE	remote job entry
rms	root-mean-square (deviation)
RO	read only; receive only
ROA	recognized operating agency
ROC	required operational capability
ROM	read-only memory
ROSE	remote operations service element protocol
rpm	revolutions per minute
RPM	rate per minute
RPOA	recognized private operating agency
rps	revolutions per second
RQ	repeat-request
RR	repetition rate
RSL	received signal level
rss	root-sum-square
R/T	real time
RTA	remote trunk arrangement
RTS	request to send
RTTY	radio teletypewriter
RTU	remote terminal unit
RTX	request to transmit
RVA	reactive volt-ampere
RVWG	Reliability and Vulnerability Working Group

RWI	radio and wire integration
RX	receive; receiver
RZ	return-to-zero

S

s	second
SCC	specialized common carrier
SCE	service creation environment
SCF	service control facility
SCP	service control point
SCPC	single channel per carrier
SCR	semiconductor-controlled rectifier; silicon-controlled rectifier
SCSR	single channel signaling rate
SDLC	synchronous data link control
SDM	space-division multiplexing
SDN	software-defined network
SECDEF	Secretary of Defense
SECORD	secure voice cord board
SECTEL	secure telephone
SETAMS	systems engineering, technical assistance, and management services
SEVAS	Secure Voice Access System
S-F	store-and-forward
SF	single-frequency (signaling)
SGDF	supergroup distribution frame
S/H	sample and hold
SHA	sidereal hour angle
SHARES	Shared Resources (SHARES) HF Radio Program
SHF	super high frequency
SI	International System of Units
SID	sudden ionospheric disturbance
SIGINT	signals intelligence
SINAD	signal-plus-noise-plus-distortion to noise-plus-distortion ratio
SLD	superluminescent diode
SLI	service logic interpreter
SLP	service logic program
SMDR	station message-detail recording
SMSA	standard metropolitan statistical area
SNR	signal-to-noise ratio
SOH	start-of-heading character
SOM	start of message
sonar	sound navigation and ranging
SONET	synchronous optical network
SOP	standard operating procedure
SOR	start of record
SOW	statement of work
(S+N)/N	signal-plus-noise-to-noise ratio
sr	steradian
S/R	send and receive

SSB	single-sideband (transmission)
SSB-SC	single-sideband suppressed carrier (transmission)
SSN	station serial number
SSP	service switching point
SSUPS	solid-state uninterruptible power system
STALO	stabilized local oscillator
STD	subscriber trunk dialing
STFS	standard time and frequency signal (service); standard time and frequency service
STL	standard telegraph level; studio-to-transmitter link
STP	standard temperature and pressure; signal transfer point
STU	secure telephone unit
STX	start-of-text character
SUB	substitute character
SWR	standing wave ratio
SX	simplex signaling
SXS	step-by-step switching system
SYN	synchronous idle character
SYSGEN	system generation

T

T	tera (10^{12})
TADIL	tactical data information link
TADIL-A	tactical data information link-A
TADIL-B	tactical data information link-B
TADS	teletypewriter automatic dispatch system
TADSS	Tactical Automatic Digital Switching System
TAI	International Atomic Time
TASI	time-assignment speech interpolation
TAT	trans-Atlantic telecommunication (cable)
TC	toll center
TCB	trusted computing base
TCC	telecommunications center
TCCF	Tactical Communications Control Facility
TCF	technical control facility
TCP	transmission control protocol
TCS	trusted computer system
TCU	teletypewriter control unit
TCVXO	temperature compensated-voltage controlled crystal oscillator
TCXO	temperature controlled crystal oscillator
TD	time delay; transmitter distributor
TDD	Telecommunications Device for the Deaf
TDM	time-division multiplexing
TDMA	time-division multiple access
TE	transverse electric [mode]
TED	trunk encryption device
TEK	traffic encryption key
TEM	transverse electric and magnetic [mode]

TEMPEST	compromising emanations
TEMS	telecommunications management system
TGM	trunk group multiplexer
THD	total harmonic distortion
THF	tremendously high frequency
THz	terahertz
TIA	Telecommunications Industry Association
TIE	time interval error
TIFF	tag image file format
TIP	terminal interface processor
T_K	response timer
TLP	transmission level point
TM	transverse magnetic [mode]
TP	toll point
TRANSEC	transmission security
TRC	transverse redundancy check
TRF	tuned radio frequency
TRI-TAC	tri-services tactical [equipment]
TSK	transmission security key
TSP	Telecommunications Service Priority [system]
TSPS	traffic service position system
TSR	telecommunications service request; terminate and stay resident
TTL	transistor-transistor logic
TTTN	tandem tie trunk network
TTY	teletypewriter
TTY/TDD	Telecommunications Device for the Deaf
TV	television
TW	traveling wave
TWT	traveling wave tube
TWTA	traveling wave tube amplifier
TWX®	teletypewriter exchange service
TX	transmit; transmitter

U

UDP	User Datagram Protocol
UHF	ultra high frequency
U/L	uplink
ULF	ultra low frequency
UPS	uninterruptible power supply
UPT	Universal Personal Telecommunications service
USB	upper sideband
USDA	U.S. Department of Agriculture
USFJ	U.S. Forces, Japan
USFK	U.S. Forces, Korea
USNO	U.S. Naval observatory
USTA	U.S. Telephone Association
UT	Universal Time
UTC	Coordinated Universal Time

uv	ultraviolet

V

V	volt
VA	volt-ampere
VAN	value-added network
VAR	value added reseller
VARISTAR	variable resistor
vars	volt-amperes reactive
VC	virtual circuit
VCO	voltage-controlled oscillator
VCXO	voltage-controlled crystal oscillator
V/D	voice/data
Vdc	volts direct current
VDU	video display unit; visual display unit
VF	voice frequency
VFCT	voice frequency carrier telegraph
VFDF	voice frequency distribution frame
VFO	variable-frequency oscillator
VFTG	voice-frequency telegraph
VHF	very high frequency
VLF	very low frequency
V/m	volts per meter
VNL	via net loss
VNLF	via net loss factor
vocoder	voice-coder
vodas	voice-operated device anti-sing
vogad	voice-operated gain-adjusting device
volcas	voice-operated loss control and echo/signaling suppression
vox	voice-operated relay circuit; voice operated transmit
VRC	vertical redundancy check
VSAT	very small aperture terminal
VSB	vestigial sideband [transmission]
VSM	vestigial sideband modulation
VSWR	voltage standing wave ratio
VT	virtual terminal
VTU	video teleconferencing unit
vu	volume unit

W

WADS	wide area data service
WAIS	Wide Area Information Servers
WAN	wide area network
WARC	World Administrative Radio Conference
WATS	Wide Area Telecommunications Service; Wide Area Telephone Service
WAWS	Washington Area Wideband System
WDM	wavelength-division multiplexing

WHSR	White House Situation Room
WIN	WWMCCS Intercomputer Network
WITS	Washington Integrated Telecommunications System
WORM	write once, read many times
wpm	words per minute
wps	words per second
wv	working voltage
WVDC	working voltage direct current
WWDSA	worldwide digital system architecture
WWMCCS	Worldwide Military Command and Control System
WWW	World Wide Web

X

XMIT	transmit
XMSN	transmission
XMTD	transmitted
XMTR	transmitter
XO	crystal oscillator
XOFF	transmitter off
XON	transmitter on
XT	crosstalk
XTAL	crystal

Z

Z	Zulu time
ZD	zero defects
Z_o	characteristic impedance
0TLP	zero transmission level point

Conversion Factors

23.1 Standard Units

Name	Symbol	Quantity
ampere	A	electric current
ampere per meter	A/m	magnetic field strength
ampere per square meter	A/m^2	current density
becquerel	Bg	activity (of a radionuclide)
candela	cd	luminous intensity
coulomb	C	electric charge
coulomb per kilogram	C/kg	exposure (x and gamma rays)
coulomb per sq. meter	C/m^2	electric flux density
cubic meter	m^3	volume
cubic meter per kilogram	m^3/kg	specific volume
degree Celsius	°C	Celsius temperature
farad	F	capacitance
farad per meter	F/m	permittivity
henry	H	inductance
henry per meter	H/m	permeability
hertz	Hz	frequency
joule	J	energy, work, quantity of heat
joule per cubic meter	J/m^3	energy density
joule per kelvin	J/K	heat capacity
joule per kilogram K	J/(kg•K)	specific heat capacity
joule per mole	J/mol	molar energy
kelvin	K	thermodynamic temperature
kilogram	kg	mass
kilogram per cubic meter	kg/m^3	density, mass density
lumen	lm	luminous flux
lux	lx	luminance
meter	m	length
meter per second	m/s	speed, velocity
meter per second sq.	m/s^2	acceleration

mole	mol	amount of substance
newton	N	force
newton per meter	N/m	surface tension
ohm	Ω	electrical resistance
pascal	Pa	pressure, stress
pascal second	Pa•s	dynamic viscosity
radian	rad	plane angle
radian per second	rad/s	angular velocity
radian per second squared	rad/s^2	angular acceleration
second	s	time
siemens	S	electrical conductance
square meter	m^2	area
steradian	sr	solid angle
tesla	T	magnetic flux density
volt	V	electrical potential
volt per meter	V/m	electric field strength
watt	W	power, radiant flux
watt per meter kelvin	W/(m•K)	thermal conductivity
watt per square meter	W/m^2	heat (power) flux density
weber	Wb	magnetic flux

23.2 Standard Prefixes

Multiple	Prefix	Symbol
10^{18}	exa	E
10^{15}	peta	P
10^{12}	tera	T
10^{9}	giga	G
10^{6}	mega	M
10^{3}	kilo	k
10^{2}	hecto	h
10	deka	da
10^{-1}	deci	d
10^{-2}	centi	c
10^{-3}	milli	m
10^{-6}	micro	μ
10^{-9}	nano	n
10^{-12}	pico	p
10^{-15}	femto	f
10^{-18}	atto	a

23.3 Common Standard Units

Unit	Symbol
centimeter	cm
cubic centimeter	cm^3
cubic meter per second	m^3/s
gigahertz	GHz
gram	g
kilohertz	kHz
kilohm	kΩ
kilojoule	kJ
kilometer	km
kilovolt	kV
kilovoltampere	kVA
kilowatt	kW
megahertz	MHz
megavolt	MV
megawatt	MW
megohm	MΩ
microampere	μA
microfarad	μF
microgram	μg
microhenry	μH
microsecond	μs
microwatt	μW
milliampere	mA
milligram	mg
millihenry	mH
millimeter	mm
millisecond	ms
millivolt	mV
milliwatt	mW
nanoampere	nA
nanofarad	nF
nanometer	nm
nanosecond	ns
nanowatt	nW
picoampere	pA
picofarad	pF
picosecond	ps
picowatt	pW

23.4 Conversion Reference Data

A

To Convert	Into	Multiply By
abcoulomb	statcoulombs	2.998×10^{10}
acre	sq. chain (Gunters)	10
acre	rods	160
acre	square links (Gunters)	1×10^5
acre	Hectare or sq. hectometer	0.4047
acre-feet	cubic feet	43,560.0
acre-feet	gallons	3.259×10^5
acres	sq. feet	43,560.0
acres	sq. meters	4,047
acres	sq. miles	1.562×10^{-3}
acres	sq. yards	4,840
ampere-hours	coulombs	3,600.0
ampere-hours	faradays	0.03731
amperes/sq. cm	amps/sq. in	6.452
amperes/sq. cm	amps/sq. meter	10^4
amperes/sq. in	amps/sq. cm	0.1550
amperes/sq. in	amps/sq. meter	1,550.0
amperes/sq. meter	amps/sq. cm	10^{-4}
amperes/sq. meter	amps/sq. in	6.452×10^{-4}
ampere-turns	gilberts	1.257
ampere-turns/cm	amp-turns/in	2.540
ampere-turns/cm	amp-turns/meter	100.0
ampere-turns/cm	gilberts/cm	1.257
ampere-turns/in	amp-turns/cm	0.3937
ampere-turns/in	amp-turns/m	39.37
ampere-turns/in	gilberts/cm	0.4950
ampere-turns/meter	amp-turns/cm	0.01
ampere-turns/meter	amp-turns/in	0.0254
ampere-turns/meter	gilberts/cm	0.01257
Angstrom unit	inch	3937×10^{-9}
Angstrom unit	meter	1×10^{-10}
Angstrom unit	micron or (Mu)	1×10^{-4}
are	acre (U.S.)	0.02471
ares	sq. yards	119.60
ares	acres	0.02471
ares	sq. meters	100.0
astronomical unit	kilometers	1.495×10^8
atmospheres	ton/sq. in	0.007348
atmospheres	cm of mercury	76.0
atmospheres	ft of water (at 4°C)	33.90
atmospheres	in of mercury (at 0°C)	29.92
atmospheres	kg/sq. cm	1.0333
atmospheres	kg/sq. m	10,332

| atmospheres | pounds/sq. in | 14.70 |
| atmospheres | tons/sq. ft | 1.058 |

B

To Convert	Into	Multiply By
barrels (U.S., dry)	cubic inches	7056
barrels (U.S., dry)	quarts (dry)	105.0
barrels (U.S., liquid)	gallons	31.5
barrels (oil)	gallons (oil)	42.0
bars	atmospheres	0.9869
bars	dynes/sq. cm	10^4
bars	kg/sq. m	1.020×10^4
bars	pounds/sq. ft	2,089
bars	pounds/sq. in	14.50
baryl	dyne/sq. cm	1.000
bolt (U.S. cloth)	meters	36.576
Btu	liter-atmosphere	10.409
Btu	ergs	1.0550×10^{10}
Btu	foot-lb	778.3
Btu	gram-calories	252.0
Btu	horsepower-hr	3.931×10^{-4}
Btu	joules	1,054.8
Btu	kilogram-calories	0.2520
Btu	kilogram-meters	107.5
Btu	kilowatt-hr	2.928×10^{-4}
Btu/hr	foot-pounds/s	0.2162
Btu/hr	gram-calories/s	0.0700
Btu/hr	horsepower-hr	3.929×10^{-4}
Btu/hr	watts	0.2931
Btu/min	foot-lbs/s	12.96
Btu/min	horsepower	0.02356
Btu/min	kilowatts	0.01757
Btu/min	watts	17.57
Btu/sq. ft/min	watts/sq. in	0.1221
bucket (br. dry)	cubic cm	1.818×10^4
bushels	cubic ft	1.2445
bushels	cubic in	2,150.4
bushels	cubic m	0.03524
bushels	liters	35.24
bushels	pecks	4.0
bushels	pints (dry)	64.0
bushels	quarts (dry)	32.0

C

To Convert	Into	Multiply By
calories, gram (mean)	Btu (mean)	3.9685×10^{-3}
candle/sq. cm	Lamberts	3.142
candle/sq. in	Lamberts	0.4870
centares (centiares)	sq. meters	1.0
Centigrade	Fahrenheit	(C° x 9/5) + 32

centigrams	grams	0.01
centiliter	ounce fluid (U.S.)	0.3382
centiliter	cubic inch	0.6103
centiliter	drams	2.705
centiliter	liters	0.01
centimeter	feet	3.281×10^{-2}
centimeter	inches	0.3937
centimeter	kilometers	10^{-5}
centimeter	meters	0.01
centimeter	miles	6.214×10^{-6}
centimeter	millimeters	10.0
centimeter	mils	393.7
centimeter	yards	1.094×10^{-2}
centimeter-dynes	cm-grams	1.020×10^{-3}
centimeter-dynes	meter-kg	1.020×10^{-8}
centimeter-dynes	pound-ft	7.376×10^{-8}
centimeter-grams	cm-dynes	980.7
centimeter-grams	meter-kg	10^{-5}
centimeter-grams	pound-ft	7.233×10^{-5}
centimeters of mercury	atmospheres	0.01316
centimeters of mercury	feet of water	0.4461
centimeters of mercury	kg/sq. meter	136.0
centimeters of mercury	pounds/sq. ft	27.85
centimeters of mercury	pounds/sq. in	0.1934
centimeters/sec	feet/min	1.9686
centimeters/sec	feet/sec	0.03281
centimeters/sec	kilometers/hr	0.036
centimeters/sec	knots	0.1943
centimeters/sec	meters/min	0.6
centimeters/sec	miles/hr	0.02237
centimeters/sec	miles/min	3.728×10^{-4}
centimeters/sec/sec	feet/sec/sec	0.03281
centimeters/sec/sec	km/hr/sec	0.036
centimeters/sec/sec	meters/sec/sec	0.01
centimeters/sec/sec	miles/hr/sec	0.02237
chain	inches	792.00
chain	meters	20.12
chains (surveyor's or Gunter's)	yards	22.00
circular mils	sq. cm	5.067×10^{-6}
circular mils	sq. mils	0.7854
circular mils	sq. inches	7.854×10^{-7}
circumference	radians	6.283
cord feet	cubic feet	16
cords	cord feet	8
coulomb	statcoulombs	2.998×10^{9}
coulombs	faradays	1.036×10^{-5}
coulombs/sq. cm	coulombs/sq. in	64.52
coulombs/sq. cm	coulombs/sq. meter	10^{4}
coulombs/sq. in	coulombs/sq. cm	0.1550

coulombs/sq. in	coulombs/sq. meter	1,550
coulombs/sq. meter	coulombs/sq. cm	10^{-4}
coulombs/sq. meter	coulombs/sq. in	6.452×10^{-4}
cubic centimeters	cubic feet	3.531×10^{-5}
cubic centimeters	cubic inches	0.06102
cubic centimeters	cubic meters	10^{-6}
cubic centimeters	cubic yards	1.308×10^{-6}
cubic centimeters	gallons (U.S. liq.)	2.642×10^{-4}
cubic centimeters	liters	0.001
cubic centimeters	pints (U.S. liq.)	2.113×10^{-3}
cubic centimeters	quarts (U.S. liq.)	1.057×10^{-3}
cubic feet	bushels (dry)	0.8036
cubic feet	cubic cm	28,320.0
cubic feet	cubic inches	1,728.0
cubic feet	cubic meters	0.02832
cubic feet	cubic yards	0.03704
cubic feet	gallons (U.S. liq.)	7.48052
cubic feet	liters	28.32
cubic feet	pints (U.S. liq.)	59.84
cubic feet	quarts (U.S. liq.)	29.92
cubic feet/min	cubic cm/sec	472.0
cubic feet/min	gallons/sec	0.1247
cubic feet/min	liters/sec	0.4720
cubic feet/min	pounds of water/min	62.43
cubic feet/sec	million gal/day	0.646317
cubic feet/sec	gallons/min	448.831
cubic inches	cubic cm	16.39
cubic inches	cubic feet	5.787×10^{-4}
cubic inches	cubic meters	1.639×10^{-5}
cubic inches	cubic yards	2.143×10^{-5}
cubic inches	gallons	4.329×10^{-3}
cubic inches	liters	0.01639
cubic inches	mil-feet	1.061×10^{5}
cubic inches	pints (U.S. liq.)	0.03463
cubic inches	quarts (U.S. liq.)	0.01732
cubic meters	bushels (dry)	28.38
cubic meters	cubic cm	10^{6}
cubic meters	cubic feet	35.31
cubic meters	cubic inches	61,023.0
cubic meters	cubic yards	1.308
cubic meters	gallons (U.S. liq.)	264.2
cubic meters	liters	1,000.0
cubic meters	pints (U.S. liq.)	2,113.0
cubic meters	quarts (U.S. liq.)	1,057.
cubic yards	cubic cm	7.646×10^{5}
cubic yards	cubic feet	27.0
cubic yards	cubic inches	46,656.0
cubic yards	cubic meters	0.7646
cubic yards	gallons (U.S. liq.)	202.0
cubic yards	liters	764.6

cubic yards	pints (U.S. liq.)	1,615.9
cubic yards	quarts (U.S. liq.)	807.9
cubic yards/min	cubic ft/sec	0.45
cubic yards/min	gallons/sec	3.367
cubic yards/min	liters/sec	12.74

D

To Convert	Into	Multiply By
Dalton	gram	1.650×10^{-24}
days	seconds	86,400.0
decigrams	grams	0.1
deciliters	liters	0.1
decimeters	meters	0.1
degrees (angle)	quadrants	0.01111
degrees (angle)	radians	0.01745
degrees (angle)	seconds	3,600.0
degrees/sec	radians/sec	0.01745
degrees/sec	revolutions/min	0.1667
degrees/sec	revolutions/sec	2.778×10^{-3}
dekagrams	grams	10.0
dekaliters	liters	10.0
dekameters	meters	10.0
drams (apothecaries or troy)	ounces (avoirdupois)	0.1371429
drams (apothecaries or troy)	ounces (troy)	0.125
drams (U.S., fluid or apothecaries)	cubic cm	3.6967
drams	grams	1.7718
drams	grains	27.3437
drams	ounces	0.0625
dyne/cm	erg/sq. millimeter	0.01
dyne/sq. cm	atmospheres	9.869×10^{-7}
dyne/sq. cm	inch of mercury at 0°C	2.953×10^{-5}
dyne/sq. cm	inch of water at 4°C	4.015×10^{-4}
dynes	grams	1.020×10^{-3}
dynes	joules/cm	10^{-7}
dynes	joules/meter (newtons)	10^{-5}
dynes	kilograms	1.020×10^{-6}
dynes	poundals	7.233×10^{-5}
dynes	pounds	2.248×10^{-6}
dynes/sq. cm	bars	10^{-6}

E

To Convert	Into	Multiply By
ell	cm	114.30
ell	inches	45
em, pica	inch	0.167
em, pica	cm	0.4233
erg/sec	Dyne-cm/sec	1.000

ergs	Btu	9.480×10^{-11}
ergs	dyne-centimeters	1.0
ergs	foot-pounds	7.367×10^{-8}
ergs	gram-calories	0.2389×10^{-7}
ergs	gram-cm	1.020×10^{-3}
ergs	horsepower-hr	3.7250×10^{-14}
ergs	joules	10^{-7}
ergs	kg-calories	2.389×10^{-11}
ergs	kg-meters	1.020×10^{-8}
ergs	kilowatt-hr	0.2778×10^{-13}
ergs	watt-hours	0.2778×10^{-10}
ergs/sec	Btu/min	$5,688 \times 10^{-9}$
ergs/sec	ft-lb/min	4.427×10^{-6}
ergs/sec	ft-lb/sec	7.3756×10^{-8}
ergs/sec	horsepower	1.341×10^{-10}
ergs/sec	kg-calories/min	1.433×10^{-9}
ergs/sec	kilowatts	10^{-10}

F

To Convert	Into	Multiply By
farad	microfarads	10^{6}
Faraday/sec	ampere (absolute)	9.6500×10^{4}
faradays	ampere-hours	26.80
faradays	coulombs	9.649×10^{4}
fathom	meter	1.828804
fathoms	feet	6.0
feet	centimeters	30.48
feet	kilometers	3.048×10^{-4}
feet	meters	0.3048
feet	miles (naut.)	1.645×10^{-4}
feet	miles (stat.)	1.894×10^{-4}
feet	millimeters	304.8
feet	mils	1.2×10^{4}
feet of water	atmospheres	0.02950
feet of water	in of mercury	0.8826
feet of water	kg/sq. cm	0.03048
feet of water	kg/sq. meter	304.8
feet of water	pounds/sq. ft	62.43
feet of water	pounds/sq. in	0.4335
feet/min	cm/sec	0.5080
feet/min	feet/sec	0.01667
feet/min	km/hr	0.01829
feet/min	meters/min	0.3048
feet/min	miles/hr	0.01136
feet/sec	cm/sec	30.48
feet/sec	km/hr	1.097
feet/sec	knots	0.5921
feet/sec	meters/min	18.29
feet/sec	miles/hr	0.6818
feet/sec	miles/min	0.01136

feet/sec/sec	cm/sec/sec	30.48
feet/sec/sec	km/hr/sec	1.097
feet/sec/sec	meters/sec/sec	0.3048
feet/sec/sec	miles/hr/sec	0.6818
feet/100 feet	per centigrade	1.0
foot-candle	lumen/sq. meter	10.764
foot-pounds	Btu	1.286×10^{-3}
foot-pounds	ergs	1.356×10^{7}
foot-pounds	gram-calories	0.3238
foot-pounds	hp-hr	5.050×10^{-7}
foot-pounds	joules	1.356
foot-pounds	kg-calories	3.24×10^{-4}
foot-pounds	kg-meters	0.1383
foot-pounds	kilowatt-hr	3.766×10^{-7}
foot-pounds/min	Btu/min	1.286×10^{-3}
foot-pounds/min	foot-pounds/sec	0.01667
foot-pounds/min	horsepower	3.030×10^{-5}
foot-pounds/min	kg-calories/min	3.24×10^{-4}
foot-pounds/min	kilowatts	2.260×10^{-5}
foot-pounds/sec	Btu/hr	4.6263
foot-pounds/sec	Btu/min	0.07717
foot-pounds/sec	horsepower	1.818×10^{-3}
foot-pounds/sec	kg-calories/min	0.01945
foot-pounds/sec	kilowatts	1.356×10^{-3}
Furlongs	miles (U.S.)	0.125
furlongs	rods	40.0
furlongs	feet	660.0

G

To Convert	Into	Multiply By
gallons	cubic cm	3,785.0
gallons	cubic feet	0.1337
gallons	cubic inches	231.0
gallons	cubic meters	3.785×10^{-3}
gallons	cubic yards	4.951×10^{-3}
gallons	liters	3.785
gallons (liq. Br. Imp.)	gallons (U.S. liq.)	1.20095
gallons (U.S.)	gallons (Imp.)	0.83267
gallons of water	pounds of water	8.3453
gallons/min	cubic ft/sec	2.228×10^{-3}
gallons/min	liters/sec	0.06308
gallons/min	cubic ft/hr	8.0208
gausses	lines/sq. in	6.452
gausses	webers/sq. cm	10^{-8}
gausses	webers/sq. in	6.452×10^{-8}
gausses	webers/sq. meter	10^{-4}
gilberts	ampere-turns	0.7958
gilberts/cm	amp-turns/cm	0.7958
gilberts/cm	amp-turns/in	2.021
gilberts/cm	amp-turns/meter	79.58

gills	liters	0.1183
gills	pints (liq.)	0.25
gills (British)	cubic cm	142.07
grade	radian	0.01571
grains	drams (avoirdupois)	0.03657143
grains (troy)	grains (avdp.)	1.0
grains (troy)	grams	0.06480
grains (troy)	ounces (avdp.)	2.0833×10^{-3}
grains (troy)	pennyweight (troy)	0.04167
grains/Imp. gal	parts/million	14.286
grains/U.S. gal	parts/million	17.118
grains/U.S. gal	pounds/million gal	142.86
gram-calories	Btu	3.9683×10^{-3}
gram-calories	ergs	4.1868×10^{7}
gram-calories	foot-pounds	3.0880
gram-calories	horsepower-hr	1.5596×10^{-6}
gram-calories	kilowatt-hr	1.1630×10^{-6}
gram-calories	watt-hr	1.1630×10^{-3}
gram-calories/sec	Btu/hr	14.286
gram-centimeters	Btu	9.297×10^{-8}
gram-centimeters	ergs	980.7
gram-centimeters	joules	9.807×10^{-5}
gram-centimeters	kg-calories	2.343×10^{-8}
gram-centimeters	kg-meters	10^{-5}
grams	dynes	980.7
grams	grains	15.43
grams	joules/cm	9.807×10^{-5}
grams	joules/meter (newtons)	9.807×10^{-3}
grams	kilograms	0.001
grams	milligrams	1,000
grams	ounces (avdp.)	0.03527
grams	ounces (troy)	0.03215
grams	poundals	0.07093
grams	pounds	2.205×10^{-3}
grams/cm	pounds/inch	5.600×10^{-3}
grams/cubic cm	pounds/cubic ft	62.43
grams/cubic cm	pounds/cubic in	0.03613
grams/cubic cm	pounds/mil-foot	3.405×10^{-7}
grams/liter	grains/gal	58.417
grams/liter	pounds/1,000 gal	8.345
grams/liter	pounds/cubic ft	0.062427
grams/liter	parts/million	1,000.0
grams/sq. cm	pounds/sq. ft	2.0481

H

To Convert	Into	Multiply By
hand	cm	10.16
hectares	acres	2.471
hectares	sq. feet	1.076×10^{5}
hectograms	grams	100.0

hectoliters	liters	100.0
hectometers	meters	100.0
hectowatts	watts	100.0
henries	millihenries	1,000.0
horsepower	Btu/min	42.44
horsepower	foot-lb/min	33,000
horsepower	foot-lb/sec	550.0
horsepower	kg-calories/min	10.68
horsepower	kilowatts	0.7457
horsepower	watts	745.7
horsepower (boiler)	Btu/hr	33.479
horsepower (boiler)	kilowatts	9.803
horsepower (metric) (542.5 ft lb./sec)	horsepower (550 ft lb./sec)	0.9863
horsepower (550 ft lb./sec)	horsepower (metric) (542.5 ft lb./sec)	1.014
horsepower-hr	Btu	2,547
horsepower-hr	ergs	2.6845×10^{13}
horsepower-hr	foot-lb	1.98×10^{6}
horsepower-hr	gram-calories	641,190
horsepower-hr	joules	2.684×10^{6}
horsepower-hr	kg-calories	641.1
horsepower-hr	kg-meters	2.737×10^{5}
horsepower-hr	kilowatt-hr	0.7457
hours	days	4.167×10^{-2}
hours	weeks	5.952×10^{-3}
hundredweights (long)	pounds	112
hundredweights (long)	tons (long)	0.05
hundredweights (short)	ounces (avoirdupois)	1,600
hundredweights (short)	pounds	100
hundredweights (short)	tons (metric)	0.0453592
hundredweights (short)	tons (long)	0.0446429

I

To Convert	Into	Multiply By
inches	centimeters	2.540
inches	meters	2.540×10^{-2}
inches	miles	1.578×10^{-5}
inches	millimeters	25.40
inches	mils	1,000.0
inches	yards	2.778×10^{-2}
inches of mercury	atmospheres	0.03342
inches of mercury	feet of water	1.133
inches of mercury	kg/sq. cm	0.03453
inches of mercury	kg/sq. meter	345.3
inches of mercury	pounds/sq. ft	70.73
inches of mercury	pounds/sq. in	0.4912
inches of water (at 4°C)	atmospheres	2.458×10^{-3}
inches of water (at 4°C)	inches of mercury	0.07355
inches of water (at 4°C)	kg/sq. cm	2.540×10^{-3}

inches of water (at 4°C)	ounces/sq. in	0.5781
inches of water (at 4°C)	pounds/sq. ft	5.204
inches of water (at 4°C)	pounds/sq. in	0.03613
international ampere	ampere (absolute)	0.9998
international Volt	volts (absolute)	1.0003
international volt	joules (absolute)	1.593×10^{-19}
international volt	joules	9.654×10^{4}

J

To Convert	Into	Multiply By
joules	Btu	9.480×10^{-4}
joules	ergs	10^{7}
joules	foot-pounds	0.7376
joules	kg-calories	2.389×10^{-4}
joules	kg-meters	0.1020
joules	watt-hr	2.778×10^{-4}
joules/cm	grams	1.020×10^{4}
joules/cm	dynes	10^{7}
joules/cm	joules/meter (newtons)	100.0
joules/cm	poundals	723.3
joules/cm	pounds	22.48

K

To Convert	Into	Multiply By
kilogram-calories	Btu	3.968
kilogram-calories	foot-pounds	3,088
kilogram-calories	hp-hr	1.560×10^{-3}
kilogram-calories	joules	4,186
kilogram-calories	kg-meters	426.9
kilogram-calories	kilojoules	4.186
kilogram-calories	kilowatt-hr	1.163×10^{-3}
kilogram meters	Btu	9.294×10^{-3}
kilogram meters	ergs	9.804×10^{7}
kilogram meters	foot-pounds	7.233
kilogram meters	joules	9.804
kilogram meters	kg-calories	2.342×10^{-3}
kilogram meters	kilowatt-hr	2.723×10^{-6}
kilograms	dynes	980,665
kilograms	grams	1,000.0
kilograms	joules/cm	0.09807
kilograms	joules/meter (newtons)	9.807
kilograms	poundals	70.93
kilograms	pounds	2.205
kilograms	tons (long)	9.842×10^{-4}
kilograms	tons (short)	1.102×10^{-3}
kilograms/cubic meter	grams/cubic cm	0.001
kilograms/cubic meter	pounds/cubic ft	0.06243
kilograms/cubic meter	pounds/cubic in	3.613×10^{-5}
kilograms/cubic meter	pounds/mil-foot	3.405×10^{-10}
kilograms/meter	pounds/ft	0.6720

kilograms/sq. cm	dynes	980,665
kilograms/sq. cm	atmospheres	0.9678
kilograms/sq. cm	feet of water	32.81
kilograms/sq. cm	inches of mercury	28.96
kilograms/sq. cm	pounds/sq. ft	2,048
kilograms/sq. cm	pounds/sq. in	14.22
kilograms/sq. meter	atmospheres	9.678×10^{-5}
kilograms/sq. meter	bars	98.07×10^{-6}
kilograms/sq. meter	feet of water	3.281×10^{-3}
kilograms/sq. meter	inches of mercury	2.896×10^{-3}
kilograms/sq. meter	pounds/sq. ft	0.2048
kilograms/sq. meter	pounds/sq. in	1.422×10^{-3}
kilograms/sq. mm	kg/sq. meter	10^{6}
kilolines	maxwells	1,000.0
kiloliters	liters	1,000.0
kilometers	centimeters	10^{5}
kilometers	feet	3,281
kilometers	inches	3.937×10^{4}
kilometers	meters	1,000.0
kilometers	miles	0.6214
kilometers	millimeters	10^{4}
kilometers	yards	1,094
kilometers/hr	cm/sec	27.78
kilometers/hr	feet/min	54.68
kilometers/hr	feet/sec	0.9113
kilometers/hr	knots	0.5396
kilometers/hr	meters/min	16.67
kilometers/hr	miles/hr	0.6214
kilometers/hr/sec	cm/sec/sec	27.78
kilometers/hr/sec	feet/sec/sec	0.9113
kilometers/hr/sec	meters/sec/sec	0.2778
kilometers/hr/sec	miles/hr/sec	0.6214
kilowatt-hr	Btu	3,413
kilowatt-hr	ergs	3.600×10^{13}
kilowatt-hr	foot-lb	2.655×10^{6}
kilowatt-hr	gram-calories	859,850
kilowatt-hr	horsepower-hr	1.341
kilowatt-hr	joules	3.6×10^{6}
kilowatt-hr	kg-calories	860.5
kilowatt-hr	kg-meters	3.671×10^{5}
kilowatt-hr	pounds of water raised from 62° to 212°F	22.75
kilowatts	Btu/min	56.92
kilowatts	foot-lb/min	4.426×10^{4}
kilowatts	foot-lb/sec	737.6
kilowatts	horsepower	1.341
kilowatts	kg-calories/min	14.34
kilowatts	watts	1,000.0
knots	feet/hr	6,080
knots	kilometers/hr	1.8532

knots	nautical miles/hr	1.0
knots	statute miles/hr	1.151
knots	yards/hr	2,027
knots	feet/sec	1.689

L

To Convert	Into	Multiply By
league	miles (approx.)	3.0
light year	miles	5.9×10^{12}
light year	kilometers	9.4637×10^{12}
lines/sq. cm	gausses	1.0
lines/sq. in	gausses	0.1550
lines/sq. in	webers/sq. cm	1.550×10^{-9}
lines/sq. in	webers/sq. in	10^{-8}
lines/sq. in	webers/sq. meter	1.550×10^{-5}
links (engineer's)	inches	12.0
links (surveyor's)	inches	7.92
liters	bushels (U.S. dry)	0.02838
liters	cubic cm	1,000.0
liters	cubic feet	0.03531
liters	cubic inches	61.02
liters	cubic meters	0.001
liters	cubic yards	1.308×10^{-3}
liters	gallons (U.S. liq.)	0.2642
liters	pints (U.S. liq.)	2.113
liters	quarts (U.S. liq.)	1.057
liters/min	cubic ft/sec	5.886×10^{-4}
liters/min	gal/sec	4.403×10^{-3}
lumen	spherical candle power	0.07958
lumen	watt	0.001496
lumens/sq. ft	foot-candles	1.0
lumens/sq. ft	lumen/sq. meter	10.76
lux	foot-candles	0.0929

M

To Convert	Into	Multiply By
maxwells	kilolines	0.001
maxwells	webers	10^{-8}
megalines	maxwells	10^{6}
megohms	microhms	10^{12}
megohms	ohms	10^{6}
meter-kilograms	cm-dynes	9.807×10^{7}
meter-kilograms	cm-grams	10^{5}
meter-kilograms	pound-feet	7.233
meters	centimeters	100.0
meters	feet	3.281
meters	inches	39.37
meters	kilometers	0.001
meters	miles (naut.)	5.396×10^{-4}
meters	miles (stat.)	6.214×10^{-4}

meters	millimeters	1,000.0
meters	yards	1.094
meters	varas	1.179
meters/min	cm/sec	1,667
meters/min	feet/min	3.281
meters/min	feet/sec	0.05468
meters/min	km/hr	0.06
meters/min	knots	0.03238
meters/min	miles/hr	0.03728
meters/sec	feet/min	196.8
meters/sec	feet/sec	3.281
meters/sec	kilometers/hr	3.6
meters/sec	kilometers/min	0.06
meters/sec	miles/hr	2.237
meters/sec	miles/min	0.03728
meters/sec/sec	cm/sec/sec	100.0
meters/sec/sec	ft/sec/sec	3.281
meters/sec/sec	km/hr/sec	3.6
meters/sec/sec	miles/hr/sec	2.237
microfarad	farads	10^{-6}
micrograms	grams	10^{-6}
microhms	megohms	10^{-12}
microhms	ohms	10^{-6}
microliters	liters	10^{-6}
microns	meters	1×10^{-6}
miles (naut.)	feet	6,080.27
miles (naut.)	kilometers	1.853
miles (naut.)	meters	1,853
miles (naut.)	miles (statute)	1.1516
miles (naut.)	yards	2,027
miles (statute)	centimeters	1.609×10^{5}
miles (statute)	feet	5,280
miles (statute)	inches	6.336×10^{4}
miles (statute)	kilometers	1.609
miles (statute)	meters	1,609
miles (statute)	miles (naut.)	0.8684
miles (statute)	yards	1,760
miles/hr	cm/sec	44.70
miles/hr	feet/min	88
miles/hr	feet/sec	1.467
miles/hr	km/hr	1.609
miles/hr	km/min	0.02682
miles/hr	knots	0.8684
miles/hr	meters/min	26.82
miles/hr	miles/min	0.1667
miles/hr/sec	cm/sec/sec	44.70
miles/hr/sec	feet/sec/sec	1.467
miles/hr/sec	km/hr/sec	1.609
miles/hr/sec	meters/sec/sec	0.4470
miles/min	cm/sec	2,682

miles/min	feet/sec	88
miles/min	km/min	1.609
miles/min	knots/min	0.8684
miles/min	miles/hr	60
mil-feet	cubic inches	9.425×10^{-6}
milliers	kilograms	1,000
milligrams	grains	0.01543236
milligrams	grams	0.001
milligrams/liter	parts/million	1.0
millihenries	henries	0.001
milliliters	liters	0.001
millimeters	centimeters	0.1
millimeters	feet	3.281×10^{-3}
millimeters	inches	0.03937
millimeters	kilometers	10^{-6}
millimeters	meters	0.001
millimeters	miles	6.214×10^{-7}
millimeters	mils	39.37
millimeters	yards	1.094×10^{-3}
millimicrons	meters	1×10^{-9}
million gal/day	cubic ft/sec	1.54723
mils	centimeters	2.540×10^{-3}
mils	feet	8.333×10^{-5}
mils	inches	0.001
mils	kilometers	2.540×10^{-8}
mils	yards	2.778×10^{-5}
miner's inches	cubic ft/min	1.5
minims (British)	cubic cm	0.059192
minims (U.S., fluid)	cubic cm	0.061612
minutes (angles)	degrees	0.01667
minutes (angles)	quadrants	1.852×10^{-4}
minutes (angles)	radians	2.909×10^{-4}
minutes (angles)	seconds	60.0
myriagrams	kilograms	10.0
myriameters	kilometers	10.0
myriawatts	kilowatts	10.0

N

To Convert	Into	Multiply By
nepers	decibels	8.686
Newton	dynes	1×105

O

To Convert	Into	Multiply By
ohm (international)	ohm (absolute)	1.0005
ohms	megohms	10^{-6}
ohms	microhms	10^{6}
ounces	drams	16.0
ounces	grains	437.5
ounces	grams	28.349527

ounces	pounds	0.0625
ounces	ounces (troy)	0.9115
ounces	tons (long)	2.790×10^{-5}
ounces	tons (metric)	2.835×10^{-5}
ounces (fluid)	cubic inches	1.805
ounces (fluid)	liters	0.02957
ounces (troy)	grains	480.0
ounces (troy)	grams	31.103481
ounces (troy)	ounces (avdp.)	1.09714
ounces (troy)	pennyweights (troy)	20.0
ounces (troy)	pounds (troy)	0.08333
ounces/sq. inch	dynes/sq. cm	4,309
ounces/sq. in	pounds/sq. in	0.0625

P

To Convert	Into	Multiply By
parsec	miles	19×10^{12}
parsec	kilometers	3.084×10^{13}
parts/million	grains/U.S. gal	0.0584
parts/million	grains/Imp. gal	0.07016
parts/million	pounds/million gal	8.345
pecks (British)	cubic inches	554.6
pecks (British)	liters	9.091901
pecks (U.S.)	bushels	0.25.
pecks (U.S.)	cubic inches	537.605
pecks (U.S.)	liters	8.809582
pecks (U.S.)	quarts (dry)	8
pennyweights (troy)	grains	24.0
pennyweights (troy)	ounces (troy)	0.05
pennyweights (troy)	grams	1.55517
pennyweights (troy)	pounds (troy)	4.1667×10^{-3}
pints (dry)	cubic inches	33.60
pints (liq.)	cubic cm	473.2
pints (liq.)	cubic feet	0.01671
pints (liq.)	cubic inches	28.87
pints (liq.)	cubic meters	4.732×10^{-4}
pints (liq.)	cubic yards	6.189×10^{-4}
pints (liq.)	gallons	0.125
pints (liq.)	liters	0.4732
pints (liq.)	quarts (liq.)	0.5
Planck's quantum	erg - second	6.624×10^{-27}
poise	gram/cm sec	1.00
poundals	dynes	13,826
poundals	grams	14.10
poundals	joules/cm	1.383×10^{-3}
poundals	joules/meter (newtons)	0.1383
poundals	kilograms	0.01410
poundals	pounds	0.03108
pound-feet	cm-dynes	1.356×10^{7}
pound-feet	cm-grams	13,825

pound-feet	meter-kg	0.1383
pounds	drams	256
pounds	dynes	44.4823 x 10^4
pounds	grains	7,000
pounds	grams	453.5924
pounds	joules/cm	0.04448
pounds	joules/meter (newtons)	4.448
pounds	kilograms	0.4536
pounds	ounces	16.0
pounds	ounces (troy)	14.5833
pounds	poundals	32.17
pounds	pounds (troy)	1.21528
pounds	tons (short)	0.0005
pounds (avoirdupois)	ounces (troy)	14.5833
pounds (troy)	grains	5,760
pounds (troy)	grams	373.24177
pounds (troy)	ounces (avdp.)	13.1657
pounds (troy)	ounces (troy)	12.0
pounds (troy)	pennyweights (troy)	240.0
pounds (troy)	pounds (avdp.)	0.822857
pounds (troy)	tons (long)	3.6735 x 10^{-4}
pounds (troy)	tons (metric)	3.7324 x 10^{-4}
pounds (troy)	tons (short)	4.1143 x 10^{-4}
pounds of water	cubic ft	0.01602
pounds of water	cubic inches	27.68
pounds of water	gallons	0.1198
pounds of water/min	cubic ft/sec	2.670 x 10^{-4}
pounds/cubic ft	grams/cubic cm	0.01602
pounds/cubic ft	kg/cubic meter	16.02
pounds/cubic ft	pounds/cubic in	5.787 x 10^{-4}
pounds/cubic ft	pounds/mil-foot	5.456 x 10^{-9}
pounds/cubic in	gm/cubic cm	27.68
pounds/cubic in	kg/cubic meter	2.768 x 10^4
pounds/cubic in	pounds/cubic ft	1,728
pounds/cubic in	pounds/mil-foot	9.425 x 10^{-6}
pounds/ft	kg/meter	1.488
pounds/in	gm/cm	178.6
pounds/mil-foot	gm/cubic cm	2.306 x 10^6
pounds/sq. ft	atmospheres	4.725 x 10^{-4}
pounds/sq. ft	feet of water	0.01602
pounds/sq. ft	inches of mercury	0.01414
pounds/sq. ft	kg/sq. meter	4.882
pounds/sq. ft	pounds/sq. in	6.944 x 10^{-3}
pounds/sq. in	atmospheres	0.06804
pounds/sq. in	feet of water	2.307
pounds/sq. in	inches of mercury	2.036
pounds/sq. in	kg/sq. meter	703.1
pounds/sq. in	pounds/sq. ft	144.0

Q

To Convert	Into	Multiply By
quadrants (angle)	degrees	90.0
quadrants (angle)	minutes	5,400.0
quadrants (angle)	radians	1.571
quadrants (angle)	seconds	3.24×10^5
quarts (dry)	cubic inches	67.20
quarts (liq.)	cubic cm	946.4
quarts (liq.)	cubic feet	0.03342
quarts (liq.)	cubic inches	57.75
quarts (liq.)	cubic meters	9.464×10^{-4}
quarts (liq.)	cubic yards	1.238×10^{-3}
quarts (liq.)	gallons	0.25
quarts (liq.)	liters	0.9463

R

To Convert	Into	Multiply By
radians	degrees	57.30
radians	minutes	3,438
radians	quadrants	0.6366
radians	seconds	2.063×10^5
radians/sec	degrees/sec	57.30
radians/sec	revolutions/min	9.549
radians/sec	revolutions/sec	0.1592
radians/sec/sec	revolutions/min/min	573.0
radians/sec/sec	revolutions/min/sec	9.549
radians/sec/sec	revolutions/sec/sec	0.1592
revolutions	degrees	360.0
revolutions	quadrants	4.0
revolutions	radians	6.283
revolutions/min	degrees/sec	6.0
revolutions/min	radians/sec	0.1047
revolutions/min	revolutions/sec	0.01667
revolutions/min/min	radians/sec/sec	1.745×10^{-3}
revolutions/min/min	revolutions/min/sec	0.01667
revolutions/min/min	revolutions/sec/sec	2.778×10^{-4}
revolutions/sec	degrees/sec	360.0
revolutions/sec	radians/sec	6.283
revolutions/sec	revolutions/min	60.0
revolutions/sec/sec	radians/sec/sec	6.283
revolutions/sec/sec	revolutions/min/min	3,600.0
revolutions/sec/sec	revolutions/min/sec	60.0
rod	chain (Gunters)	0.25
rod	meters	5.029
rods	feet	16.5
rods (surveyors' meas.)	yards	5.5

S

To Convert	Into	Multiply By
scruples	grains	20
seconds (angle)	degrees	2.778×10^{-4}
seconds (angle)	minutes	0.01667
seconds (angle)	quadrants	3.087×10^{-6}
seconds (angle)	radians	4.848×10^{-6}
slug	kilogram	14.59
slug	pounds	32.17
sphere	steradians	12.57
square centimeters	circular mils	1.973×10^{5}
square centimeters	sq. feet	1.076×10^{-3}
square centimeters	sq. inches	0.1550
square centimeters	sq. meters	0.0001
square centimeters	sq. miles	3.861×10^{-11}
square centimeters	sq. millimeters	100.0
square centimeters	sq. yards	1.196×10^{-4}
square feet	acres	2.296×10^{-5}
square feet	circular mils	1.833×10^{8}
square feet	sq. cm	929.0
square feet	sq. inches	144.0
square feet	sq. meters	0.09290
square feet	sq. miles	3.587×10^{-8}
square feet	sq. millimeters	9.290×10^{4}
square feet	sq. yards	0.1111
square inches	circular mils	1.273×10^{6}
square inches	sq. cm	6.452
square inches	sq. feet	6.944×10^{-3}
square inches	sq. millimeters	645.2
square inches	sq. mils	10^{6}
square inches	sq. yards	7.716×10^{-4}
square kilometers	acres	247.1
square kilometers	sq. cm	10^{10}
square kilometers	sq. ft	10.76×10^{6}
square kilometers	sq. inches	1.550×10^{9}
square kilometers	sq. meters	10^{6}
square kilometers	sq. miles	0.3861
square kilometers	sq. yards	1.196×10^{6}
square meters	acres	2.471×10^{-4}
square meters	sq. cm	10^{4}
square meters	sq. feet	10.76
square meters	sq. inches	1,550
square meters	sq. miles	3.861×10^{-7}
square meters	sq. millimeters	10^{6}
square meters	sq. yards	1.196
square miles	acres	640.0
square miles	sq. feet	27.88×10^{6}
square miles	sq. km	2.590
square miles	sq. meters	2.590×10^{6}
square miles	sq. yards	3.098×10^{6}

square millimeters	circular mils	1,973
square millimeters	sq. cm	0.01
square millimeters	sq. feet	1.076×10^{-5}
square millimeters	sq. inches	1.550×10^{-3}
square mils	circular mils	1.273
square mils	sq. cm	6.452×10^{-6}
square mils	sq. inches	10^{-6}
square yards	acres	2.066×10^{-4}
square yards	sq. cm	8,361
square yards	sq. feet	9.0
square yards	sq. inches	1,296
square yards	sq. meters	0.8361
square yards	sq. miles	3.228×10^{-7}
square yards	sq. millimeters	8.361×10^{5}

T

To Convert	Into	Multiply By
temperature (°C)+273	absolute temperature (°C)	1.0
temperature (°C)+17.78	temperature (°F)	1.8
temperature (°F)+460	absolute temperature (°F)	1.0
temperature (°F)−32	temperature (°C)	5/9
tons (long)	kilograms	1,016
tons (long)	pounds	2,240
tons (long)	tons (short)	1.120
tons (metric)	kilograms	1,000
tons (metric)	pounds	2,205
tons (short)	kilograms	907.1848
tons (short)	ounces	32,000
tons (short)	ounces (troy)	29,166.66
tons (short)	pounds	2,000
tons (short)	pounds (troy)	2,430.56
tons (short)	tons (long)	0.89287
tons (short)	tons (metric)	0.9078
tons (short)/sq. ft	kg/sq. meter	9,765
tons (short)/sq. ft	pounds/sq. in	2,000
tons of water/24 hr	pounds of water/hr	83.333
tons of water/24 hr	gallons/min	0.16643
tons of water/24 hr	cubic ft/hr	1.3349

V

To Convert	Into	Multiply By
volt (absolute)	statvolts	0.003336
volt/inch	volt/cm	0.39370

W

To Convert	Into	Multiply By
watt-hours	Btu	3.413
watt-hours	ergs	3.60×10^{10}
watt-hours	foot-pounds	2,656
watt-hours	gram-calories	859.85

watt-hours	horsepower-hr	1.341×10^{-3}
watt-hours	kilogram-calories	0.8605
watt-hours	kilogram-meters	367.2
watt-hours	kilowatt-hr	0.001
watt (international)	watt (absolute)	1.0002
watts	Btu/hr	3.4129
watts	Btu/min	0.05688
watts	ergs/sec	107
watts	foot-lb/min	44.27
watts	foot-lb/sec	0.7378
watts	horsepower	1.341×10^{-3}
watts	horsepower (metric)	1.360×10^{-3}
watts	kg-calories/min	0.01433
watts	kilowatts	0.001
watts (Abs.)	Btu (mean)/min	0.056884
watts (Abs.)	joules/sec	1
webers	maxwells	10^{8}
webers	kilolines	10^{5}
webers/sq. in	gausses	1.550×10^{7}
webers/sq. in	lines/sq. in	10^{8}
webers/sq. in	webers/sq. cm	0.1550
webers/sq. in	webers/sq. meter	1,550
webers/sq. meter	gausses	10^{4}
webers/sq. meter	lines/sq. in	6.452×10^{4}
webers/sq. meter	webers/sq. cm	10^{-4}
webers/sq. meter	webers/sq. in	6.452×10^{-4}

Y

To Convert	**Into**	**Multiply By**
yards	centimeters	91.44
yards	kilometers	9.144×10^{-4}
yards	meters	0.9144
yards	miles (naut.)	4.934×10^{-4}
yards	miles (stat.)	5.682×10^{-4}
yards	millimeters	914.4

23.5 Reference Tables

Table 23.1 Power Conversion Factors (decibels to watts)

dBm	dBw	Watts	Multiple	Prefix
+150	+120	1,000,000,000,000	10^{12}	1 Terawatt
+140	+110	100,000,000,000	10^{11}	100 Gigawatts
+130	+100	10,000,000,000	10^{10}	10 Gigawatts
+120	+90	1,000,000,000	10^{9}	1 Gigawatt
+110	+80	100,000,000	10^{8}	100 Megawatts
+100	+70	10,000,000	10^{7}	10 Megawatts
+90	+60	1,000,000	10^{6}	1 Megawatt
+80	+50	100,000	10^{5}	100 Kilowatts
+70	+40	10,000	10^{4}	10 Kilowatts
+60	+30	1,000	10^{3}	1 Kilowatt
+50	+20	100	10^{2}	1 Hectrowatt
+40	+10	10	10	1 Decawatt
+30	0	1	1	1 Watt
+20	−10	0.1	10^{-1}	1 Deciwatt
+10	−20	0.01	10^{-2}	1 Centiwatt
0	−30	0.001	10^{-3}	1 Milliwatt
−10	−40	0.0001	10^{-4}	100 Microwatts
−20	−50	0.00001	10^{-5}	10 Microwatts
−30	−60	0.000,001	10^{-6}	1 Microwatt
−40	−70	0.0,000,001	10^{-7}	100 Nanowatts
−50	−80	0.00,000,001	10^{-8}	10 Nanowatts
−60	−90	0.000,000,001	10^{-9}	1 Nanowatt
−70	−100	0.0,000,000,001	10^{-10}	100 Picowatts
−80	−110	0.00,000,000,001	10^{-11}	10 Picowatts
−90	−120	0.000,000,000,001	10^{-12}	1 Picowatt

Table 23.2 Specifications of Standard Copper Wire Sizes

Wire Size AWG	Diam. in mils	Circular mil Area	Turns per Linear Inch[1] Enam.	SCE	DCC	Ohms per 100 ft[2]	Current Carrying Capacity[3]	Diam. in mm
1	289.3	83810	-	-	-	0.1239	119.6	7.348
2	257.6	05370	-	-	-	0.1563	94.8	6.544
3	229.4	62640	-	-	-	0.1970	75.2	5.827
4	204.3	41740	-	-	-	0.2485	59.6	5.189
5	181.9	33100	-	-	-	0.3133	47.3	4.621
6	162.0	26250	-	-	-	0.3951	37.5	4.115
7	144.3	20820	-	-	-	0.4982	29.7	3.665
8	128.5	16510	7.6	-	7.1	0.6282	23.6	3.264
9	114.4	13090	8.6	-	7.8	0.7921	18.7	2.906
10	101.9	10380	9.6	9.1	8.9	0.9989	14.8	2.588
11	90.7	8234	10.7	-	9.8	1.26	11.8	2.305
12	80.8	6530	12.0	11.3	10.9	1.588	9.33	2.063
13	72.0	5178	13.5	-	12.8	2.003	7.40	1.828
14	64.1	4107	15.0	14.0	13.8	2.525	5.87	1.628
15	57.1	3257	16.8	-	14.7	3.184	4.65	1.450
16	50.8	2583	18.9	17.3	16.4	4.016	3.69	1.291
17	45.3	2048	21.2	-	18.1	5.064	2.93	1.150
18	40.3	1624	23.6	21.2	19.8	6.386	2.32	1.024
19	35.9	1288	26.4	-	21.8	8.051	1.84	0.912
20	32.0	1022	29.4	25.8	23.8	10.15	1.46	0.812
21	28.5	810	33.1	-	26.0	12.8	1.16	0.723
22	25.3	642	37.0	31.3	30.0	16.14	0.918	0.644
23	22.6	510	41.3	-	37.6	20.36	0.728	0.573
24	20.1	404	46.3	37.6	35.6	25.67	0.577	0.511
25	17.9	320	51.7	-	38.6	32.37	0.458	0.455
26	15.9	254	58.0	46.1	41.8	40.81	0.363	0.406
27	14.2	202	64.9	-	45.0	51.47	0.288	0.361
28	12.6	160	72.7	54.6	48.5	64.9	0.228	0.321
29	11.3	127	81.6	-	51.8	81.83	0.181	0.286
30	10.0	101	90.5	64.1	55.5	103.2	0.144	0.255
31	8.9	50	101	-	59.2	130.1	0.114	0.227
32	8.0	63	113	74.1	61.6	164.1	0.090	0.202
33	7.1	50	127	-	66.3	206.9	0.072	0.180
34	6.3	40	143	86.2	70.0	260.9	0.057	0.160
35	5.6	32	158	-	73.5	329.0	0.045	0.143
36	5.0	25	175	103.1	T7.0	414.8	0.036	0.127
37	4.5	20	198	-	80.3	523.1	0.028	0.113
38	4.0	16	224	116.3	83.6	659.6	0.022	0.101
39	3.5	12	248	-	86.6	831.8	0.018	0.090

[1] Based on 25.4 mm.
[2] Ohms per 1,000 ft measured at 20°C.
[3] Current-carrying capacity at 700 cm/amp.

Table 23.3 Celsius-to-Fahrenheit Conversion Table

°Celsius	°Fahrenheit	°Celsius	°Fahrenheit
−50	−58	125	257
−45	−49	130	266
−40	−40	135	275
−35	−31	140	284
−30	−22	145	293
−25	−13	150	302
−20	4	155	311
−15	5	160	320
−10	14	165	329
−5	23	170	338
0	32	175	347
5	41	180	356
10	50	185	365
15	59	190	374
20	68	195	383
25	77	200	392
30	86	205	401
35	95	210	410
40	104	215	419
45	113	220	428
50	122	225	437
55	131	230	446
60	140	235	455
65	149	240	464
70	158	245	473
75	167	250	482
80	176	255	491
85	185	260	500
90	194	265	509
95	203	270	518
100	212	275	527
105	221	280	536
110	230	285	545
115	239	290	554
120	248	295	563

Table 23.4 Inch-to-Millimeter Conversion Table

Inch	0	1/8	1/4	3/8	1/2	5/8	3/4	7/8	Inch
0	0.0	3.18	6.35	9.52	12.70	15.88	19.05	22.22	0
1	25.40	28.58	31.75	34.92	38.10	41.28	44.45	47.62	1
2	50.80	53.98	57.15	60.32	63.50	66.68	69.85	73.02	2
3	76.20	79.38	82.55	85.72	88.90	92.08	95.25	98.42	3
4	101.6	104.8	108.0	111.1	114.3	117.5	120.6	123.8	4
5	127.0	130.2	133.4	136.5	139.7	142.9	146.0	149.2	5
6	152.4	155.6	158.8	161.9	165.1	168.3	171.4	174.6	6
7	177.8	181.0	184.2	187.3	190.5	193.7	196.8	200.0	7
8	203.2	206.4	209.6	212.7	215.9	219.1	222.2	225.4	8
9	228.6	231.8	235.0	238.1	241.3	244.5	247.6	250.8	9
10	254.0	257.2	260.4	263.5	266.7	269.9	273.0	276.2	10
11	279	283	286	289	292	295	298	302	11
12	305	308	311	314	317	321	324	327	12
13	330	333	337	340	343	346	349	352	13
14	356	359	362	365	368	371	375	378	14
15	381	384	387	391	394	397	400	403	15
16	406	410	413	416	419	422	425	429	16
17	432	435	438	441	445	448	451	454	17
18	457	460	464	467	470	473	476	479	18
19	483	486	489	492	495	498	502	505	19
20	508	511	514	518	521	524	527	530	20

Table 23.5 Conversion of Millimeters to Decimal Inches

mm	Inches	mm	Inches	mm	Inches
1	0.039370	46	1.811020	91	3.582670
2	0.078740	47	1.850390	92	3.622040
3	0.118110	48	1.889760	93	3.661410
4	0.157480	49	1.929130	94	3.700780
5	0.196850	50	1.968500	95	3.740150
6	0.236220	51	2.007870	96	3.779520
7	0.275590	52	2.047240	97	3.818890
8	0.314960	53	2.086610	98	3.858260
9	0.354330	54	2.125980	99	3.897630
10	0.393700	55	2.165350	100	3.937000
11	0.433070	56	2.204720	105	4.133848
12	0.472440	57	2.244090	110	4.330700
13	0.511810	58	2.283460	115	4.527550
14	0.551180	59	2.322830	120	4.724400
15	0.590550	60	2.362200	125	4.921250
16	0.629920	61	2.401570	210	8.267700
17	0.669290	62	2.440940	220	8.661400
18	0.708660	63	2.480310	230	9.055100
19	0.748030	64	2.519680	240	9.448800
20	0.787400	65	2.559050	250	9.842500
21	0.826770	66	2.598420	260	10.236200
22	0.866140	67	2.637790	270	10.629900
23	0.905510	68	2.677160	280	11.032600
24	0.944880	69	2.716530	290	11.417300
25	0.984250	70	2.755900	300	11.811000
26	1.023620	71	2.795270	310	12.204700
27	1.062990	72	2.834640	320	12.598400
28	1.102360	73	2.874010	330	12.992100
29	1.141730	74	2.913380	340	13.385800
30	1.181100	75	2.952750	350	13.779500
31	1.220470	76	2.992120	360	14.173200
32	1.259840	77	3.031490	370	14.566900
33	1.299210	78	3.070860	380	14.960600
34	1.338580	79	3.110230	390	15.354300
35	1.377949	80	3.149600	400	15.748000
36	1.417319	81	3.188970	500	19.685000
37	1.456689	82	3.228340	600	23.622000
38	1.496050	83	3.267710	700	27.559000
39	1.535430	84	3.307080	800	31.496000
40	1.574800	85	3.346450	900	35.433000
41	1.614170	86	3.385820	1000	39.370000
42	1.653540	87	3.425190	2000	78.740000
43	1.692910	88	3.464560	3000	118.110000
44	1.732280	89	3.503903	4000	157.480000
45	1.771650	90	3.543300	5000	196.850000

Table 23.6 Conversion of Common Fractions to Decimal and Millimeter Units

Common Fractions	Decimal Fractions	mm (approx.)	Common Fractions	Decimal Fractions	mm (approx.)
1/128	0.008	0.20	1/2	0.500	12.70
1/64	0.016	0.40	33/64	0.516	13.10
1/32	0.031	0.79	17/32	0.531	13.49
3/64	0.047	1.19	35/64	0.547	13.89
1/16	0.063	1.59	9/16	0.563	14.29
5/64	0.078	1.98	37/64	0.578	14.68
3/32	0.094	2.38	19/32	0.594	15.08
7/64	0.109	2.78	39/64	0.609	15.48
1/8	0.125	3.18	5/8	0.625	15.88
9/64	0.141	3.57	41/64	0.641	16.27
5/32	0.156	3.97	21/32	0.656	16.67
11/64	0.172	4.37	43/64	0.672	17.07
3/16	0.188	4.76	11/16	0.688	17.46
13/64	0.203	5.16	45/64	0.703	17.86
7/32	0.219	5.56	23/32	0.719	18.26
15/64	0.234	5.95	47/64	0.734	18.65
1/4	0.250	6.35	3/4	0.750	19.05
17/64	0.266	6.75	49/64	0.766	19.45
9/32	0.281	7.14	25/32	0.781	19.84
19/64	0.297	7.54	51/64	0.797	20.24
5/16	0.313	7.94	13/16	0.813	20.64
21/64	0.328	8.33	53/64	0.828	21.03
11/32	0.344	8.73	27/32	0.844	21.43
23/64	0.359	9.13	55/64	0.859	21.83
3/8	0.375	9.53	7/8	0.875	22.23
25/64	0.391	9.92	57/64	0.891	22.62
13/32	0.406	10.32	29/32	0.906	23.02
27/64	0.422	10.72	59/64	0.922	23.42
7/16	0.438	11.11	15/16	0.938	23.81
29/64	0.453	11.51	61/64	0.953	24.21
15/32	0.469	11.91	31/32	0.969	24.61
31/64	0.484	12.30	63/64	0.984	25.00

Table 23.7 Decimal Equivalent Size of Drill Numbers

Drill no.	Decimal Equiv.	Drill no.	Decimal Equiv.	Drill no.	Decimal Equiv.
80	0.0135	53	0.0595	26	0.1470
79	0.0145	52	0.0635	25	0.1495
78	0.0160	51	0.0670	24	0.1520
77	0.0180	50	0.0700	23	0.1540
76	0.0200	49	0.0730	22	0.1570
75	0.0210	48	0.0760	21	0.1590
74	0.0225	47	0.0785	20	0.1610
73	0.0240	46	0.0810	19	0.1660
72	0.0250	45	0.0820	18	0.1695
71	0.0260	44	0.0860	17	0.1730
70	0.0280	43	0.0890	16	0.1770
69	0.0292	42	0.0935	15	0.1800
68	0.0310	41	0.0960	14	0.1820
67	0.0320	40	0.0980	13	0.1850
66	0.0330	39	0.0995	12	0.1890
65	0.0350	38	0.1015	11	0.1910
64	0.0360	37	0.1040	10	0.1935
63	0.0370	36	0.1065	9	0.1960
62	0.0380	35	0.1100	8	0.1990
61	0.0390	34	0.1110	7	0.2010
60	0.0400	33	0.1130	6	0.2040
59	0.0410	32	0.1160	5	0.2055
58	0.0420	31	0.1200	4	0.2090
57	0.0430	30	0.1285	3	0.2130
56	0.0465	29	0.1360	2	0.2210
55	0.0520	28	0.1405	1	0.2280
54	0.0550	27	0.1440		

Table 23.8 Decimal Equivalent Size of Drill Letters

Letter Drill	Decimal Equiv.	Letter Drill	Decimal Equiv.	Letter Drill	Decimal Equiv.
A	0.234	J	0.277	S	0.348
B	0.238	K	0.281	T	0.358
C	0.242	L	0.290	U	0.368
D	0.246	M	0.295	V	0.377
E	0.250	N	0.302	W	0.386
F	0.257	O	0.316	X	0.397
G	0.261	P	0.323	Y	0.404
H	0.266	Q	0.332	Z	0.413
I	0.272	R	0.339		

Table 23.9 Conversion Ratios for Length

Known Quantity	Multiply by	Quantity to Find
inches (in)	2.54	centimeters (cm)
feet (ft)	30	centimeters (cm)
yards (yd)	0.9	meters (m)
miles (mi)	1.6	kilometers (km)
millimeters (mm)	0.04	inches (in)
centimeters (cm)	0.4	inches (in)
meters (m)	3.3	feet (ft)
meters (m)	1.1	yards (yd)
kilometers (km)	0.6	miles (mi)
centimeters (cm)	10	millimeters (mm)
decimeters (dm)	10	centimeters (cm)
decimeters (dm)	100	millimeters (mm)
meters (m)	10	decimeters (dm)
meters (m)	1000	millimeters (mm)
dekameters (dam)	10	meters (m)
hectometers (hm)	10	dekameters (dam)
hectometers (hm)	100	meters (m)
kilometers (km)	10	hectometers (hm)
kilometers (km)	1000	meters (m)

Table 23.10 Conversion Ratios for Area

Known Quantity	Multiply by	Quantity to Find
square inches (in^2)	6.5	square centimeters (cm^2)
square feet (ft^2)	0.09	square meters (m^2)
square yards (yd^2)	0.8	square meters (m^2)
square miles (mi^2)	2.6	square kilometers (km^2)
acres	0.4	hectares (ha)
square centimeters (cm^2)	0.16	square inches (in^2)
square meters (m^2)	1.2	square yards (yd^2)
square kilometers (km^2)	0.4	square miles (mi^2)
hectares (ha)	2.5	acres
square centimeters (cm^2)	100	square millimeters (mm^2)
square meters (m^2)	10,000	square centimeters (cm^2)
square meters (m^2)	1,000,000	square millimeters (mm^2)
ares (a)	100	square meters (m^2)
hectares (ha)	100	ares (a)
hectares (ha)	10,000	square meters (m^2)
square kilometers (km^2)	100	hectares (ha)
square kilometers (km^2)	1,000	square meters (m^2)

Table 23.11 Conversion Ratios for Mass

Known Quantity	Multiply by	Quantity to Find
ounces (oz)	28	grams (g)
pounds (lb)	0.45	kilograms (kg)
tons	0.9	tonnes (t)
grams (g)	0.035	ounces (oz)
kilograms (kg)	2.2	pounds (lb)
tonnes (t)	100	kilograms (kg)
tonnes (t)	1.1	tons
centigrams (cg)	10	milligrams (mg)
decigrams (dg)	10	centigrams (cg)
decigrams (dg)	100	milligrams (mg)
grams (g)	10	decigrams (dg)
grams (g)	1000	milligrams (mg)
dekagram (dag)	10	grams (g)
hectogram (hg)	10	dekagrams (dag)
hectogram (hg)	100	grams (g)
kilograms (kg)	10	hectograms (hg)
kilograms (kg)	1000	grams (g)

Table 23.12 Conversion Ratios for Volume

Known Quantity	Multiply by	Quantity to Find
milliliters (mL)	0.03	fluid ounces (fl oz)
liters (L)	2.1	pints (pt)
liters (L)	1.06	quarts (qt)
liters (L)	0.26	gallons (gal)
gallons (gal)	3.8	liters (L)
quarts (qt)	0.95	liters (L)
pints (pt)	0.47	liters (L)
cups (c)	0.24	liters (L)
fluid ounces (fl oz)	30	milliliters (mL)
teaspoons (tsp)	5	milliliters (mL)
tablespoons (tbsp)	15	milliliters (mL)
liters (L)	100	milliliters (mL)

Table 23.13 Conversion Ratios for Cubic Measure

Known Quantity	Multiply by	Quantity to Find
cubic meters (m^3)	35	cubic feet (ft^3)
cubic meters (m^3)	1.3	cubic yards (yd^3)
cubic yards (yd^3)	0.76	cubic meters (m^3)
cubic feet (ft^3)	0.028	cubic meters (m^3)
cubic centimeters (cm^3)	1000	cubic millimeters (mm^3)
cubic decimeters (dm^3)	1000	cubic centimeters (cm^3)
cubic decimeters (dm^3)	1,000,000	cubic millimeters (mm^3)
cubic meters (m^3)	1000	cubic decimeters (dm^3)
cubic meters (m^3)	1	steres
cubic feet (ft^3)	1728	cubic inches (in^3)
cubic feet (ft^3)	28.32	liters (L)
cubic inches (in^3)	16.39	cubic centimeters (cm^3)
cubic meters (m^3	264	gallons (gal)
cubic yards (yd^3)	27	cubic feet (ft^3)
cubic yards (yd^3)	202	gallons (gal)
gallons (gal)	231	cubic inches (in^3)

Table 23.14 Conversion Ratios for Electrical Quantities

Known Quantity	Multiply by	Quantity to Find
Btu per minute	0.024	horsepower (hp)
Btu per minute	17.57	watts (W)
horsepower (hp)	33,000	foot-pounds per min (ft-lb/min)
horsepower (hp)	746	watts (W)
kilowatts (kW)	57	Btu per minute
kilowatts (kW)	1.34	horsepower (hp)

Index

A

A/D converter, 271
active, 171
active component, 181, 259
active hub, 324
active set input, 275
adjustable resistor networks, 186
air-core inductor, 214
aliasing, 271
all-pole filters, 225
alternating current, 3
aluminum electrolytic capacitor, 200, 205
amplifier stage, 259
amplitude modulation, 289
apparent power, 173
application layer, 326
ARCnet network, 323
atomic theory, 1
atoms, 1
attenuation band, 223
avalanche voltage, 246

B

bandgap reference diodes, 247
base, 271
base-collector junction, 247
base-emitter junction, 247
bathtub curve, 352, 354
Bessel function, 302
binary frequency-shift keying, 316
binary on-off keying, 316
binary phase-shift keying, 316
binary-coded decimal, 283
biphase coding, 320
bipolar transistor, 247
blackbody, 235
Boolean algebra, 275
boundary layer, 235
branch, 172

brightness, 157, 161
British Thermal Units, 233
broadband network, 323
buffer coating, 333
buffer tube, 333
build specifications, 325
burn-in failure, 346
burn-in testing, 354
byte, 283

C

calefaction, 238
calorie, 233
campus area network, 326
capacitance, 199 - 200
capacitor, 199 - 200
carbon composition resistors, 185
carbon film resistors, 184
carrier wave, 292
Cartesian form, 174 - 175
cascade connection, 263
catastrophic failure, 238
CCIR, 40
ceramic capacitors, 204
cermet resistor, 184
check sum, 285
circuit Q, 178
circuit resistance, 178
closed loop, 172
CMOS, 281
cold-soak, 353
combinational, 275
common-base, 262
common-collector, 264
common-drain, 264
common-emitter, 249, 259
common-gate, 262
common-mode failure, 362
common-mode signals, 266
common-source, 259